DISPELLING
THE DARKNESS

Voyage in the Malay Archipelago and the
Discovery of Evolution by Wallace and Darwin

DISPELLING
THE DARKNESS

Voyage in the Malay Archipelago and the
Discovery of Evolution by Wallace and Darwin

John van Wyhe

National University of Singapore

 **World Scientific**

NEW JERSEY · LONDON · SINGAPORE · BEIJING · SHANGHAI · HONG KONG · TAIPEI · CHENNAI

Published by

World Scientific Publishing Co. Pte. Ltd.

5 Toh Tuck Link, Singapore 596224

USA office: 27 Warren Street, Suite 401-402, Hackensack, NJ 07601

UK office: 57 Shelton Street, Covent Garden, London WC2H 9HE

Library of Congress Cataloging-in-Publication Data
van Wyhe, John, 1971–
   Dispelling the darkness : voyage in the malay archipelago and the discovery of evolution by Wallace and Darwin / John van Wyhe, National University of Singapore, Singapore.
   pages cm
   Includes bibliographical references and index.
   ISBN 978-9814458795 (hardcover : alk. paper)
   ISBN 978-9814458801 (softcover : alk. paper)
   1. Wallace, Alfred Russel, 1823–1913. 2. Naturalists--Great Britain--Biography.
3. Natural selection. I. Title.
   QH31.W2V36 2013
   576.8'2092--dc23
   [B]
                                                        2012051671

British Library Cataloguing-in-Publication Data
A catalogue record for this book is available from the British Library.

In-house Editor: Veronica Low

Typeset by Stallion Press
Email: enquiries@stallionpress.com

Printed in Singapore by Fulsland Offset Printing (S) Pte Ltd Singapore.

Für Alexander

CONTENTS

ACKNOWLEDGEMENTS

I am very grateful to those who have helped me, in so many ways, with the research and writing of this book. The National University of Singapore and in particular the Department of Biological Sciences and Tembusu College have generously supported me throughout my research and writing of this book. My students have been a refreshing source of insightful questions and new perspectives. My assistant Kees Rookmaaker has helped me on countless occasions and this book would have been much the poorer without his expertise, assistance and suggestions. I am grateful to Michael B. Pearson who generously sent me an electronic version of his transcription of Wallace's *Journal* which I used constantly during the early stages of researching and writing this book before I was able to access the original documents to create my own transcriptions with Kees Rookmaaker.

I am indebted to the anonymous donor who funded the *Darwin Online* and *Wallace Online* projects between 2009 and 2013. Both projects and this research would have been impossible otherwise. I am as ever deeply indebted and grateful to my parents Richard and Donna van Wyhe. Thanks are also due to my agent Bill Hamilton for his helpful suggestions and feedback on early drafts though the book was sadly delayed through a long period of inactivity. I am grateful to Ambassador Tommy Koh and Ho Yi Kai for recommending and inviting me to publish with World Scientific Publishing as the Wallace anniversary year drew nigh. Thanks also to my editor Veronica Low for her painstaking efforts to improve the book and my inconsistent punctuation. Antranig Basman read and commented on the entire manuscript and offered much valuable feedback which was of enormous help to me. Jon Hodge also read the entire manuscript and gave wonderful encouragement and much thoughtful and helpful feedback, particularly on the Ternate essay discussion as the book neared completion.

Judith Magee and Lorraine Portch sent images from the Natural History Museum (NHM) collection with unparallelled kindness and promptness. I thank the Wallace Correspondence Project at the NHM for their generous assistance with Wallace letters. Gerrell M. Drawhorn generously shared many details from his Wallace research. Daniel Tham kindly showed me the William Farquhar drawings at the National Museum of Singapore. I am also indebted to the workers behind online text collections such as Google Books, the Internet Archive, JSTOR, the Biodiversity Heritage Library and many others. The Rev. Rene Nicolas of Missions Étrangères de Paris answered my queries about the Bukit Timah mission. I am very grateful to Muhammad Abdullah Sinus 'Dillah' and Andi Muhammad Yusuf 'Uccu' for showing me around Macassar and Maros. I am also grateful to Chor Lin Lee, Director of the National Museum of Singapore, for permission to reproduce the images from that collection.

I am also grateful for help that contributed to this book by Cordula van Wyhe, Anna Mayer, George Beccaloni, Erik Holmberg, Christine Garwood, David Clifford, Jim Secord, Gordon Chancellor, Richard Corlett, Janet Browne, Daniel Tay, Tim Barnard, Jan Van Der Putten, Peter C. Kjaergaard, Paul Clark, Duncan Porter, Frank Sulloway, Greg Radick, Tony Larkum, Ian Singleton, Shannon Bohle, Andrew Wee, Brenda Yeoh, John S. Ascher, Gerda and Gesa Bleckwedel, Anna Ng, Yuchen Ang, James Tann, Umiaty, Cathryn Hindle, Pat Rich and George Quinn. At the Department of Biological Sciences at the National University of Singapore, I thank Rudolf Meier, Paul Matsudaira, Theo Evans, Al Davis, Dee Dupuy, Lisa Lau Li-Cheng, Joan Choo Beng Goon, Yee Ngoh Chan, Nursyidah Binti Mansor, Ann Nee Yong, Lai Soh Fun, Soong Beng Ching, Laurence Gwee and Hew Choy Leong. At the Raffles Museum of Biodiversity Research, thanks are due to Peter Ng, Leo Tan and Kelvin Lim. At Tembusu College, I thank the Fellows and students and especially the Master Greg Clancey and Danielle Henricus. Lucille Yap of the Philatelic Museum was helpful with several queries. Thanks also to the staff at the National Museum of Singapore and the National Archives of Singapore. At the Library of the Linnean Society of London, Lynda Brooks, Gina Douglas, Leonie Berwick and Andrea Deneau kindly gave me crucial assistance with the Wallace manuscripts in their collection.

Between 2003 and 2005, I was a Research Fellow on an earlier Wallace correspondence project at the Open University (apparently now defunct) and, by cataloguing and reading thousands of Wallace's extant letters, gained my first in-depth understanding of Wallace. In 2005, I then moved my *Darwin Online* project (first launched while I was a Senior Research Fellow at the National University of Singapore in 2002) to the University of Cambridge with the help of Jim Secord and Janet Browne. So now, after ten years studying Darwin and publishing the first complete edition of his published works and the largest collection of his edited manuscripts, it has been delightful to get back to Wallace. This book began when I was invited to contribute a biographical chapter on Wallace to the *Cambridge Encyclopedia of Darwin*. I became intrigued with the Ternate letter mystery. It was thought impossible to know when Wallace sent it and when Darwin really received it, a loophole that has fuelled a small but very vocal conspiracy industry. With so many conflicting claims, I wanted to see just how far the surviving evidence allowed us to go. As I dug deeper, a detective-story investigation unfolded, in which I was eventually cheerfully joined by Kees Rookmaaker, to track down how Wallace's fateful essay travelled across the globe to Darwin. The full story is told below. Inspired by this solution to an intractable mystery, I thought more of Wallace's story might also turn out to be different than conventionally told if subjected to intensive historical research and not just taken on faith.

Portions of Chapter 10 are reproduced, though heavily revised and supplemented by subsequent research, from my 2007 article "Mind the gap: did Darwin avoid publishing his theory for many years?" courtesy of the Royal Society. I am grateful to the Department of Special Collections and University Archives, W.E.B. Du Bois Library, University of Massachusetts Amherst for permission to reproduce Woodbury photographs. For permission to quote from unpublished manuscripts in their collections, I thank the Wallace Literary Estate, the Syndics of Cambridge University Library, the Linnean Society of London and the Trustees of the Natural History Museum (London).

John van Wyhe
Singapore, December 2012

View of Ternate. J.C. Rappard, c. 1883–9. Tropenmuseum.

INTRODUCTION

It's February 1858. In the shadow of a slumbering volcano on the small East Indies island of Ternate, an English naturalist named Alfred Russel Wallace lay on his bamboo bed racked with tropical fever. He shivered under blankets despite the perpetual tropical heat. A few hours later, the fever swung towards heat and he was drenched with sweat. Wallace had been privately speculating about evolution for years. Unable to venture out in the midst of this fever, he thought again of the history of life on Earth. It had changed so much over time.

He was sure that the only way to explain the fossil record was by evolution of some kind. He flipped through his notebooks and recent mail and thought about some of the puzzling creatures he had recently collected. Then something just clicked. He had an exhilarating idea. There is a struggle for existence in nature. Many are born, only a few survive. Those that fit their environments survive best. Only those few that survive leave descendants. Surely this is the key! The world would later know it by Charles Darwin's phrase "natural selection".

It is one of the most romantic beginnings of a scientific theory ever recorded. Archimedes was inspired in the bath, the structure of benzene came from a day dream about a snake biting its tail and the atomic bomb was first imagined at a London traffic light. When Wallace recovered from his fever, he took up his pen and wrote out, in his exemplary rounded handwriting, an essay outlining how local varieties could become new species. He later mailed it, in one of the greatest ironies in history, to Darwin, more than 9,000 miles away in England.

After two months in transit, Wallace's essay arrived at Darwin's house. Darwin had been working on a similar theory of evolution for twenty years but had not yet published it. The astonished Darwin exclaimed to a friend, "I never saw a more striking coincidence, if Wallace had my M.S. sketch written out in 1842 he could not have made a better short abstract! Even his terms now stand as Heads of my Chapters."[1] To avoid a priority dispute, Darwin's colleagues arranged to have Wallace's essay, and two shorter pieces by Darwin, read together at a meeting of the Linnean Society of London on 1 July 1858.

The shockwaves from that day still reverberate. It was the beginning of the greatest scientific revolution in history — nothing short of an apocalypse for traditional ideas about nature, philosophy and religion. Finally we would be able to unlock the secrets of life on Earth and of our true origins, not only where we come from but what we actually are.

The acerbic naturalist Thomas Huxley later wrote, "The facts of variability, of the struggle for existence, of adaptation to conditions, were notorious enough; but none of us had suspected that the road to the heart of the species problem lay through them, until Darwin and Wallace dispelled the darkness."[2] Even 150 years later, the "unexpected thunderclap" of 1 July 1858 is said to have marked "the beginning of the 'Darwinian Revolution', 'the beginning of Modern Biology', 'the beginning of a new era in scientific thinking', and many other such labels that indicate that from this date, dynamic new ideas, concepts and forces were turned loose in biology, philosophy and theology".[3]

And no one had seen it coming, least of all the principal players — Charles Darwin and Alfred Wallace. This story has been told a thousand times before. But, amazingly, much about the traditional story is wrong. That we should have bookshelves full of legends and myths about one of the most important turning points in history is just not good enough. It is time to set the record straight.

There's Something About Wallace

Wallace is unlike any other figure in the history of science. He is usually described as neglected, obscure or forgotten. If he is not an "elusive Victorian" or a "forgotten naturalist", he is "Darwin's moon", "in Darwin's shadow" or even a "heretic in Darwin's court" — all titles of Wallace biographies. Relegating Wallace's name to the subtitle of his own biography hardly helps. Yet, scores of articles and books have been written on Wallace in the past forty years and continue to appear. He is apparently famous for being forgotten!

But he has not always been seen as forgotten or overshadowed. Until the late 1960s, no one seems to have thought so.[4] But of course it is undeniable that compared to his illustrious co-discoverer Darwin, Wallace is little known to the general public. But then, so are all the other naturalists of the Victorian era.

Nowadays, Wallace has many admirers who are incredibly affectionate, passionate and sometimes downright angry. For some, Wallace is an icon of conservation and exploration or the father of biogeography. The science journalist David Quammen called Wallace "the greatest field biologist of the nineteenth century".[5] Everyone points out that Wallace was the co-discoverer of natural selection with Darwin. But if Darwin and Wallace published natural selection jointly, why is Darwin so famous and Wallace so unknown? Many Wallace fans are convinced that something is wrong. And like many admirers of a figure who deserves far more than he gets, they have a tendency to emphasise and sometimes exaggerate his accomplishments and originality. Generations of scholars have studied Darwin's publications and papers in great detail. We now understand Darwin's work remarkably well. But Wallace has never received anything like the same intensity of research. Overshadowed by the towering Darwin, in more ways than one, a century after his death Wallace has become more myth than man.

It is astonishing how many myths and misconceptions about Wallace are at large. But given how frequently legends are published, even in best-selling books that reach millions of readers, it at least makes sense. For example, virtually every writer claims that Wallace first declared his belief in evolution in his 1855 Sarawak law paper. Yet the paper does not mention evolution at all, and Wallace very carefully refrained from doing so. Modern readers nevertheless read evolution between the lines. The ambiguity of language often allows for multiple readings. Such errors have serious consequences. Wallace repeatedly refrained from espousing evolution in print, yet no writer on him seems to have realised it. This helps to explain one of the biggest of all conundrums: why did he not send his 1858 evolution essay directly for publication but instead privately to Darwin?[6]

Some people believe that Wallace was really first to the finish line with natural selection and that he was shamefully relegated to an ignominious second-place by Darwin or his friends. Did Darwin and Wallace not come up with the identical theory and so deserve equal credit? Simon Winchester in his best-selling *Krakatoa* and biologist Jerry Coyne in his celebrated *Why evolution is true* state that Wallace and Darwin thought of natural selection "at the very same time".[7] But Wallace's breakthrough came *twenty years* after Darwin's (1838). Knowing this, Arnold Brackman and John G. Wilson claimed that Wallace may not have thought of natural selection first, but was still the first to "write down" the "full" or "complete" theory of evolution.[8]

Most of all, some Wallace fans bemoan his lack of fame or recognition vis-à-vis Darwin.[9] Was Wallace cheated of his rightful priority or at least equal fame with Darwin? Some claim that Wallace's essay ought to have come first rather than last in the 1858 presentation or that it was not acceptable to publish the paper without his explicit prior consent. In the BBC book *Genius of Britain* (2010), we read that, "If he hadn't been so decent and respectful of Darwin, Wallace would have been credited for the discovery of a theory that would now be known as Wallaceism, not Darwinism." According to historian Ian McCalman, "Thanks to the way in which Darwin and Wallace's papers were announced to the world of science… 'Darwin is today remembered as the genius discoverer of the theory, while Wallace… is famous for being forgotten.'"[10]

Another popular explanation is that Wallace was working class and this or his lack of wealth led to his obscurity in favour of the scientifically respected patrician Darwin. In A. S. Byatt's novella *Morpho Eugenia* (1992) and the film version *Angels and insects* (1996), Wallace was recast as naturalist William Adamson who speaks with a working-class accent and is taunted by his betters — "you're not one of us". Was Wallace the victim of a Victorian class-based glass ceiling? I have found no evidence to support it. In fact, the number of contemporary men from humble, even *humbler*, origins who achieved towering reputations, from David Livingstone, Michael Faraday, William Whewell, Richard Owen, Herbert Spencer, Thomas Huxley to Charles Dickens, is simply too great to accept that it explains Wallace's reputation.

One of the most intractable mysteries is just when did Darwin receive Wallace's essay? The original doesn't survive. Did Darwin lie about its receipt and secretly keep it for weeks or even months? A letter from Wallace to another correspondent does survive with postmarks showing its receipt in London on a different date. It is reproduced in many books as if it were a smoking gun proving that Darwin lied.[11] Did Darwin then borrow elements of the theory from or even plagiarise Wallace? Many, many writers repeat these stories.

The lawyer David Hallmark made the news by making such accusations against Darwin: "[Hallmark] has hired a specialist in plagiarism software — the kind used to catch deceitful college students." Hallmark claimed in an interview, "The descent of Wallace from equality to relative invisibility is the direct result of the unlawful conduct of Charles Darwin by suppressing the true worth of Wallace as the author of the theory."[12]

To some, the entire Linnean Society affair was a "delicate arrangement", a shabby stitch-up and even a "scientific crime". Were some of the documents relating to this story selectively or intentionally destroyed?[13] The former BBC producer Roy Davies claims to have uncovered "a deliberate and iniquitous case of intellectual theft, deceit and lies perpetrated by Charles Darwin".[6] Few are as extreme as this, but a remarkable number of people believe that something improper or unfair happened. Would Darwin never have published without being "forced" into the open by Wallace? And why had Darwin not published by then? Most books say that he kept his theory secret for twenty years because he was afraid. Is that true?

So part of what divides many over Darwin and Wallace is sympathy for the apparent have-not, the outsider versus the insider, the self-educated versus the Cambridge man, the lower class versus the upper, the poor man versus the rich man and so forth. These stereotypical dichotomies feel so natural that they help to reinforce the image of a legendary Wallace who is very different from the historical Wallace.

This book shows why these romantic tales are just plain wrong. Some are just sloppy mistakes. But mistakes matter because they form a legendary story with a mythical Wallace who never existed. The real Wallace's breakthrough cannot be understood without a more accurate historical reconstruction.

Wallace's voyage has never received anything like the in-depth study of Darwin's *Beagle* voyage or his path to evolutionary theory. The fragmentary remains of Wallace's thoughts and experiences from those extraordinary years are like fossils. An extinct creature is gone forever, its substance vanished, but remarkable and highly specific traces sometimes remain. Like fossils, the remnants of Wallace's thoughts are tiny specks amidst vast gaps in the record. It is difficult to reconstruct a more complete picture.

Other evidence from his context is needed to reconstruct the story of what follows. There were hundreds of other publications from the time that fill great voids in the old story. The research for this book included the additional book-length projects of editing Wallace's notebooks and letters to modern scholarly standards — transcribing handwritten scrawls, dating passages, ink types, identifying his sources and so forth. This allowed his itinerary to be reconstructed. All previous ones, including his own, were highly inaccurate. The results allow Wallace's voyage and his path to natural selection to be revealed for the first time.

This book aims to completely revise our understanding of Wallace's voyage from the ground up. As Wallace travelled from island to island and collected insect after insect, his ideas about species gradually evolved. His voyage is as close as we will ever get to a historical experiment. What if there was no Darwin?

The orthodox explanation of Darwin's theory by historians of science is arguably *social* — that is that the social, political and economic context of Victorian Britain explains or is reflected in Darwin's *Origin of species*. As Wallace is supposed to be from the other end of the social and political spectrum, and yet came up with such similar views about evolution, there is an incentive to show that Wallace does not contradict but actually supports the same theoretical approach to the history of science.[14] But Wallace's odyssey may show we have been hunting in the wrong haystack.

Evolution and Science Amnesia

The Galápagos Islands are famous all over the world as the site of the discovery of evolution. Yet Darwin did not discover evolution there. Instead, material collected from the Islands became one of three main kinds of evidence that convinced Darwin after returning to England in 1836.[15] Another region does have a bona fide claim as the *field site* of the discovery of the theory of evolution by natural selection, and that is Southeast Asia, where Wallace worked between 1854 and 1862. Whereas Darwin's evolutionary conclusions were formalised and written down on reflection after his travels, Wallace's independent discovery of natural selection happened in the field. This is not entirely surprising considering how comparatively barren the Galápagos are in contrast to the rich biodiversity and striking patterns of zoological distribution in Southeast Asia.

There is an unfolding curtain of forgetfulness about scientific knowledge and discoveries that closes ever after us as time and newer discoveries ensue. Since the day of Darwin and Wallace, X-rays, penicillin, antibiotics, genetics, jet airplanes, nuclear physics, DNA, plate tectonics, space exploration and digital technology have all in their turn filled newspapers, popular science books and the attention of the public. Thus, the cutting-edge science of the generation before is largely forgotten. There simply isn't room for all the old to stay as prominent as so much new science appears.

This inevitable amnesia is one of the main problems with the public understanding of evolution today. The main outlines of what we know about the story of life on Earth are now unclear. Instead, we acquire shorthand histories. One of these could be called the "Richard Dawkins history of science". According to popular writers and broadcasters like Dawkins, everyone before Darwin and Wallace believed that the world was 6,000 years old, and that all species on Earth were created by God just as they are now. Imagined in this way, the revolutionary proposals of Darwin and Wallace must have caused a titanic clash between science and religion.

This science amnesia is also one of the principal reasons for a glaring anomaly. Compared to today, evolution was originally presented to a far more religious world and with far less evidence. Nevertheless, evolution came to be accepted as fact by the international scientific community and much of the educated public within twenty years of its announcement by Wallace and Darwin. Yet now, we have vastly more complete evidence and a vastly more secular world, but evolution is widely perceived to be controversial! How can this be?

The essential background for understanding evolution, such as the age of the earth and the progressive nature of the fossil record, were new and familiar to the

Victorians. But these are widely unknown and unappreciated now. Hence, the modern public is often confronted with evolution in a vacuum, as it were. Or it is shrouded with details of the latest understandings of DNA and microbiology. No wonder an uninformed or sceptical audience is so often not convinced, and indeed utterly confused. A definition of evolution in terms of changing gene frequencies may be useful amongst biologists, but it is worse than useless for the general public. Darwin and Wallace did not know about genes, and they certainly understood evolution. If the general reader today could understand and appreciate evolution to the general extent that the Victorians did, that would be preferable to the widespread misunderstandings that now cause so much trouble.

The year 2013 is the centenary of Wallace's death. Many new publications and commemorations can be expected. The majority of his private papers are now wonderfully catalogued and curated at London's Natural History Museum and the Linnean Society of London. There is a new correspondence project devoted to him; Charles H. Smith's *Wallace Page* and my own *Wallace Online* have made access to Wallace's writings easier than ever before. There has never been a better time to reveal the true story of Wallace and his independent path to evolution.

Chapter 1

GREAT EXPECTATIONS

O n the morning of 19 April 1854, a giant iron steamship dominated the crowded harbour of Singapore. As pungent black-grey coal smoke curled out of the paddle steamer's twin funnels and mixed with the humid morning air, small native boats rowed to and fro ferrying passengers and baggage in a continuous conveyor belt.

In more senses than one, this was as far from London as one could get. It was a strange land. The trees, birds, insects and the people — all were foreign and different. Even in the morning it was already quite hot and the humidity was near eighty percent. The region was usually called the Eastern Archipelago, now Malaysia, Singapore and Indonesia.[16]

About a dozen Westerners disembarked. One of them clambered from a small boat onto the jetty. This was a thirty-one-year-old Englishman named Alfred Russel Wallace. He was six feet tall, thin and good-looking, with long legs, brown hair and sparkling blue eyes. With his pale clothes, wire-rimmed spectacles, Panama hat and dark whiskers, he looked like any other European in this bustling Asian entrepôt. However, when his teenage English servant and the Malay boatmen began unloading his baggage, which included two double-barrelled guns, long-handled insect nets and collecting boxes, it was obvious that this was not the usual merchant or colonial official. The locals probably paid no attention to another eccentric-looking Westerner. Wallace had no precise route or plan; he just hoped to collect natural history specimens for sale in London. But his adventures and discoveries would change the world in ways no one could have predicted.

Years later, when writing his great book of travels in the East, Wallace remarked to Darwin, "Like every other traveller, I suppose, I feel dreadfully the want of copious notes on common everyday objects, sights and sounds and incidents, which I imagined I could never forget but which I now find it impossible to recall with any accuracy."[17] The present book supplies what Wallace did not record, and corrects much that he did inaccurately.

But one cannot make sense of Wallace's explorations, and the new science that emerged as a result, without understanding his background and context. So we must begin before his arrival in the Eastern Archipelago with Wallace's upbringing in Britain.

Bringing Up Wallace

The 1820s marked the end of the Napoleonic Wars and the pax Britannica that followed saw Britain expand industrially, economically and militarily. These were decades of prosperity, unparalleled growth and dominance in commerce, technology and science though often with dramatic swings in social distress. In the midst of this post-Jane Austen and pre-Charles Dickens England, Wallace was born on 8 January 1823 at Kensington Cottage on the bank of the River Usk, about half a mile from the market town of the same name in Monmouthshire (now Gwent) on the Welsh borders, 140 miles northwest of London. Usk had a market every Monday. On a hill overlooking the town were the remains of a Norman castle. The water of the river was so clear one could see the pebbles on the bottom.[18] Wallace had two elder brothers, William and John, and two elder sisters, Elizabeth and Frances (known as Fanny).

Some modern writers refer to Wallace as working class.[19] But this does not at all fit how social class was defined in the 19th century. Class was not just wealth.[20] His parents, Thomas Vere (1771–1843) and Mary Anne Greenell Wallace (1788–1868) were gentlefolk, and would fit seamlessly into a Jane Austen novel. Wallace's baptism record in the little church at the nearby village of Llanbadoc reads "Feb[ry]. 16th, Alfred Russell [sic] Son of Thomas Vere and Mary Ann Wallace Lanbadock Gentleman".[21] Thomas Vere was a solicitor by training, though he inherited property sufficient to generate an income of £500 per annum.[22] Thus, Wallace's father lived the independent life of a gentleman. Their genteel status is obvious from their wedding portraits. We would today better describe the Wallaces as lower middle class.

Straight out of Jane Austen. Wallace's parents. ML1:16.

Kensington Cottage. ML1:20.

Charles Darwin, born in February 1809, was already fourteen and attending his local Shrewsbury school when Wallace was born. Darwin's family was not so different from the Wallaces in terms of their social class or distance from the aristocracy on the one hand and the working classes on the other. But the Darwins, heavily intermarried with the wealthy pottery magnate Wedgwoods, were very rich indeed. Darwin's family wealth enabled him to live as an independent gentleman all his life. He never had a job, unlike Wallace who would have so many.

The Wallace's financial circumstances rapidly declined, partly from the growing family size and a series of bad investments by Thomas Vere who had lived the idle life of an uninspired gentleman, unremarkably gracing the social scene in London and Bath. So the family moved from London to the Welsh borders, where Wallace was born. As far as Wallace could later remember, the family kept only one servant.

Because of his birthplace, Wallace is sometimes referred to as Welsh, but his parents were English. As a small boy in Usk, Wallace could remember, because of his blonde hair, "I was generally spoken of among the Welsh-speaking country people as the little Saxon."[23] Wallace referred to himself as "English" and an "English naturalist" many times in his publications.[24]

Public school boy

In 1828, the family moved to Hertford, twenty-one miles north of London. The very day after arriving at their house, Wallace met a neighbour boy about his age, George Silk. They would remain friends for the rest of their lives. At the same time, Charles Darwin was just going up to Cambridge to study for a B.A. degree with the intention of becoming a country clergyman.[25] During the next eight or nine years, the Wallaces lived in five different houses in Hertford. In June 1829, the last child, Herbert Edward, was born.

Wallace attended, perhaps from October 1829, Hertford Grammar School, a "gent.'s boarding" school, under headmaster Clement Henry Crutwell.[26] There were about forty students. Boarders were charged 25–30 guineas a year. There were seven scholarships to send students to Peterhouse, the oldest Cambridge college. Wallace followed a classical education, not unlike Darwin's at Shrewsbury School, including lessons on Latin and English grammar, classical geography, writing and "some Euclid and algebra".[27] It would be Wallace's only formal education. Many modern writers claim that Wallace suffered from an unusual hardship in having no formal education beyond grammar school. But this was more than most people acquired. Universal state education would not be introduced until the 1870s. Despite the family's finances, Wallace was sent to the

Hertford Grammar School c. 1830. ML1:49.

gentleman's school, not to one of the large inexpensive schools in the town for the poor such as the National School with 120 boys.

On 7 June 1832, the Reform Act became law, widening the eligible electorate in England and Wales and sweeping away much entrenched political corruption.

Although hailed as a great reform at the time, the electorate by one estimate only increased from c. 400,000 to c. 650,000.[28] Wallace and his family went to see a banquet held in Hertford's main street to mark the occasion. Sadly a week later, his sister Elizabeth died from consumption. In 1832 or 1833, sister Fanny went to Hoddesdon as governess to a gentleman's family. Around the same time, Darwin was setting out on the voyage of HMS *Beagle* as naturalist (not the captain's gentleman companion as fashion now has it), an event that would change Darwin's life, and so much else, forever.

A year or two later, John Wallace was apprenticed to a builder in London. To make ends meet, Wallace's father became librarian for a subscription library and took a few pupils.[29] At least this meant Wallace was exposed to many books as a child. During his last year in Hertford, the family's finances declined even further, so much so that Wallace was obliged to tutor other students to pay his school fees. He was deeply conscious of this fall in status before his peers. He later described the shame of this and other cost-saving measures imposed by his parents as a "cruel disgrace", "exceedingly distasteful" and perhaps "the severest punishment I ever endured".[30] Wallace left school aged fourteen in March 1837 (not Christmas 1836 as in his autobiography), five months after Darwin returned from the *Beagle* voyage.[31] The Wallaces moved to Hoddesdon, a coaching town on the road between London and Cambridge where Fanny opened an ill-fated ladies' boarding school.

In London town

In early 1837, the year Queen Victoria ascended the throne on the death of her uncle William IV, Wallace joined his elder brother John in London. John was apprenticed to a builder, a Mr. Webster, on Robert Street, Hampstead Road.[32] Here, Wallace lived amongst working-class men or artisans for the first time. It is clear from his careful recollections of their language, behaviour and dress that he saw them as a different kind of people. His association with working-class people adds to the modern misconception that Wallace was working class. Like other Victorians of his generation, Wallace described a society composed variously of "the higher classes", "the middle classes", "tradesmen and labourers", "peasantry" and the "lowest class of manufacturing operatives".[33]

Not fitting easily into the usual categories, Wallace seems to have gone through life with the impression of watching all "classes" from the outside, though he clearly felt the greatest affinity with his middle-class peers. This, in addition to his formative experiences in a radical working-class context, left him with a sense that the social arrangements of his country were not fair. This also gave him an outsider's perspective, something that often correlates with innovation and a greater readiness to adopt unconventional ideas.

Wallace spent his evenings in a "hall of science" or mechanics' institute. This was an institution for the adult education of working men, normally having a library, reading room and sometimes a museum, and offering courses of lectures. Wallace recollected attending the John Street Institution, just off Tottenham Court Road. It was founded and supported by a philanthropic wine merchant named William Devonshire Saull. Saull, a Fellow of the Geological and Royal Antiquarian Societies, amassed a large geological, antiquarian and phrenological collection which was open to the public, particularly working people.

Wallace recalled that at the John Street Institution, he first encountered the writings of religious sceptics and early socialists, even attending a lecture by the famous Welsh industrialist, reformer and proto-socialist Robert Owen.[34] Wallace, according to his later recollections, was deeply impressed by Owen's utopian social ideals, with a stress on environment determining character and behaviour. Hence, people who suffered from bad social arrangements did not simply deserve all they got. If the social environment was improved, so would the morals and well-being of the workers.

However, according to historian of science James Secord, Wallace's recollection cannot be correct as the John Street Institution was occupied by Owenites only three years later.[35] Similarly, Wallace recalled reading a pamphlet in 1837 by Owen's son, Robert Dale Owen, condemning the Christian doctrine of eternal punishment. This pamphlet was not published until 1840 or 1841.[36] So Wallace's recollections in his later life mixed some events and experiences from different times (he lived in London again in early 1844). It is an important lesson we will encounter again. Later recollections should not be treated like contemporary records. In fact, Wallace may not have been much of a socialist until his later life, if we take the evidence of his own writings.

While at the mechanics' institute, Wallace read a classic philosophical dilemma on the origin of evil: "'Is God able to prevent evil but not willing? Then he is not benevolent. Is he willing but not able? Then he is not omnipotent. Is he both able and willing? Whence then is evil?' This struck me very much, and it seemed quite unanswerable."[37]

This dilemma dates at least to the ancient Greek philosopher Epicurus, but had been more recently popularised by the Scottish philosopher David Hume. By Wallace's time, the dilemma was found in many freethinking tracts and articles.[38] One of their favourite targets was the old-fashioned but still largely respectable natural theology of writers like the Rev. William Paley. Natural theologians claimed that studying nature provided evidence for the existence and beneficence of God. This was supposedly seen in the apparent design of living things. But such facile assertions were easy to contradict since there could be other explanations for the complexity of living things. At Cambridge, students like Darwin were still taught these outmoded but safe doctrines. But for Wallace, the experience of reading radical freethinkers and Owenite socialists "laid the foundation of my religious scepticism".[39] Hence, Wallace never experienced the stereotypical Victorian crisis of faith, as he lost what little he had at an early age.

But something else must have happened at this time which has left no trace in the historical record. At exactly the same time, just a mile south of where Wallace was staying, Darwin took genteel lodgings at 36 Great Marlborough Street. He was just back from his round-the-world *Beagle* voyage and needed to attend scientific meetings and work on his collections. So, at the very moment that Darwin began his evolutionary theorising, he and Wallace unknowingly crossed paths for the first time. It would be twenty years before they would finally be thrust inextricably together.

Origins and Species

It is important not to fall into the tempting but incorrect story that "the origin of species" was a mystery that investigators had pondered for centuries or millennia before Darwin and Wallace cracked it. The question of where new species come from is not a timeless one that has always been around. There are no timeless questions. The things people wonder about and believe are very specific to their time and place. Reading things from completely different times and contexts as if they were in conversation with one another leads to an imaginary history of disembodied ideas that is as unreal as it is inaccurate.

Curiosity about the origin of species in the modern sense arose at the end of the 18th century when the great French anatomist Georges Cuvier finally demonstrated that some species had gone extinct. It became universally accepted by these mostly Christian men of science that the earth was immeasurably, incalculably ancient. Geologists across Europe found that the world had seen many, many eras of life come and go, and that each of these was characterised by different fossils.

The many layers of the earth's rocks showed clearly that species vanished after a time and never reappeared. In the rocks above them new species appeared. Where did the new ones come from? What was their origin? One possibility was that new species had somehow been created after great catastrophes or floods. The prevailing view, but by no means the only one, was that some form of divine creation was needed to explain how the new species had appeared.

Perhaps the most profound realisation that gradually emerged was that the fossil record was progressive. The most primitive creatures were found in the oldest rocks, at the bottom. During the ensuing history of the earth, more "advanced" creatures appeared — first primitive shells, then bizarre-looking armoured fish, then amphibians and reptiles, then birds and then mammals.

But the mammals were of unfamiliar extinct types. Only in the very newest rocks were there fossil mammals and birds closely resembling those alive today. Nowhere in the record were any fossil humans or traces of human existence to be found. So, there was little difficulty accepting that these ancient ages of the earth existed before the creation of man. Perhaps, some supposed, that is why these ancient eras and extinct creatures were not mentioned in the Bible. Perhaps the Bible was only concerned with the last of a series of creations, the one in which human beings appeared.

The other incentive for curiosity about species origins came from the living world and from the daunting puzzle of how to organise and categorise it all. The binomial system of the great Swedish naturalist Carl Linnaeus was a good start. But how might an arrangement according to common features show how species were *really* related?

There were many proposals. As the number of extinct and living species known increased vastly beyond what could have been imagined a generation before, the questions became more specific and detailed to particular groups and then sub-groups or particular regions of the earth. A big picture was quite naturally lost in the dazzling array of detail coming in.

One big-picture system was notoriously put forward by a colleague of Cuvier, Jean-Baptiste Lamarck (1744–1829), curator of invertebrates at the Paris

Museum of Natural History. Ever since Darwin's day, Lamarck's views have been known more in a caricatured form than anything resembling what he actually wrote.[40] Today, the situation is even worse because since about 1900, the term "Lamarckian" has come to mean "inheritance of acquired characteristics". The classic example is the giraffe's long neck. How did it get like that? Supposedly by stretching to reach leaves on the highest branches, making its neck a little longer. And this acquired characteristic was inherited by its offspring. And so on. This definition of Lamarckism is found in countless textbooks.

In fact, "inheritance of acquired characteristics" is an entirely misleading shorthand for Lamarck's theory. First of all, it was not invented by or unique to Lamarck. Virtually all naturalists in the 19th century believed in some form of "inheritance of acquired characteristics", including Darwin. Lamarck's real target was Cuvier's demonstration of extinction.

According to Lamarck's theory, fossils are not truly extinct because they will come again. For Lamarck, there was an inherent tendency for life to progress towards greater complexity. Starting from the spontaneous generation of life from mud or slime, this generation would slowly evolve up the great chain of being. In response to conditions in the local environment however, to adapt, the "inheritance of acquired characteristics" would act. The reason there are no mastodons in the world today is that their lineage has moved on beyond that stage, to modern elephants.

How to explain that there are still primitive creatures in the world? These were the result of more recent spontaneous generations of life. It was still going on. Hence, life on Earth was not a branching tree as Darwin and Wallace would later describe it, but a series of parallel lines at different stages of upward advance. Human beings were the highest point yet reached, and probably the culmination of life on Earth.

But taking on Cuvier was dangerous business. Over the course of his distinguished career, Cuvier became perhaps the most eminent man of science in the world. And he also left crushed rivals in his wake like the strangled snakes of Hercules. Cuvier's point in demolishing Lamarck was not so much to attack evolution (and certainly not to defend special creation), but to defend the fact of extinction and what he saw as proper science. Nevertheless, the unscientific smear of Lamarck's theory emanating from that greatest man of science tarred any even apparently similar views for decades.[41] This was not Richard Dawkins' version of history in which religion or the Bible somehow made these questions forbidden.

Darwin Gets to Work

When Darwin set out on the voyage of the *Beagle,* he accepted that species were permanent like almost everyone else. But he was repeatedly intrigued by a number of puzzling findings such as unearthing extinct creatures and some curious patterns of animal and plant distribution. It was hard to explain them. There was no direct line of discoveries or insights of genius that led to "the solution". Darwin, as we will see with Wallace, went through a series of hypotheses, false starts and interim theories before formulating natural selection.

Contrary to popular legend, Darwin experienced no eureka moment on the Galápagos Islands and did not become an evolutionist there.[42] About a year later, as the *Beagle* was sailing back to England, Darwin began to prepare a list of notes on his bird collection for an expert ornithologist to describe. The Galápagos birds obviously resembled those of South America, although the environment was totally different. As a geologist, Darwin understood that the islands had erupted out of the sea in the geologically recent past. He noted how the mockingbirds on different islands seemed to be of different types. But the environments were the same. Why weren't the birds the same? He penned his first doubt that species were unchangeable: "If there is the slightest foundation for these remarks [that the birds really were different types on different islands], such facts would undermine the stability of Species."[43]

Back in London, Darwin scribbled away in his lodgings. He collated his materials to generate a theory for life on the model of a great geologist of the day. Darwin swung back and forth between examining the long term and short term. He started by asking himself some very basic and fundamental questions about living things. Why don't organisms just live forever? He concluded that if they did, injuries or mutilations would accumulate and be passed on to offspring. Any species that did that would become spoiled with imperfections and die out. Therefore, short lifespans act as a filter to remove the accumulation of inherited injuries. From what he had read and heard from others, he was convinced that a species, like an individual, had some sort of an inbuilt lifespan. But if it or some of its members changed into a *new* species, this would start a new clock and so life could continue indefinitely.

He had collected two species of rhea in Patagonia whose ranges abutted. Maybe they descended from a common parent. Darwin first thought that "if one species does change into another it must be per saltum" or "at one blow" — what is now called saltational speciation. Saltum is Latin for "jump".[44]

How have living things developed in such a way as to form the patterns that taxonomists had found for the past century? All groups and species fit together in nesting relationships, groups with sub-groups, and each sub-group having its own sub-groups. Some had close similarities, others had big gaps between them. Why not just lots of different types that didn't fit together in this way?

If the process of life was always parent to offspring, maybe geographical isolation sometimes made some diverge away from ancestors. This would make sub-types from a common starting point. The irregular extinctions of an ever-changing world would make the gaps. Then, Darwin sketched the first tree of life diagram in his notebook. This branching process mapped one-to-one on the nested taxonomic system like a superimposed drawing from a camera lucida (see p. 213, Chapter 8).

But how did organisms come to fit their environments and way of life? It was not until 28 September 1838 that Darwin read the Rev. Thomas Malthus' *Essay on the principle of population* (1826) "for amusement" as he later recalled.[45] But, as Janet Browne has written, he was "clearly following up lines of inquiry relating to individual variation, averages, and chance, as well as seeking information on human population statistics".[46]

Malthus (1766–1834) was the first professor of history and political economy at the East India Company College, Haileybury (1805–1834). Malthus argued that the utopian philosophies of Nicolas de Condorcet and William Godwin were wrong because humanity could not inevitably progress. The population could grow geometrically, so it would necessarily outstrip food production. For example, in four generations there would be an increase from 2 to 4 to 24 to 96 and so forth, whereas food production was believed to increase only at an arithmetic rate of 1, 2, 3, 4, 5, etc. Hence, in the absence of severe checks to population growth such as war, pestilence, famine and so forth, disaster loomed.[47]

The argument of Malthus ignited Darwin's imagination to boil down his many ideas into a more succinct principle, what he later called "natural selection". Why is the world not overrun by one type of orchid, insect or frog given how many seeds or eggs they produce? Their numbers must be kept down by checks — "the warring of species". Countless thousands did not make it, only a few survive long enough to reproduce. But which ones?

Darwin imagined the convulsing cosmos of living things all over the globe — all reproducing at a fantastic rate, and almost all being ruthlessly pruned back — devoured, starved or destroyed. The numbers proved this had to be true. The bursting outward force of reproduction was checked by the carnage of ingestion and death. These two opposing processes were like a war that never ends. Yet, those with the right stuff to slip through the gauntlet and survive would pass on their characteristics to their own offspring. Thus, small changes could gradually accumulate to become larger changes over many generations. The result would be the change of species over time. Adaptations that would help suit a new environment would emerge from this otherwise destructive feature of nature.

Every part of every organism was likely to vary — if one examined individual organisms carefully enough. Hence, every feature was varying hither and yon constantly. There was an endless supply of variety. If circumstances were such that one of these happened to benefit its possessor, then it would get through the filtering process of natural selection and be preferentially passed on.

Darwin outlined this process in a famous passage in which he compared the possibility space of organisms to a surface covered with sharp wedges. "One may say there is a force like a hundred thousand wedges trying [to] force into every kind of adapted structure into the gaps in the oeconomy of Nature, or rather forming gaps by thrusting out weaker ones. The final cause of all this wedgings, must be to sort out proper structure & adapt it to change." But adaptation was relative; "chance & unfavourable conditions to parent may be become favourable to offspring".[48]

In October 1838, he opened another notebook and summarised his theories to date:

> Three principles will account for all
> (1)   Grandchildren like grandfathers
> (2)   Tendency to small change especially with physical change
> (3)   Great fertility in proportion to support of parents[49]

At the same time as his work on the *Beagle* collections, Darwin also "thought much upon religion". It was during these years of work and intense reading and theorising that he came to disbelieve in Christianity and divine revelation. There was simply no evidence. He downgraded to a deist. He still believed in a supernatural creator that had established the laws of nature in the first place, but as far as Darwin was concerned, nature worked according to natural laws.

Eventually, his notes began to resemble his final theories, "there is a contest & a grain of sand turns the balance", "if a seed were produced with infinitesimal advantage it would have better chance of being propagated". Just as animal breeders chose only some animals to breed from, and thereby drastically transformed plant and animal breeds, so in nature the fact that only a few could survive could transform species, "domesticated races...are made by precisely same means as species".[50]

His notebooks end full of questions and lines of new enquiry. A vast research programme would be needed to answer them. As he later wrote, "I had at last got a theory by which to work."[51] He mentioned it to his cousin William Darwin Fox, "It is my prime hobby & I really think some day, I shall be able to do something on that most intricate subject species & varieties."[52] This is the earliest known statement that Darwin intended to publish his species theory someday.

A. R. Wallace, Surveyor

In the summer of 1837, as Darwin scribbled about species, Wallace joined his freethinking brother William as an apprentice land surveyor, first in Bedfordshire. The Tithe Commutation Act was passed the year before. It replaced the ancient system of the payment of tithes in kind with monetary payments based on the average value of tithable produce and productivity of the land. The valuation process required accurate maps. It was an excellent time to be a surveyor. Over the next six and a half years, the Wallace brothers moved from place to place carrying out surveying jobs, staying mostly in inns and lodging houses. Wallace liked the instruments of surveying and the mathematics involved was intellectually stimulating. He began to read about mechanics and optics. His days in the open air of the countryside led him to an interest in natural history.

In January 1839, there was a drop in demand for surveyors so William sent Wallace to live with another surveyor named Matthews who had a watch- and clock-making business at Leighton Buzzard, Bedfordshire. Wallace, now sixteen, did not like quiet indoor watch repair. Fortunately after nine months, Matthews moved on to a job in London and Wallace returned to the care of his brother William. Wallace was later grateful that "circumstances allowed me to continue in the more varied, more interesting, and more healthy occupation of a land-surveyor".[53]

In the autumn of 1839, the brothers made their base at Kington, Herefordshire, where they worked under contract for a firm of surveyors and estate agents. The Kington Mechanic's Institution had opened earlier that year. As in London, Wallace became an active participant. The historian of science Jack Morrell pointed out how provincial societies like this were "an opportunity for publication, social legitimation, and the satisfactions of power, vanity, and emulation were all to be savored". After all, science was one option for those "aspiring to the status of the cultivated English gentleman". A magazine of the time recommended "without hesitation" "the study of Natural History" to expand one's mind and to elevate oneself socially.[54] It was considered morally uplifting and socially safe.

Still only eighteen, Wallace wrote a five-page article recommending science as one of the best means of conducting the institution. The essay contains a list of recommended works which may offer a clue to some of his earliest scientific reading, by all accounts an impressive list for an eighteen-year-old: "the Natural History and Natural Philosophy, Volumes of the Cabinet Cyclopædia, Loudon's Encyclopædia of Agriculture, Plants, &c., Lyell's Geology, Murchison's Silurian System, Kirby and Spence's Entomology, Lindley's Natural System of Botany, Brande's Chemistry, George Coombe's Constitution of Man, and other works— Dr. Channing's works, Humboldt's Personal Narrative, &c. &c."[55]

In late autumn 1841, the Wallace brothers left Kington for the town of Neath on Swansea Bay. For Wallace, it was a "turning-point of my life" because the freedom from steady work that followed allowed him to read more widely about science.[56] He took up an amateur pursuit of botany although he had no one to guide or encourage his nascent interests. He bought his first work on natural history, a shilling paperback by the Society for the Diffusion of Useful Knowledge. He could not recall the name of the book, but it may have been Lindley's *Vegetable physiology*. Wallace carried this thin booklet about with him to identify plants and flowers and learnt how to identify and classify where a plant belonged in the system of classification. He felt a tingle of excitement as he realised there was some order underlying the apparent chaotic diversity of plants.

But many plants could not be found in his introductory pamphlet so he looked for something more detailed. He recalled finding one in advertisements. If his booklet was by Lindley, this may explain why Wallace chose the latest edition of Lindley's 292-page textbook *Elements of botany* at the high price of 10s. 6d.[57] As so often in later life, Wallace took the plunge without checking first. He was disappointed to find that the book was a systematic textbook of world botany and not a British field guide. He supplemented the book with notes copied from Loudon's "Encyclopædia of Plants".[58] Wallace began a collection of dried plants so that he could accurately record which ones he had already collected. "I first named the species as nearly as I could do so, and then laid them out to be pressed and dried. At such times I experienced the joy which every discovery of a new form of life gives to the lover of nature."[59]

In Kington and Leicester, Wallace read some very influential works for his future life. Among these were two famous books which introduced him to the exciting prospect of scientific travel: Darwin's *Journal of researches* and the great German naturalist Alexander von Humboldt's *Personal narrative of travels to the equinoctial regions of the New Continent* (1819–1829). Humboldt referred to "the system of Malthus" several times. So it may have been through this or other connections that Wallace also read Malthus around this time.

The most influential work for Wallace's budding scientific mind was the geologist Charles Lyell's (1797–1875) *Principles of geology* (1830–3).[60] Lyell offered not just a new geology but a new way of understanding nature. And Lyell was everything Wallace admired, a gentleman naturalist and a respected member of the scientific community who debated the great questions of the history of the earth, natural causes and species.

Born to a wealthy family that owned an estate on the edge of the Scottish Highlands, Lyell was as socially respectable as could be. From 1816 to 1819, he attended Exeter College, Oxford, where he took the courses on geology of the great William Buckland. Lyell soon became part of British geology's inner circle and member of the Geological Society of London while still a student. He travelled widely, studying in particular the great Italian volcanoes of Vesuvius and Etna. He found evidence of gradual and successive formation where lava flows had overflowed one another. In stratigraphy, he showed how some of the most recent layers could be differentiated by their distinct fossil marine shells.

His *Principles of geology* was a synthesis of the already sophisticated geology and palaeontology of the day and added his own ambitious spin now characterised as "uniformitarianism". Textbook histories represent Lyell as a "uniformitarian" in debate with another school called "catastrophists" — those who supposedly thought that the history of the earth was shaped only by great eruptions and cataclysms instead of the more mundane or uniform natural causes proposed by Lyell. In fact, the issue that distinguished Lyell, whom the historian of geology Martin Rudwick describes as the only "uniformitarian", was scale. Other geologists already accepted observable causes like erosion, earthquakes and volcanoes. Lyell's radical point was that the *intensity* of modern events must have been the same throughout the history of the world. His contemporaries were highly critical of this blanket assertion.

But it was an influential part of Lyell's scheme. Lyell felt that uniformity of law had to be assumed to make geology truly scientific.[61] Small, slow, gradual and cumulative effects over immense periods of time could produce large change to the surface of the earth. Forces in the real world such as earthquakes, running

Frontispiece to Lyell's *Priniciples of geology*, 6th edn., 1840.

water and volcanoes must have always acted just as they do now. Natural, visible and non-miraculous causes should be sought to explain natural phenomena.

The frontispiece to Lyell's book was a perfect example. It depicted the ruins of an ancient Roman temple near Naples (actually it was a marketplace). When it was excavated, lava rock covered the base up to a line of holes that can be seen as a dark band in the middle of the columns. The holes were made by marine organisms. The ruins showed that since Roman times, the land on which this structure stands had slowly subsided so that the columns were partly submerged. Later, the land had risen again so that the ruins were now above the sea. All of this had happened gradually enough that the columns had not fallen down — and all since Roman times.

The other way in which Lyell's book was controversial was his denial that the earth's history was directional and that life had progressively appeared. Instead, Lyell argued for a steady-state model of an essentially eternal earth. The imperfect state of the evidence only made it look as if the earth's and life's history was progressive.

Through a masterful exegesis of the palaeontological evidence, Lyell argued for a "gradual birth and death of species". Instead of a series of Cuvierian large-scale creations and destructions, there was a continuous state of change. As the world slowly changed, the species living in a particular area would gradually become unsuited to their environment. Since they could not change beyond fixed limits, their numbers would dwindle and the species would die out. New species were then somehow formed to fit the new environment. This happened piecemeal throughout the history of the world. But because the time scale was so great, it was unlikely that we would ever witness the formation of a new species.

Lyell spent considerable space reviewing and critiquing Lamarck's theory of evolution. Lyell could not accept evolution, particularly as it threatened, in his view, to degrade and bestialise mankind. Although Lyell was more hostile to evolution in his book than in his private letters, he was most obviously a figure of liberal modern science. He stood for the power of natural causes and the rule of natural laws as opposed to the old-fashioned biblical interpretations of previous generations of geologists. Here, Wallace's mechanics' institute education overlapped with Lyell. Wallace probably placed Lyell above all other scientific writers in the genteel pantheon of contemporary science. *Principles of geology* was highly respected and admired. Lyell was a thorough insider who made, within the realm of the acceptable, innovative pushes towards new ideas. He was knighted in 1848.

As Wallace worked in the outdoors as a land surveyor, Darwin continued his *Beagle* publishing projects and species theory in his London lodgings. Lyell and

Darwin became good friends. Darwin admired Lyell as his scientific inspiration and Lyell enjoyed the substantial support of the up-and-coming Darwin, his most sincere convert. Darwin's unpretentious enthusiasm for natural puzzles, such as the way that the mysterious ring-shaped coral atolls were formed in the Pacific Ocean, enthralled Lyell. The gradual upward growth of the reef encircling an island kept pace as the ocean bed slowly subsided. Eventually, the island disappeared beneath the waves, leaving only the encircling reef. Lyell had proposed that coral atolls were growths on top of slightly submerged volcanoes.

Lyell would get carried away during these animated conversations. "On such occasions, while absorbed in thought, he would throw himself into the strangest attitudes, often resting his head on the seat of a chair, while standing up." This gives new meaning to the old metaphor of turning a thinker's theories on their head. The freedom from embarrassment showed how close the two men had become, possible partly because of their almost equal social rank. Otherwise, Lyell "had a strong sense of humour and often told amusing anecdotes. He was very fond of society, especially of eminent men, and of persons high in rank; and this over-estimation of a man's position in the world, seemed to me his chief foible. He used to discuss with Lady Lyell as a most serious question, whether or not they should accept some particular invitation."[62]

Darwin married his cousin Emma Wedgwood in 1839 and by 1842, after finishing his narrative of the *Beagle* voyage and most of his specimens were safely in the hands of experts, he was able to move with his growing family to a former parsonage in the more congenial village of Down in Kent. He would live there for the rest of his life.

Wallace in Leicester

In 1843, Wallace's father died. He was buried in the family vault at Saint Andrew's Church, Hertford. With a decline in the demand for surveyors, William no longer had enough work for his younger brother. His apprenticeship ended in December. After a brief period of unemployment in early 1844, Wallace, although barely qualified, was employed as a teacher or "English and Drawing Master" at the Collegiate School in the industrial city of Leicester. His fine grammar school education paid off. The job left him with plenty of free time for private study.

It is hardly surprising that, as a young man interested in science and reading works on the most intriguing scientific questions of the day at the Leicester town library, Wallace met another budding young naturalist, an enthusiastic entomologist named Henry Walter Bates. Just three years his junior, Bates was from a family in the hosiery business and attended school to the age of thirteen. Bates introduced Wallace to his next scientific pursuit: insect collecting, particularly beetles. Wallace's first scientific publication therefore was, like Darwin's, a record of a beetle capture.[63]

Other lifelong interests Wallace encountered in Leicester were two of the great Victorian movements usually dismissed today as pseudosciences: phrenology and mesmerism.[64] Phrenology began as the doctrine of an eccentric Viennese physician named Franz Joseph Gall in the 1790s.[65] Gall was an early theorist of cerebral localisation. Rather than seeing the brain as a largely undifferentiated single organ, Gall believed that the various faculties of the immaterial mind were tied to specific physical areas or organs of the brain. The physical size of each organ should be an indicator of its power or preponderance in the individual's personality and abilities. Furthermore, as the shape of the skull was believed to take its shape from the underlying brain, Gall concluded that the shape of the skull was an accurate indicator of the relative powers of the faculties of the mind. Although the personal divination of head-reading was only one facet of Gall's system, it was the facet that attracted the most attention.

Gall's assistant was a young German physician named Johann Gaspar Spurzheim. Spurzheim left Gall in 1813 and set out to imitate his former master's success with a lecture tour of Britain. But Spurzheim blended Gall's system with an enlightenment philosophy of natural laws to create quite a new doctrine. Spurzheim's modified version became phrenology in the English-speaking world. Phrenology developed most actively in Edinburgh amongst middle-class professionals personally converted by Spurzheim such as lawyers and medical men from the 1800s to 1820s.[66] This context included an eighteen-year-old medical student named Charles Darwin. Darwin picked up a bit of phrenology there too, although his "very little belief of it" was later "entirely battered down".[67] But Darwin, like Wallace, used phrenological language for the rest of his life; "rational organs" or the "moral" and "intellectual" "faculties" are found throughout the *Descent of man* (1871) and *Expression of the emotions* (1872).

Mesmerism was founded by another German physician, Franz Anton Mesmer.[68] This doctrine was also very popular in the 1840s and involved influencing other people by means of an invisible magnetic fluid. Hence the phrase "animal magnetism". Some of what mesmerism involved is similar to modern hypnosis. Sometimes patients could be mesmerised and endure surgery — apparently an astonishing confirmation of the reality of the phenomena in the era before anaesthesia. Wallace and Bates attended a presentation on mesmerism by the itinerant lecturer Spencer Hall.[69] The German industrialist and social theorist Frederick Engels also attended Hall's lectures and left an account that has been previously overlooked:

> [Hall] was a very mediocre charlatan…The lady was sent into a magnetico-sleep and then, as soon as the operator touched any part of the skull corresponding to one of Gall's organs, she gave a bountiful display of theatrical, demonstrative gestures and poses representing the activity of the organ concerned; for instance, for the organ of philoprogenitiveness she fondled and kissed an imaginary baby, etc.…The effect on me and one of my acquaintances was exactly the same as on Mr. Wallace; the phenomena interested us and we tried to find out how far we could reproduce them.[70]

Wallace experimented by mesmerising some of his Leicester students, to cause rigidity of the limbs, a trance state, suggestion, as well as phrenomesmerism — eliciting the behaviour of a phrenological organ by stimulating it mesmerically with passes of the hands. As Wallace wrote in his autobiography: "The importance of these experiments to me was that they convinced me, once for all, that the antecedently incredible may nevertheless be true; and, further, that the accusations of imposture by scientific men should have no weight whatever against the detailed observations and statements of other men, presumably as sane and sensible as their opponents, who had witnessed and tested the phenomena."[71]

This was perhaps the earliest instance of Wallace's lifelong characteristic of becoming convinced by a few coincidences that something was true and then never again doubting it or losing his belief. The fact that his mesmerised subjects were familiar with the phrenological map of the head, for example, never entered his consideration as an alternative explanation for their actions. Engels, who later read Wallace's account, concluded:

> While we with our frivolous scepticism thus found that the basis of magnetico-phrenological charlatanry lay in a series of phenomena

which for the most part differ only in degree from those of the waking state and require no mystical interpretation, Mr. Wallace's "ardour" led him into a series of self-deceptions, in virtue of which he confirmed Gall's map of the skull in all its details and noted a mysterious relation between operator and patient. Everywhere in Mr. Wallace's account, the sincerity of which reaches the degree of naiveté, it becomes apparent that he was much less concerned in investigating the factual background of charlatanry than in reproducing all the phenomena at all costs. Only this frame of mind is needed for the man who was originally a scientist to be quickly converted into an "adept" by means of simple and facile self-deception. Mr. Wallace ended up with faith in magnetico-phrenological miracles and so already stood with one foot in the world of spirits.[72]

Wallace in Neath

Wallace's brother William died suddenly of pneumonia in March 1845. Wallace left the school at Leicester to take over William's surveying firm in Neath, together with his brother John.[73] The business did not succeed. Wallace next worked for a few months as a surveyor for a proposed rail line. Then, he and John attempted to establish an architectural firm which produced a few projects such as the building for the Mechanics' Institute of Neath which still stands. One of the directors of the Institute invited Wallace to give lectures on science and engineering. In late 1846, Wallace and his brother John bought a cottage near Neath where they lived with their mother and sister Fanny. Historian R. E. Hughes discovered that Wallace was a curator of the museum library at Neath, something Wallace strangely omitted from his autobiography.[74]

Two itinerant phrenological lecturers came through Neath from 1845–6. The lecturers, the pro-mesmerism "Professor of Phrenology" Edwin Thomas Hicks (bankrupt in December 1846) and his rival, the more up-market and ambitious James Quilter Rumball, used different systems. Hicks used George Combe's orthodox list of thirty-five phrenological organs, but Rumball had made his own "discoveries" and therefore counted thirty-nine. Wallace attended their evening lectures on the amazing power of the new science of the mind. Afterwards, he queued up with a handful of others to have his head read or "delineated". It is a common misconception that phrenological readings required head shaving.[75] Although a few earnest individuals such as the novelist George Eliot did so, almost all phrenological readings were done without shaving. Phrenologists, eager to encourage as many (often paying) people as possible to have their heads read, were adamant that shaving was unnecessary. The skilled hands of a phrenologist could discover all the bumps and dents through hair.

Wallace could not afford a ten-shilling essay-length character analysis, so he made do with the five-shilling character sketch.[76] Both Hicks and Rumball agreed verbatim on one thing in their readings of Wallace, his lack of "self-confidence". No doubt this was as apparent in Wallace's demeanour as in his cranial bumps of Self-Esteem and Love of Approbation. For Wallace, this accuracy "confirmed me in the belief that the science is a true and important one".[77] Wallace later continued his phrenomesmerism in Brazil, but there is no trace of any of it throughout the Eastern Archipelago voyage. Only after his voyage does further interest in phrenology appear in Wallace's writings.

In 1845, Wallace read a recent controversial work, the mysteriously anonymous *Vestiges of the natural history of creation* (1844).[78] Even as an old man, he could still remember "the excitement caused by the publication of the *Vestiges*, and the eagerness and delight with which I read it".[79] Hughes uncovered a lecture at the Swansea Literary and Scientific Society and subsequent discussion in the area that seems very likely to have been the context in which Wallace first heard of and read *Vestiges*.

Vestiges convinced Wallace of evolution — of a sort anyway. It was not the branching genealogical tree of life that Darwin and Wallace would announce thirteen years later. *Vestiges* grew out of a very different background from Lyell's *Principles* or Lamarck's philosophical transmutation. *Vestiges* sprang from the same soil as another book which "greatly interested" Wallace in these years, the cantankerous Scottish lawyer and phrenologist George Combe's *Constitution of man* (1828).[80]

Cheap, mass-produced works like Combe's were another way in which steam technology was revolutionising the age. Just as steam transformed the major industries of textiles and iron and the railways connected cities and ports, steam-powered paper manufacture and printing made the mass production and dissemination of books and magazines one of the most important innovations of the Victorian era. Although now forgotten, Combe's *Constitution of man* was produced in vast numbers in cheap editions and became one of the best-selling works of its genre of the century.

Nowadays, the *Constitution of man* is usually described as a phrenology book, but this is a mistake. True, Combe was the leading phrenologist in Britain, and *Constitution of man* included a chapter on phrenology, but the book is actually about a "doctrine of natural laws". Combe's phrenological laws of mind were described as the most recently discovered laws of nature. Combe elaborated a system of hierarchically arranged natural laws: physical, organic and moral. These three classes mapped onto man's constitution as described by phrenology. By combining these with a "law of hereditary descent", Combe argued that the human race would ascend the scale of improvement in organic and mental spheres.[81] *Constitution of man* was thus a self-help guide on how to live a good life. The trick was to study and submit to the laws of nature, rather than religion. Combe's book was hugely controversial, on one occasion it was even publicly burnt!

The evolutionary *Vestiges* was secretly written by a Scottish publisher and friend of Combe named Robert Chambers. *Vestiges* was an attempt to supply a scientific big picture for how the world worked according to the findings of modern science. Established men of science were not providing one. Following *Constitution of man*, *Vestiges* argued that nature operated according to natural laws, and that a fundamental outcome of the way the laws worked was progress. But *Vestiges* called it "development", meaning part of what we would now call evolution, not just living things, but all of nature. Solar systems developed from swirling clouds of interstellar dust or "fire-mist", the geological processes of the earth had developed, preparing the world for life and finally living things arose, these too undergoing a process of progressive development — new features in organisms appeared as a reaction to environmental changes. Organisms produce offspring like themselves. But sometimes, according to an even higher natural law, they would produce an offspring of a higher type. Over time, this led to life progressing ever upwards. Invertebrates, fish, reptiles, mammals and finally humans had all followed in succession, and *Vestiges* hinted that something higher would follow us.

Natural progress was just as applicable to human mental faculties as organic ones. Therefore, a "doctrine of natural laws", rather than religion, would lead to future scientific and social progress. By religion, Combe and *Vestiges* were mostly concerned with their conservative evangelical opponents in Edinburgh. These progressive themes appeared again and again in Wallace's later writings as he adopted much of the rationalist, sceptical and naturalistic outlook of his Owenite working-class environment with an optimistic faith in physical and social progress through the unimpeded operation of beneficent natural laws.[82] Even at the end of his life, he said in an interview, "Give the people good conditions, improve their environment, and all will tend towards the highest type."[83]

All of this contributed to Wallace accepting that species were not fixed but could change, from 1845 onwards. However, there is and was no homogeneous idea of evolution. There were many very different ideas of biological change. The genealogical descent and branching pattern does not appear in Wallace's writings until the mid-1850s.[84] Wallace was clearly open to new and radical ideas: phrenology, mesmerism and evolution. If the expectations that might have been predicated from Wallace's family background had financially evaporated, intellectual and scientific expectations beckoned instead.

Darwin's Drafts

By the time Wallace became interested in the evolutionary theories of *Vestiges*, Darwin had developed his own theories in a series of notebooks and draft essays. The first short sketch was written after he completed his book on coral reefs in 1842. By February of 1844 when Darwin finished one of his three volumes on the geology of the *Beagle* voyage, and before starting the next book, Darwin enlarged and improved his 1842 pencil sketch into a 230-page essay in ink. He then wrote the famous memorandum to his wife on 5 July 1844 asking her to have the sketch published in case of his sudden death.

Countless writers have claimed that Darwin wanted to publish his theory *only* after his death. Historian James Moore even claimed that the essay *was* written for publication, but that "it was only over his dead body that [Darwin] was going to allow these things out in 1844". Bill Bryson and Rebecca Stott wrote that Darwin "locked it away in a drawer". And now it is even claimed that Darwin hid the essay under the stairs.[85] But one has merely to read the memorandum to see how wrong these are.

> I have just finished my sketch of my species theory. If, as I believe, my theory in time be accepted even by one competent judge, it will be a considerable step in science. I therefore write this in case of my sudden death, as my most solemn and last request…that you will

devote 400 pounds to its publication....I wish that my sketch be given to some competent person, with this sum to induce him to take trouble in its improvement and enlargement.

He went on to remark that if necessary, another hundred pounds should be offered "as the correcting and enlarging and altering my sketch will also take considerable time" as it was written "from memory without consulting any works, and with no intention of publication in its present form".[86] Darwin went out of his way to say that the essay was not intended for publication, and only in case of his "sudden death" was it to be published — not after he died as an old man in many years. And yet this is the memorandum that is endlessly referred to as Darwin asking his wife to publish his theory only after his death!

This 1844 essay was sent to the village schoolmaster to be copied out in finer handwriting.[87] We are also endlessly told that Darwin put aside the essay because of the fiery reception the evolutionary *Vestiges* received about this time. But *Vestiges* was published in October 1844, and Darwin had already continued with his *Beagle* publications in September. This was the final reference to his species work in the journal where he recorded his work for ten years, until he had finished his next projects on marine invertebrates, especially barnacles, in 1854.[88]

So in July 1844, Darwin had written a lengthy essay outlining his theory of evolution as it then stood, "with no intention of publication in its present form". And publication, intended at some unspecified future date, would first require "considerable time" for "correcting and enlarging and altering".[89] The fair copy was intentionally left with plenty of margins and empty pages for further revision. This was not a finished work.[90]

As he ploughed on with his main occupation, the *Beagle* publishing programme, species remained a "far distant work".[91] In February 1845, he wrote to his old friend, the clergyman naturalist Leonard Jenyns, "It will be years before I publish [on species], so that I shall have plenty of time to think of better words."[92] There was no rush.

In early September 1845, Darwin's correspondent, the status-conscious and prickly botanist Joseph Dalton Hooker (later to become Darwin's best friend), criticised a speculative work on species by a French botanist: "I am not inclined to take much for granted from anyone [who] treats the subject in his way and who does not know what it is to be a specific Naturalist himself."[93] Darwin, aware of Hooker's "peppery" temper, responded that this remark could also include himself.

How painfully (to me) true is your remark that no one has hardly a right to examine the question of species who has not minutely described many. I was, however, pleased to hear from Owen (who is vehemently opposed to any mutability in species) that he thought it was a very fair subject & that there was a mass of facts to be brought to bear on the question, not hitherto collected. My only comfort is, (as I mean to attempt the subject) that I have dabbled in several branches of Nat. Hist: & seen good specific men work out my species & know something of geology; (an indispensable union) & though I shall get more kicks than half-pennies, I will, life serving, attempt my work.[94]

Passages like this are often interpreted by those looking for reasons for Darwin's purported postponement as confessions that fear of critical reaction must have held him back. Yet, Darwin explicitly stated *twice* that he would publish despite anticipated criticism. "All which you so kindly say about my species work does not alter one iota my long self-acknowledged presumption in accumulating facts & speculating on the subject of variation, without having worked out my due share of species. But now for nine years it has been anyhow the greatest amusement to me."[95] In another letter of November 1845, Darwin wrote, "I hope this next summer to finish my S. American geology; then to get out a little Zoology & hurrah for my species-work, in which, according to every law of probability, I shall stick & be confounded in the mud."[96] Darwin actually finished *Geology* in October 1846. If his "little Zoology" took him another year, this estimate might mean he expected (in late 1845) that he would commence species by the beginning of 1848. If he then spent five years working on it as he supposed, publication would result in 1853 — just before Wallace would set out for the Eastern Archipelago.

The Amazon Expedition, 1848–1852

In 1848, Wallace and Bates were inspired by the recent book by American traveller William Henry Edwards, *A voyage up the River Amazon*. If they could make their way cheaply to Brazil, they could make a good living from natural history collecting. They engaged a man named Samuel Stevens to act as their agent.

Stevens would become one of Wallace's most important friends. Born in 1817 in Middlesex, Stevens' poor health in childhood meant his parents encouraged his interests in natural history collecting.[97] He soon specialised in insects,

Samuel Stevens. Royal Entomological Society.

then a growing craze which Darwin was pursuing at Cambridge. Stevens became a pioneer of collecting insects at night and encouraged others to do so.[98] He became a member of the Entomological Society of London in 1837 and served as treasurer and member of its council. In 1840, he joined his brother John Stevens as a partner in the Stevens Auction Rooms on King Street, Covent Garden, London. In 1848, Samuel opened his own Natural History Agency at 24 Bloomsbury Street, Bedford Square, London, across from the British Museum. Stevens' agency sold natural history specimens to collectors and institutions such as the British Museum. He advertised his items for sale in the 1850s and 1860s in *Transactions of the Entomological Society*, *Hooker's journal of botany*, *Zoologist*, *Athenaeum*, *Lancet*, *Notes and queries* and *Household words*.

The offerings were incredibly varied and as eclectic as a cluttered Victorian drawing room. They ranged from sets of microscope slides to eggs from the Bismarck Archipelago, plants from the Seychelles and Ceylon, antelope heads from Central Africa and insects from Britain, Africa, Rangoon, China, Japan, Venezuela, Brazil, Peru, Mexico and the Eastern Archipelago. He also sold the full range of collecting equipment and supplies. As an agent for collectors, Stevens charged a twenty-percent commission on the sales of specimens and another five percent on remittances of money and despatch of freight.[99]

So while Darwin was burying himself in barnacle studies, Wallace and Bates embarked in April 1848 on an adventure that would set the course for their future lives and careers. There would be no turning back. Wallace later described it as "a wild scheme, a journey to the almost unknown forests of the Amazon in order to observe nature and make a living by collecting".[100] Wallace's brother John emigrated to California in the spring of 1849 to chase the gold rush. His descendants still live there.

Wallace and Bates initially stayed in Para (Belém). After nine months of collecting Amazonian specimens together and the arrival of Wallace's brother Herbert, Wallace and Bates set out separately to cover different territory. Several writers have suggested some sort of dissension in the split of Wallace and Bates.[101] Sadly, Herbert died of yellow fever on his way back home in 1851. Collecting was not for him.

Wallace focused particularly on the Upper Rio Negro. He used his surveying skills to map the river and tributaries upstream; he claimed further inland than any previous European. During his time in Brazil, he published four pieces in periodicals. The principal results were great experience in collecting and an appreciation of the biogeographical boundaries, particularly broad rivers, that separated different species, such as monkeys. He also became proficient in Portuguese.

By 1852, Wallace had accumulated sufficient materials to set himself up in a modest rural home where he could spend years working on his collections. What he imagined was very similar to the middle-class rural ideal Darwin had enjoyed for the past ten years. But it was not to be. As Wallace was returning home aboard a very dilapidated sailing ship called the *Helen*, disaster struck. The ship caught fire in the mid-Atlantic and sank, destroying almost the entirety of Wallace's notes and personal collection. Fortunately, the collection had been insured by Stevens for £200. If Wallace collected any notes or material for his interest in the origin of species, none has survived, and he never referred to any in his later writings nor indeed to any work or progress on this subject. He returned to London on 5 October 1852.

Wallace's subsequent publications on the Amazon suffered from the dearth of data that survived the shipwreck. His first book, *Palm trees of the Amazon and their uses* (1853), described the distribution and uses of the palms he had observed and was illustrated from his own beautiful sketches — some of the few documents he saved from the *Helen*. Because of its scanty detail, inaccuracies in some of the drawings and sometimes amateurish descriptions, all resulting from his lack of experience as a botanist, the book was criticised by some contemporaries such as botanist Sir William Hooker.

Wallace's other book fared better. *A narrative of travels on the Amazon and Rio Negro* (1853), although also criticised for its scanty data, was better received and sold more copies. Wallace employed a somewhat overly genteel style, perhaps to emphasise that he was a gentleman of science — at least in style if not in means. He also read papers before scientific societies and made important connections in the London scientific community. This networking would prove absolutely crucial for his future.

Science for Sale

There were few opportunities to make a living from science in 19th-century Britain.[102] There were almost no university or industrial positions and government support was scandalously lacking. Specimen hunting was clearly appealing. The Victorian craze for collecting fossils, animals, insects, shells, plants and other natural history specimens fed and fuelled the trade.

It is hard to appreciate now just how exciting natural history was for the Victorians. Before the invention of colour photography, television or the Internet, the grandest colours and most intricately beautiful things human eyes could behold were objects of natural history. The exquisite wings of colourful butterflies or the bright screen-like landscapes visible under the microscope were dazzling and endlessly fascinating in a grey and dreary world. The expansion of European trade and exploration was funnelling unknown natural objects back to Europe at an unprecedented rate.

Collecting in the Amazon allowed Wallace to earn many times more money than he had ever had before. And it was a way of life that he enjoyed immensely. It also increased his stature. As he had lost much of his earnings, not to say his collections, from South America, it is little wonder that he decided to try again to establish himself financially and scientifically through another commercial collecting expedition overseas.

As Bates and others remained in the Amazon basin, Wallace wanted to explore new territory. He was advised that British cabinets were particularly

Ida Pfeiffer collecting in Asia. Lithograph by A. Dauthage 1840.

lacking in specimens from the Eastern Archipelago and hence it would be profit-able collecting ground. Even today, there are many websites offering specimens of the insects from this part of the world for sale, often at very high prices. But perhaps the most specific hint came from a forgotten figure who had recently travelled through the region. Stevens had recently heard of some unusually valu-able specimens from a lady traveller named Ida Pfeiffer.[103]

Ida Laura Pfeiffer might be called the dark lady of Wallace's Eastern Archipelago as she has been lost from the story as the biophysicist Rosalind Franklin, overshadowed by Watson and Crick, has been called the dark lady of DNA. Pfeiffer was born in Vienna in 1797, the only daughter amongst six sons of a wealthy merchant. Her father died in 1806 as the founder of phre-nology, Dr. Gall, was lecturing around Europe. In 1820, her mother con-vinced her to marry Dr. Pfeiffer, a Lemberg-based lawyer twenty-four years her senior. In 1835, Ida Pfeiffer and her two sons returned to live in Vienna while her husband remained in Lemberg where he died in 1838. In 1842, after her sons were grown and employed, Pfeiffer began to travel. She went first to the Holy Land and on her return in 1843 published an account of her travels that earned enough money to enable her to travel further afield, this time to Scandinavia and Iceland. Another book on this journey appeared in 1846.

In the same year, she undertook her first round-the-world journey via Brazil, Chile, Tahiti, China, Singapore, Penang, Ceylon, Persia and Greece. She returned home in 1848 and published a book about her travels in 1850. The novelty of a widow without fortune travelling around the world alone was intense and her book sold well.

In March 1851, she set out on her second circumnavigation, first to England, South Africa and then the Eastern Archipelago, visiting many of the islands Wallace later visited. At Sarawak, she was described as of "middle height, active for her age, with an open countenance and a very pleasant smile".[104] All the while Pfeiffer collected insects in Singapore, Sarawak, Java, Sumatra, Celebes, Amboyna and Ceram, some of which were sold by Stevens.[105] She returned via Australia, the United States, Peru, Ecuador, then the Great Lakes and was home in 1854, the year Wallace set out. Her second book of world travel was published in German and English in 1855.

Presumably, Stevens also introduced Wallace to his most lucrative connection, the wealthy insurance broker, insect collector and founding member and twice president of the Entomological Society, William Wilson Saunders. He contracted to buy Wallace's insects of all orders except Coleoptera and Lepidoptera. In the end, Saunders probably purchased well over 10,000 of Wallace's specimens.[106]

An expedition to the Eastern Archipelago was far more expensive than Brazil. So Wallace did some serious networking in preparation. He wrote to Sir James Brooke in March 1853 seeking assistance to visit Sarawak in Borneo. Remembered as "the White Rajah of Sarawak", Brooke (1803–1868) was one of the most colourful figures in the Eastern Archipelago.[107] He is supposedly the model for Rudyard Kipling's story *The man who would be king* and Joseph Conrad's *Lord Jim*. Born to British parents in India, Brooke spent time in the army serving in the first Anglo–Burmese war before resigning his commission after injury. When his father died in 1833, Brooke inherited £30,000. He bought a 142-tonne schooner, the *Royalist*, and sailed for Borneo in 1838.

Brooke and his crew helped the Sultan of Brunei to defeat an uprising and in 1842, Brooke was given the title of Rajah of Sarawak. Brooke established a smaller Singapore-style free trade port at Sarawak and sought to suppress rampant piracy. In 1847, Brooke was the British Consul General in Borneo and created a Knight Commander of the Order of the Bath (KCB). In answer to Wallace's appeal, Sir James forwarded instructions to receive and assist Wallace at Sarawak.[108]

Mrs Wallace, Fanny and Alfred, photograph by Thomas Sims c. 1853. Natural History Museum, London.

On 21 June, Wallace appeared before the prestigious Linnean Society of London to display his drawings of palm trees.[109] According to Wallace's later recollection, "I was introduced to Darwin in the Insect-room of the British Museum, and had a few minutes' conversation with him."[110] This was the second time Wallace and Darwin crossed paths.

After writing to Sir James Brooke, Wallace sought financial help to reach the eastern side of the globe. He submitted a paper on the Rio Negro to the Royal Geographical Society. He also presented the Society with a large map of the river based on his own survey. On 27 February 1854, Wallace was elected a Fellow. It was the only time he asked to be elected to a society. All of this allowed Wallace to approach the great geologist Sir Roderick Murchison, recently president of the Society, to ask for financial assistance for a passage to the East for Wallace and an assistant. The Geographical Society appealed on his behalf to the Admiralty who agreed to provide free passage on a navy ship. The Society also petitioned the Dutch government for permission to allow Wallace to travel through territories of the Dutch East Indies. This was granted.[111]

After a tedious run-around with Admiralty ships which were in confusion due to the outbreak of the Crimean War, Wallace was eventually given a first-class ticket to Singapore with the Peninsular & Oriental Steam Navigation Co., known as the P&O.

So after only eighteen months back in England, Wallace was about to set off again for the tropics as a specimen collector. He engaged a young assistant to accompany him, Charles Allen, "a London boy, the son of a carpenter who had done a little work for my sister, and whose parents were willing for him to go with me to learn to be a collector. He was sixteen years old, but quite undersized for his age, so that no one would have taken him for more than thirteen or four-teen."[112] Wallace had the age wrong. Charles Martin Allen was born in London in June 1839, so he was actually fourteen.[113] We know Allen could read and write, but little else about his background. As far as is known, he was Wallace's first European servant. He called Wallace "sir". Mr. Wallace was at last beginning to fill the role of the middle-class gentleman for which he had been born.

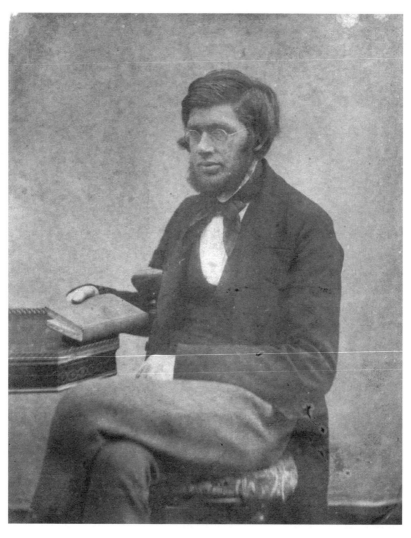

Colotype of Wallace c. 1853. Presumably by Thomas Sims. National Portrait Gallery, London.

Chapter 2

EMPIRES OF STEAM

Accounts of Wallace's journey to the Eastern Archipelago tend to pass quickly over how he went there. But this overlooks the fundamental technology that made possible Wallace's travels, correspondence, supplies and access to scientific literature — the steamship. Steam was the buzzword of the age. It was a synonym for speed, power, reliability and modernity. Advertisements in the *Times* for passenger vessels to America or Europe were headed with "STEAM!", perhaps a resonance only shared by "jet" in the 1950s and 1960s. Steam trains crisscrossed Britain and Europe, making transportation of passengers, goods and mail faster and more reliable than ever before. By 1854, there were about 8,000 miles of railways in Britain.

The 1840s and 1850s were also the beginning of the age of the steamship. Nowadays, we tend to think of steamships as a late Victorian era technology. In later life, Wallace reflected that the younger generation had no idea "how great and fundamental that change" from animal power to steam power had been.[114] Steamships were much more novel and awe-inspiring in the early decades of the 1840s and 1850s. They were by far the largest and most complex moving objects ever seen on Earth. This was all the more awesome to a generation that had known only animal power and sailing vessels. With their size, speed and independence from the winds, steamships provided unprecedented opportunities for international travel, commerce and communication. They were called "one of the greatest triumphs of art, and perhaps in its future consequences the most important invention of any age".[115] Wallace called the steamship "that highest triumph of human ingenuity, with its little floating epitome of European civilization".[116] His remark was based on his travel to the East on P&O steamers.

The first steamships were driven by paddlewheels attached to the sides of the ship near the centre and enclosed in large boxes to contain the spray they would otherwise throw on deck. By the 1850s, the more efficient screw- or propeller-driven steamers were replacing paddlewheels. Popular press reports described the latest steamships in terms that remind a modern reader of the *Titanic*, built half a century later. Writers dwelt not only on the ships' size, but also the luxury of their furnishings, the provision for every comfort from wines and the dining saloon to the mahogany of the first-class cabins. And of course there was their speed. The novelist William Makepeace Thackeray wrote of a tour of the Mediterranean on P&O steamers in 1844 "by which, in the space of a couple of months, as many men and cities were to be seen as Ulysses surveyed and noted in ten years".[117]

The P&O was first established in 1837 as the Peninsular Steam Navigation Company. The name was derived from the company's first routes from Britain to the Iberian Peninsula. In 1840, the company was awarded government mail contracts and a Royal Charter to carry the mails to the Far East. "Oriental" was added to the company's name. By the beginning of 1854, the P&O operated twenty-four paddlewheel steamers and fifteen screw steamers.

These commercial and technological developments were fuelled, or rather financed, by government mail contracts. The contracts were in turn fed by the government's need to have rapid communication with India and eastern commercial concerns after the Napoleonic Wars. The steamer companies could not have existed without substantial government contracts. Carrying passengers alone was not enough. An eight-year contract from 1853 guaranteed the P&O £199,600 per annum to carry mail twice monthly between Southampton and Hong Kong as well as bimonthly between Singapore and Sydney.[118] Postage repaid two-thirds of the government contract. So, the ability of passengers and goods to travel so quickly, regularly and reliably was derived from governmental patronage of the latest technologies to improve communications over far-flung imperial, military and commercial concerns.

Britain was already a hub of international mail networks that flowed in and outwards from America, Europe, India, South Africa, the Eastern Archipelago, China, Australia and New Zealand. By 1842, a regular mail service was extended to India "overland" via Egypt and by 1845 to China via Singapore. Fed by government mail contracts, the P&O became the largest private fleet in the world. The company had not only a line of steamers stretching halfway around the globe, but a vast support system to make these journeys possible. Coal had to be shipped ahead via sailing ships to keep the coaling stations supplied. Across the overland stretch through the Egyptian desert, there was no drinking water or food. The P&O established wells and farms to support passengers during the journey. A string of seven rest stations were established in the desert between Cairo and Suez. Hundreds of camels and horses were kept to carry the mails, baggage and passengers across the desert. The animals too had to be fed and watered. One writer in 1861 considered the establishment of the overland route to be one of the "valuable achievements of the present century".[119] The journey to India before this, around the Cape of Good Hope, had taken four to five months. It would take a year to write a letter and get a response. By 1861, this was reduced to an average of thirty-three days each way.

Chambers's papers for the people noted the arrival of letters at Hong Kong from New York in 1854 after fifty-eight days, remarking with admiration, "Thus, in less than ten years from its first establishment, [the P&O] which originally sent its steamers no farther than Gibraltar, was navigating the Mediterranean, the Red Sea, and the Indian Ocean, connecting the European shore of the Atlantic with the Asiatic shore of the Pacific, and conducting a constant communication between England and China. Such feats of science and energy will soon teach us to regard without wonder deeds of which even the glowing imagination of Eastern story-tellers did not dare to dream."[120] By 1860, British mail steamers covered over two and a half million miles a year.[121]

In 1854, P&O ships left Southampton on the 4th and 20th of each month bound for Gibraltar, Malta and Alexandria. A second line then extended from Suez, Aden, Ceylon, Madras and Calcutta. A third line ran from Bombay, Ceylon, Singapore and Hong Kong. A fourth ran between Singapore and Australia.

Passengers took the convenient South Western Railway from London's Waterloo station to the port at Southampton. Detailed guidebooks for the route were in print from the mid-1840s. *The popular overland guide* of 1861 described the scene at Southampton where Wallace's steamer waited to depart.

> The deck of the steamer is thronged with passengers and their friends; shipping agents rushing frantically to and fro, accompanied by porters bearing heaps of baggage which are scattered broadcast

wherever a vacant spot can be found. Officers and crew are busily engaged taking in cargo from barges alongside....passengers and crew impatiently await the arrival of the mail, which can now be seen approaching in several waggons escorted by Post-office officials, driven rapidly to the steamer. A few minutes suffice for its reception, and for obtaining the necessary signatures of its safe receipt, and the band strikes up a cheerful strain, while the last embrace and leave taking is made. Presently a ringing voice gives the order, "Cast off the ropes,"—a momentary pause—a vibration—one turn of the screw, or paddles—and the floating castle, like the ark of old, instinct with life, moves out majestically amid the cheers, prayers, and blessings of the crowd on shore.

At 2 pm on Saturday 4 March 1854, Wallace, Allen and about seventy other passengers embarked on the P&O steamer *Euxine*.[122] Wallace and Allen were the only passengers bound for Singapore, two were for Hong Kong and others for India. The *Euxine* was fully laden including 473 packages of gold and silver valued at £155,668, and 200 or 300 bags of mail under the care of an Admiralty agent.[123]

The *Euxine* was a paddle steamer built at Greenock, Scotland in 1847. Her name was derived from the Latin of the Black Sea (*Pontus Euxinus*), her original route. She was 222 feet long and twenty-four wide. Built of iron, she weighed 1,100 tonnes and was powered by a direct-acting two-cylinder oscillating steam engine providing 410 horsepower. She could cruise at twelve knots, almost fourteen miles per hour. There was accommodation for eighty first-class and eighteen second-class passengers and 443 tonnes of cargo. In comparison, Darwin's 1831–6 voyage on HMS *Beagle* used technology of a different era. The *Beagle* was made of wood, about ninety feet long, twenty-four wide, weighed 242 tonnes and could make about seven to eight knots depending on wind, the sails set and other factors.[124]

The *Euxine* eventually suffered a grim fate in the South Atlantic in 1874. Carrying a cargo of coal bound for Aden, she caught fire and sank. Six of the surviving crew were lost at sea in a small boat. Thinking that all would starve, they drew lots and killed and ate one of their number before being rescued by a passing Dutch ship.

By 1854, the *Euxine* was one of the smaller steamers, and no longer state-of-the-art. The larger P&O ships were being diverted as troop transports for the Crimean War.

The *Euxine* in 1850. Reproduced with kind permission from P&O Heritage collection.

Wallace later noted "the journey to the East was an expensive one".[125] Indeed it was. A gentleman's first-class ticket to Singapore was £142 and a male European servant's was £55.[126] So the combined tickets for Wallace and Allen were £197 — a princely sum, equal to the entire insured amount of Wallace's four years of collecting from Brazil. A well-to-do gentleman's annual income at the time was about £300. Wallace could never have afforded this journey without the financial help of government solicited by Murchison and the Royal Geographical Society.

While on board, Wallace wrote to his boyhood friend George Silk, "This ship is crowded, we have four berths in our cabin, 3 occupied & there is not room for two to dress in it at once. I am in the lower tier, they are luckily well ventilated and so are pretty comfortable."[127]

The steamship really was a floating town as the *Popular overland guide* put it. The deck at the front part of the ship was for the sailors and second-class passengers. The aft part was reserved for first-class passengers. The motley collection of different types of deck chairs would strike a modern person as very odd — the reason being that deck chairs were provided by the passengers themselves for their own use, and were taken on to India or other final destination. Farther astern was the large cylindrical capstan, used for hoisting the anchor. At the very back was the wheel for steering the ship, manned by two sailors with compasses. The author of the *Popular overland guide* praised the modern "power" which the science of navigation gave to the mariner, "enabling him by the aid of the

An 1870 steamship berth. The ocean steamer. *Harper's New Monthly Magazine* 41(242):193, July 1870.

wonderful properties of the loadstone, and the observation of the stars, to find his path across the trackless ocean with unerring certainty and precision, and permitting him to calculate the exact hour at which, under certain circumstances, he will arrive at a given place, is one of the greatest triumphs of science".[128]

Below deck was a long hall, the saloon, with a massive dining table in the centre, bolted to the floor to resist the motions of the sea. In the middle of the room, the mast of the ship often passed through, though ornately decorated with paint and gilt so that this utilitarian structure did not seem out of place with the rest of the elegant space. Along the side of the saloon, the doors of passengers' cabins, painted with landscapes, opened inwards. Between the doors were cushioned benches.

The engine rooms, like the rest of the ship, were open and could be toured by passengers. One could see "where the wonderful and resistless power which is

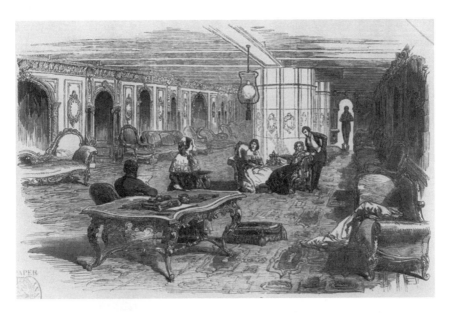

Grand saloon of the steamship *Atlantic*. *The Illustrated London News*, 16:368, May 1850.

bearing a small town upon the surface of the deep at a speed of from ten to twelve knots (or miles) per hour, can be seen pursuing its unceasing labour".[129] This ceaseless clanking of metal was itself fuelled by the labour of half-naked stokers shovelling coal into the furnaces.

Nearby was another large compartment that looked like a scene from Noah's Ark, because here, inside the great metal ship, was a menagerie of hundreds of animals, outnumbering the human inhabitants. There were three or four cows to provide fresh milk (before refrigeration, milk could not last more than a day) and coops with hundreds of chickens, ducks, turkeys, sheep and pigs. Fodder for all these animals also had to be stowed. Nearby was the butcher who gradually reduced this floating farmyard to feed the passengers and crew as the ship ploughed her way across the waves.

Wallace described the meals to Silk, "We sit down 60 to dinner, everything is generally cold & the only way to get a good dinner is to seize on the nearest dish to you and stick to it till exhausted nature is replenished. Our hours are breakfast 9 lunch 12 soda water ale & bread & cheese wine & spirits ad lib. Dinner at 4 claret port & sherry with champagne twice a week. Tea at 7 (Band plays) 8 to 9." It seems the *Euxine*, as an older replacement vessel, was not serving at the usual

P&O standard. One group of passengers on the *Euxine* in January 1854 wrote a letter to the *Times* to complain about the hurried loading of the ship moments before departure, leaving the decks and even some of the cabins and saloon stuffed with "large guns, anchors, chaincables, &c. for Constantinople" and the Crimean War.

The *Popular overland guide* noted that some passengers were wont to complain too much. "The daily bill of fare includes soups, mutton, poultry of all kinds, pork, side dishes and vegetables without number, pastry of every description, with dessert, wines, and every requisite to an excellent repast, and when I add that the bakery supplies hot rolls and buns, and the pastry-cook occasionally, ices, it will be admitted things are not so bad as have been represented....lunch, consisting of bread and biscuits, butter and cheese, with ad libitum accompaniments of bottled beer." A near contemporary passenger on the same route noted, "Ale, porter, claret, port and sherry are furnished freely at all times without extra charge, and on Wednesday and Saturdays we have champagne besides."[130] An 1854 guidebook declared, "In truth, meals are so numerous, and blended into each other so felicitously, that life on board a Peninsular and Oriental Company's steamer is one vast *monstre* refection."[131]

P&O ships strove to offer a high level of passenger service; beyond a full and select menu and free drinks, the passengers were treated to music while they dined. The waiters doubled as a band.[132] Wallace was finally travelling in style. And as much as he must have felt entitled and at home amongst his fellow first-class passengers, he probably still felt somewhat different. The social etiquette and expectations to dine at certain times constrained him. He could fit in, but it wasn't quite second nature to him. He would later feel more comfortable in a native prau where he could do as he liked when he liked.

Wallace made no mention of Allen during the voyage. The *Popular overland guide* remarked, "As the servants and the second-class passengers mess together, and the latter are not admitted on the quarter-deck, their social status is lowered in public estimation, consequently few avail themselves of the lower rates, except under very peculiar circumstances, where time and expense are equally of importance; and in that case it is always done *sub rosa* [confidentially]. I believe, however, the dietary is very good, and the quarters comfortable."[133]

On the 10th, the *Euxine* touched at Gibraltar to exchange mail. Here, they were delayed twenty-four hours because of cholera quarantine. Sailing on, she passed a sister ship on the evening of the 14th, the P&O paddle steamer *Sultan*, sailing the same route in the opposite direction. As the steam engine throbbed in the background, churning the paddles within their great boxes on the sides of the ship, Wallace wrote to Silk on 19 March.

> Our company consists of a few officers and about 20 cadets for India, 3 or 4 Scotch clerks for Calcutta, same number of business men for Australia. A Chinese interpreter and two or three others for China, Frenchmen. A Portuguese officer for Goa with whom I converse, 3 Spaniards going to the Philippines (very grave) a gentleman and two ladies (Dutch) going to Batavia and some officers & miscellaneous for Alexandria & others we have left at Malta. There are two or three chess players & I have so much improved that I think I should have some chance with you....A parson came on board at Malta going to Jerusalem, Mr. Hayward. My namesake also came on board there, he goes to Bombay, where he has been before. He is a neat figure, sharp face and very respectable, not at all like me!
>
> I have found no acquaintance on board who exactly suits me, so shall have less to regret at parting with them. One of my cabin mates is going to Australia, reads "How to make money", seems to be always thinking of it and is very dull unsociable, the other is a young cadet, very aristocratic great in Dressing case & Jewellery. Takes an hour to dress & reads the Hindostany Grammar. The Frenchman, the Portuguese & the Scotchmen I find most amusing.[134]

Egypt

After steaming 2,951 miles from Southampton, the *Euxine* arrived at Alexandria a few days late on 20 March 1854. The giant steamer *Himalaya* and ten Turkish steamers were in the harbour. Once the *Euxine* was at anchor, passengers and cargo were loaded onto smaller boats and taken to the dock where porters jostled and pressed in competition to carry things for a fee. Passengers and their baggage were cursorily examined at the customs house before heading for a hotel.

Passengers and the mail were next transported overland to Suez on the Red Sea. It was this short land crossing that gave this route the name "overland" (as compared to by sea via the Cape of Good Hope). Wallace just missed the new railway, still under construction, the following month. The Suez Canal did not open until 1869. Passengers and the mails took different routes. The mails, which

Arrival of the Indian and Australian mails at Alexandria. *The Illustrated London News* 22, 1853.

were the bulk of the P&O's revenue, and contractually obliged to meet their connections, were transported on the backs of hundreds of camels. Passengers' baggage was also sent on camels. Wallace described what happened next.

> Imagine my feelings when coming out of the Hotel (whither I had been convey'd in an omnibus) for the purpose of taking a quiet stroll through the City, I found myself in the midst of a vast crowd of donkey's & their drivers all thoroughly determined to appropriate my person to their own use and interest, without in the least consulting my inclinations. In vain with rapid strides and waving arms I endeavored to clear a way and move forward, arms and legs were seized upon, and even the Christian coat-tails were not sacred from the profane Mahometans. One would hold together two donkeys by their tails while I was struggling between them, & another, forcing together their heads, would thus hope to compel me to mount upon one or both of them; and one fellow more impudent than the rest I laid flat upon the ground, and sending the donkey staggering after him, I escaped a moment midst hideous yells and most unearthly cries. I now beckoned to a fellow more sensible-looking than the rest, and told him that I wished to walk and would take him for a guide, & hoped now to be at

rest; but vain thought! I was in the hands of the Philistines, and getting us up against a wall, they formed an impenetrable phalanx of men and Brutes thoroughly determined that I should only get away from the spot on the legs of a donkey. Bethinking myself now that donkey-riding was a national institution, and seeing a fat Yankee (very like my Paris friend) mounted, being like myself hopeless of any other means of escape, I seized upon a bridle in hopes that I should then be left in peace. But this was the signal for a more furious onset, for seeing that I would at length ride, each one was determined that he alone should profit by the transaction, and a dozen animals were forced suddenly upon me and a dozen hands tried to lift me upon their respective beasts. But now my patience was exhausted, so keeping firm hold of the bridle I had first taken with one hand, I hit right & left with the other, and calling upon my guide to do the same, we succeeded in clearing a little space around us. Now then behold your friend mounted upon a jackass in the streets of Alexandria, a boy behind holding by his tail and whipping him up. Charles (who had been lost sight of in the crowd) upon another, and my guide upon a third, and off we go among a crowd of Jews and Greeks, Turks and Arabs, and veiled women and yelling donkey-boys to see the City....You may think this account is exaggerated, but it is not; the pertinacity, vigour and screams of the Alexandrian donkey-drivers no description can do justice to.

The *Popular overland guide* forewarned passengers that they would "have to engage in a scuffle with the donkey-drivers" and hence the best advice was, as Wallace found, "securing a speedy mount". Wallace noted seeing "Pompey's Pillar", a 20.4-metre Roman triumphal column, not in fact from Pompey but from the time of Diocletian in 197 AD. The pedestal was covered by graffiti from recent European visitors.[135]

A line of small steamers took passengers from Alexandria to Cairo via the Mahmoudieh Canal. Another traveller noted, "It is a sign of the times that the turbaned Turk issues his directions in English, 'Ease her,' 'Stop her,' 'Back her,' 'Go easy,' 'Full power'—the engineers being either English or Scotch."[136] Indeed, the times were modern and convenient compared to what passengers experienced a few years before when "the canal boat, crowded to excess, affording scarcely room for every one even to sit down— the Nile steamer, filled with vermin, crawling over their victims when extended on the floor or table of the small cabin, trying to steal a few minutes sleep".[137]

The canal terminated at Atfeh, the boats proceeded through a lock into the Nile. Wallace observed along the way: "Mud villages, palm trees, camels & irrigating wheels turned by-buffaloes form the staple of the landscape with a perfectly flat country often beautifully green with crops of corn & lintels, numerous boats with immense triangular sails. Here the Pyramids came in sight, looking very large, then a handsome castillated bridge for the Alexandria & Cairo railway."[138]

Wallace and Allen then transferred into another steamer to Boolak, the port of Cairo. P&O vans then took passengers on to Cairo itself. Wallace probably stayed at the Hotel des Anglais (later Shepheard's Hotel) on Cairo's grand square. "We took a guide & walked in the city, very picturesque, and dirty, got to a quiet English hotel where mussulman waiters rejoicing in the name of 'Ale Baba' gave me some splendid tea, brown bread & fresh butter." Wallace was reminded of Thackeray's account of his first day in the East: "After there is nothing. The wonder is gone, and the thrill of that delightful shock, which so seldom touches the nerves of plain men of the world, though they seek for it everywhere."[139]

The next stage was the desert crossing to Suez. Forty vans, painted green, transported P&O passengers who sat "omnibus-fashion [facing each other], at the sides of the vehicles, entering at the back; an arrangement that is unavoidable, from the height of the wheels (a single pair), which reach half-way up the body of the van. The seats are carpeted, and the whole thing is roomy and commodious."[140] Every ten miles, the horses were changed at purpose-built roadside stations. Some of these provided meals for passengers. There was a hotel at the halfway point. "Thanks to the exertions of the British agents and associations, who make it their business to promote the intercourse with India, there is little difference now between travelling seventy miles over a post-road in England, and going over the same space of ground on the isthmus of Suez."[141]

Wallace continued his letter to Silk.

> In the morning at 7 we started for Suez in small two wheeled omnibuses…the desert is undulating, covered with a coarse volcanic gravel. The road is excellent, hundreds of camels skeletons lay all along, vultures & a few sand grouse were seen, also some small sand larks &c. We saw the mirage frequently near the middle station the Pasha has a hunting lodge like a palace. The Indian & Australian mail about 600 boxes & all the parcels & passengers luggage came over on camels which we passed on the way — a few odoriferous small plants grew here & there in the hollows, I made a small collection in my pocket book and got a few landshells. We enjoyed the ride exceedingly

Map of the overland route across Egypt. J. Barber, *Overland guide book*, 1845.

A rest station (number 4 in the map above) between Cairo and Suez. [Roberts], 1850.

and reached Suez at midnight. Suez is a miserable little town & the
Bazar is extraordinarily small, dark & dirty, no water or any green
thing exist within two miles.

In the afternoon, Wallace and Allen were transported in a small steam tender
to the giant steamship anchored about four miles out. Wallace described the
steamer as "a splendid vessel where the cabins are large & comfortable and everything very superior to the Euxine. It is a perfect calm & at length hot & pleasant."
This splendid vessel was the much larger and more luxurious *Bengal*. Built in

The *Bengal* which carried Wallace from Suez to Ceylon. Here at Malta in 1877. Reproduced with kind permission from P&O Heritage collection.

Glasgow only two years before at a cost of £68,300, at 2,300 gross tonnes, the *Bengal* was 295.9 feet long and 38.2 feet wide. A 500-horsepower geared beam steam engine and a single screw allowed her to cruise at ten knots. She began steaming the route between Southampton and Alexandria in February 1853. She could carry 135 first-class passengers and had a crew of 115. She boasted a magnificent saloon seating 100 passengers.

The *Bengal* travelled for six days down the Red Sea to Aden. In his autobiography, Wallace remarked only, "We stayed a day at desolate, volcanic Aden."[142] His letter to Silk was posted home and he wrote no more until he arrived at Singapore.

The *Bengal* then steamed 2,134 miles southeast through the Arabian Sea to the port of Galle on the southern tip of British Ceylon (Sri Lanka). All Wallace said of his first glimpse of the Asian tropics was a brief mention of the "groves of cocoa-nut palms, and crowds of natives offering for sale the precious stones of the country".[143] As the *Bengal* was heading to Bombay, Wallace and Allen transferred to the *Pottinger*, bound for Hong Kong via Singapore. She left Galle a day late on 10 April 1854. A "pleasant vessel", the *Pottinger* was another paddlewheel steamer of 1,300 tonnes and 450 horsepower. One writer described the exotic and motley apparel of the international passengers on board, "Look at the number and the variety of costumes and countenances of the passengers assembled upon deck! There are red coats, and blue coats, and black coats, and no coats, and white jackets, and green jackets, and such a variety in shape and colour of hats, that one would imagine they were contending for a prize—the owner of the ugliest hat to win the day."[144]

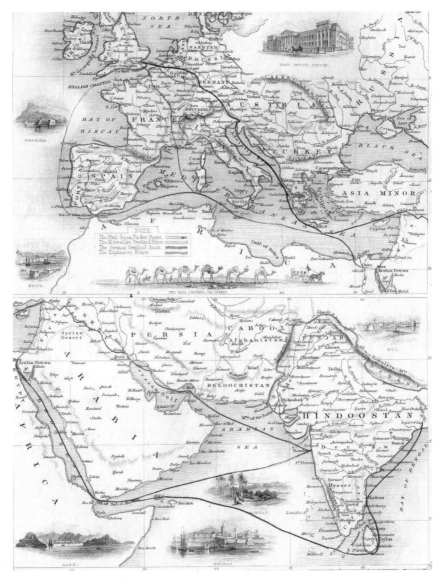

Map of the "Overland Route". J. Rapkin, *The Illustrated Atlas*, 1851.

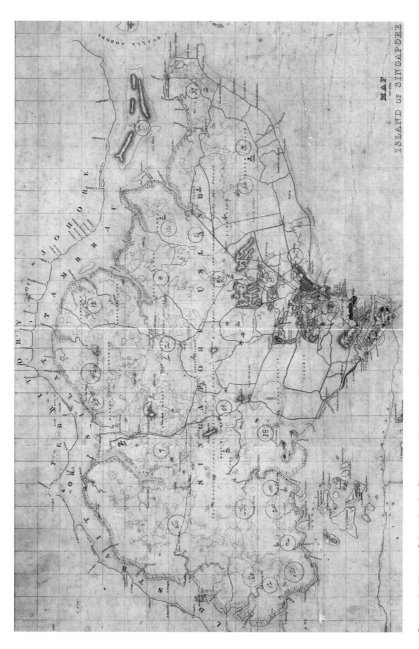

Part of an 1863 "Map of the Island of Singapore". Courtesy of National Museum of Singapore, National Heritage Board. Pulau Ubin is marked "41". The Bukit Timah mission is located near "58".

Chapter 3

SINGAPORE

T he *Pottinger* steamed east 1,286 miles across the Bay of Bengal to the island of Penang at the mouth of the Strait of Malacca between Sumatra and the Malay Peninsula, arriving on 17 April 1854. Penang had been a British possession since 1786. Wallace mentioned only Penang's famous "pictur-esque mountain, its spice-trees, and its waterfall".[145] Staying only about six hours, it was less than forty-eight more down the "richly-wooded shores" of the Malacca Straits to Wallace's final destination, the island of Singapore where he wrote, "I was to begin the eight years of wandering throughout the Malay Archipelago, which constituted the central and controlling incident of my life."[146]

The *Pottinger* steamed up to the southeastern end of Singapore and dropped anchor at 8 pm on Tuesday 18 April 1854.[147] It was already dark. The town clung along the southern coast and extended about a mile inland. Then, as now, the sea around was crowded with vessels. There were about thirty-five steamers and sailing ships and twenty-five "native craft" from many nations and countless colourful praus, junks and sampans. About 300 Chinese junks, 200 praus and 1,000 western vessels called at the port each year. As the *Pottinger* arrived after sunset, passengers and cargo were disembarked the following morning.

Small boats scurried up alongside the *Pottinger* like bees around a hive. The Malay boatmen earned their livings from this service. A German traveller recalled how a ship that anchored at Singapore was soon swarmed by boats selling every possible ware and service, even a haircut, by boatmen who all spoke a few words of English.[148]

The town of Singapore from government hill c. 1864. Bickmore 1869, p. 536.

Some of the boats carried the all important boxes of European mails up to 8 March. These were taken immediately across a narrow wooden bridge to the nearby post office on the north side of the river. Under the direction of two postmasters and a clerk, the Singapore post office processed an amazing 33,000 letters every month.[149] The international mails were as important in Singapore as in London as *The Straits Times* noted the day Wallace arrived, "The possession of the nearest and authentic intelligence from all quarters is now-a-days indispensable to guide the operations of the mercantile and shipping interest, and must become doubly so in the present moment, when the actual political complications in Europe may lead every day to events which must affect materially the money markets, as well as those of produce and goods of every kind."[150]

A young English visitor observed in 1856:

> When the mail boat arrives, which occurs twice a month, a flag is hoisted on the top of [Flagstaff] hill which can be seen all over the town, and like wildfire the news is spread. It might be thought that something serious had happened as pedestrians are on their way from all parts of the town to the post office-two or three scores Malays, or Bengalees, with each a sack over their shoulder with the name of the firm they represent printed on, and after being shown inside the office they patiently wait till the name of their firm is called out. To see the sacks of newspapers and letters handed out it may well be wondered when they will be read.[151]

Wallace, as a first-class passenger, was named in *The Straits Times* the following week. Charles Allen, as a servant and second-class passenger, was not mentioned. Wallace later wrote to his mother and Stevens that he arrived on 20 April, a date repeated by modern biographers. But the *Pottinger* arrived on the evening of the 18th and departed on the 20th for Hong Kong.[152] Wallace was not very good with dates. We will see again and again that the dates he gave were inaccurate. Presumably close enough was sufficient for his purposes, since his attention was bent more acutely to other matters. But sometimes dates make all the difference in this story.

The island of Singapore was twenty-seven miles long and eleven wide, or about 224 square miles.[153] Since independence in 1965, land reclamation has increased its size to 272 square miles, considerably changing the original outline. It continues to grow with an endless stream of dump trucks still trundling out to

ARRIVAL OF THE MAIL.

———oo———

Wallace arrives first-class. *The Straits Times*, 19 April 1854, p. 9.

the western edge of the island depositing sand to increase the small but extremely valuable amount of available real estate.

The island is separated by a narrow strait of the sea a quarter of a mile wide from the Malay Peninsula to the north. Located eighty-five miles north of the equator, Singapore's tropical climate is much the same all year round with temperatures ranging from 23–33°C (73–90°F). The humidity is high, between seventy and ninety percent. When the modern traveller arrives at Singapore, one is transferred from the airplane directly into the immaculate and heavily air-conditioned airport. Although a slight whiff of tropical air is detectable when leaving the plane, it is only when one exits the airport through large sliding glass doors that the full force of Singapore's heat and humidity envelops one like an invisible muggy blanket. Wallace, having travelled so much more gradually, and not in a pressurised container, never experienced such a shocking introduction to the atmosphere of Southeast Asia. And he never seems to have complained about the heat. The climate of Singapore was very similar to that of all the areas Wallace would visit during his voyage.

Frequent tropical downpours provide ninety-two inches of rainfall per year which keep the vegetation lush and green. There are no real seasons as it is essentially equatorial — April and May are the hottest months, just when Wallace arrived — but colonial Indian hands said Singapore's climate was very mild and healthy. The only weather fluctuations are the monsoons with their change of prevailing winds and heavier rains. The annual shift of monsoon winds from east to west powered the traditional transit of junks to and from China and the Malay praus back and forth through the archipelago.

The name Singapore derives from the Sanskrit word *Singapura* or "lion city". According to legend, a travelling Srivijayan prince, Sang Nila Utama, from Palembang, Sumatra, saw a large orange animal on the island around the year 1300 and was told it was a lion. Of course, it was actually a tiger since lions have never inhabited Singapore and tigers did.

The British had made expeditions to the East since Sir Francis Drake's voyage on the *Golden Hinde* in 1579. Expanding eastwards from India, in 1771, British trading posts were established in Penang as a way station to China. During the Napoleonic Wars between 1811 and 1815, Dutch possessions like Sumatra and Java were occupied by the British.

After the fall of Napoleon, the Dutch were again in control of their eastern possessions. Some figures in the British East India Company sought to establish a new trading settlement in the Malacca Straits between Sumatra and the Malay Peninsula. In 1819, a treaty was signed by the polymathic Sir Thomas Stamford Raffles, acting, he claimed, on behalf of the British East India Company to establish a trading post on the mostly uninhabited island of Singapore with the nominal Sultan of Johore, Hussein Shah. In 1824, the entire island was purchased by the East India Company under a further treaty with the Sultan and the local authority known as the Temenggong. In the same year, the Dutch gave Malacca to the British as part of a treaty.

Raffles spent only nine months in total in Singapore, but his influence was considerable. He laid out the design and streets of the town, with a business district around a commercial square and residential quarters for Westerners, Chinese, Malays and so forth. The modern financial district, with its forest of skyscrapers, still stands on Raffles' original commercial district. But of far greater importance for Singapore was Raffles' decision to implement the fashionable new philosophy of free trade, derived from the writings of Adam Smith and David Ricardo. The Dutch taxed vessels from other countries using their ports. At Singapore, ships, from whatever nation, were not charged import, export, tonnage or other duties to use the harbour or trade. There was no favouritism for British ships or merchants. All nations could trade on equal terms.

An article entitled "Free trade and foul trade" was printed in *The Straits Times* in February 1854. The article compared the system of the Dutch East Indies unfavourably with the *free* British system.[154] The Dutch, with all their national favouritism in trade, land and trade taxes and monopolies on the most expensive spices were said to generate £3,300,000 per year from their East Indies possessions. The imports of Singapore from 1852–3 amounted to £3,487,695 — an increase of £600,000 over the previous year. Exports totalled £3,025,980. Despite the free trade system, Singapore cleared £50,000 a year, enough to cover government and public expenses.

Under the liberal free trade policy, the new port of Singapore flourished and grew at an astonishing speed. The population of the island swelled as Chinese, Malay and Indian immigrants from the straits, the archipelago and further afield settled in ever greater numbers to take advantage of the new opportunities and freedom from tax. There was also freedom of religion. The land in the interior of the island was even free for the taking to settlers and farmers. Four years later in 1823, the population had dramatically increased to 10,000. By 1845, it had grown to 38,000. In 1854, when Wallace arrived, the population was about 65,000, of which 30,000 were Chinese, 10,000 Malays and 400 Westerners, with smaller numbers of Armenians, Arabs, Jews, Bugis, Javanese and Indians.[155]

A Forgotten Singapore

Singapore is now such a bustling, urbanised and cement and steel island with thousands of high-rise buildings and container cranes tirelessly in motion twenty-four hours a day that it has been forgotten that it was once a beautiful forested island. The Government Surveyor John Turnbull Thomson described Singapore as one approached by sea in the late 1830s:

> In the foreground, busy canoes, sampans, and tongkangs bore their noisy and laughing native crews about the harbour....Hundreds of Chinese junks, and Malay prows, lay further in shore. Behind these, stretched a sandy beach, glistening in the sun, and overhung by the graceful palm trees, the glory of Singapore planters. In the centre of the landscape was Government Hill, with its verdant lawns and snug bungalow; and at its base were the warehouses and mansions of the merchant princes. Behind these was to be seen the comely undulating

background, alternately covered with the mighty forest trees, and gambier and pepper gardens.

The tallness of the forest trees standing alone, or in clumps on the half-cleared hills and islands, gave a majesty to the scenery that I have observed nowhere else. The forest trees in thus rising into the atmosphere upright and in full verdure on the very tops of the hills, proclaimed that they nourished in a clime characterized by serenity and repose.[156]

The town was laid out on a neat grid, bisected by the Singapore River. By 1854, there were about 4,700 buildings in Singapore, mostly brick covered in whitewashed stucco with red tile roofs. The European houses were typically two storeys. The windows were made of wooden blinds rather than glass because of the heat.

Singapore was referred to in Britain as a "colony" since the 1820s. The English word colony derives from the Latin *colonia* from *colonus*, meaning farmer, cultivator, planter or settler in a new country. The word is preserved in the names of places such as Cologne in Germany. It is hard not to read the word colony today without the overtones prevailing after the Second World War — exploitation, condescension and even racism. The European residents of Singapore referred to it most often as a "settlement".

The settlement of Singapore was different from a colony. First of all, it was founded by the British, not an existing city or country conquered or taken from local peoples. The inhabitants came to Singapore of their own accord and were never a captive population. The historian Erik Holmberg has argued convincingly that the social elite of colonial Singapore was as much Asian as European. The modern image of the "natives" (themselves financial immigrants as much as the Westerners) as only coolies or rickshaw pullers conceals the fact that the majority of merchant wealth in colonial Singapore was in the hands of Asians, not Westerners.[157] Thus, early- and mid-19th-century Singapore was not a place of oppression, exploitation or racist snobbery but of commerce. Snobbery would come near the end of the colonial experience, but we should not read it back into earlier times.

Singapore became one of the British Straits Settlements in 1826. In 1854, it was still governed by the East India Company. Together with Penang and the district of Malacca, it formed the Straits Settlements under a governor with the Pickwickian name of Colonel William John Butterworth, based in Singapore in a residence on Flagstaff or Government Hill (now Fort Canning Hill) overlooking the town and harbour. From 1830 to 1867, the Straits Settlements was a residency, or subdivision, of the Presidency of Bengal, British India. The Straits Settlements became a separate Crown colony, overseen by the Colonial Office in London, in 1867.

Most English-speaking readers still find the name Singapore evocative of exoticism and the East. It is remembered for its fall to the Japanese, as important an event if not more so than the attack on Pearl Harbor. More prosaically, Singapore is known as the country that banned chewing gum and spitting and canes criminals. Singaporeans often regret that their country is so small and has so little history compared to other countries. But there is far more history than most people realise.

Even most Singaporeans today are totally unaware of what Singapore was like when Wallace arrived. The narrative familiar to most people starts with the foundation by Raffles, a colonial period (usually imagined as the 1890s–1930s), followed by the fall of the island to the Japanese in 1942, independence in 1965 and the country's subsequent unparalleled success.

Yet, the Singapore Wallace found was far more complex and developed than is imagined today. He was a stranger in a strange land to be sure, but it was still based on home. There was a complex and efficient administration which included a neoclassical law court with three judges and nine Justices of the Peace, a court for the relief of insolvent debtors and a court of commissioners for the recovery of small debts and notaries public. Public roads, bridges and sanitation were maintained by the government. The police department under its magistrate, Superintendent and Deputy Commissioner had five European constables and about 250 "native" assistants.[158]

The Singapore Library, located in the grounds of the Singapore Institution, was open from Monday to Saturday. Visitors were welcome to use it if they could provide a letter of introduction and a small fee. The 1860 catalogue listed 174 biographies, twenty-six works on divinity and ecclesiastical history, 135 of general literature, 313 of history, voyages and travels, 452 novels and fifty-eight works of poetry and drama. There were also magazines, reviews, newspapers and new books sent out from London every month.[159] The Singapore News Room on Commercial Square was open to the public and offered journals and newspapers from around the world. Singapore already had four weekly newspapers and three bimonthlies.

The Assembly Rooms at Hill Street (on the site now occupied by the Old Hill Street Police Station) offered readings, singing, violin and piano performances and theatre. The day after Wallace came ashore, Mrs. Deacle appeared as "Louisa Lovetrick" "assisted by Gentlemen Amateurs" in the "laughable Farces of 'Hunting A Turtle', and 'The Dead Shot', with Singing".[160] On Wednesday, Friday and Saturday, there were horse races north of the town. The Singapore Cricket Club was founded in 1852 (the current pavilion dates to 1859).

Commerce

Singapore existed as a commercial centre. There were no taxes on ships using the harbour, just a trivial charge from the larger ones to defray the expenses of the Horsburgh Lighthouse. Government ships, of all nations, were exempt. Shipping and commerce statistics were recorded by a trade department with two registrars of imports and exports and three clerks. The advent of steamships shifted the main trading arrangement away from Calcutta to London, and enabled the British to suppress the pirates of the region who had only sailing vessels.

Despite being a British settlement, dollars were used in Singapore then as they are now. Many assume the Singapore dollar comes from America but in fact it is Spanish. The Spanish dollar was in widespread use throughout the region since the Spanish colonisation of the Philippines in the 16th century. The Singapore government, as a sub-department of British India, used Indian rupees but all trade was conducted with Spanish dollars and cents.[161]

The trading seasons revolved around the annual monsoon winds. The northeast monsoon winds brought ships from China, Siam and Cochinchina from January to March. When the winds reversed direction, the ships left Singapore from early May. The other trading season was east–west and began in September with the Malay praus from the archipelago.

In addition to a Chamber of Commerce, there were many insurance companies thriving on the commerce including eight for life, ten for fire and fifteen for goods. There were representatives of seven international banks and colonial agencies as well as the main bank of the town, the Oriental Bank Corporation. There were Consuls of eleven countries, eight auctioneers, a book binder, printers, copperplate and lithographic printers, five boarding houses, three blacksmiths, butchers, chemists, a coach maker, an engineering firm, provisioners and ships' victualers, sail-makers, four surveyors, two soda water manufacturers, five taverns, five watchmakers and two undertakers. Amongst the many agencies were Wallace's shipping agents Hamilton, Gray and Co., established in 1832.

Seemingly every visitor to Singapore marvelled at the speed at which the settlement had expanded from a forested island inhabited by a few Malay fishermen to a thriving and bustling international entrepôt, a hub of trade and communication ranked as the fourth busiest commercial port in all of Asia.[162] On 6 February 1854, the wealthy elite of Singapore gathered at the Assembly Rooms to celebrate the thirty-fifth anniversary of the settlement. John Purvis, Singapore's oldest merchant, gave a speech which demonstrates a remarkable continuity with the government rhetoric of today:

> We have had no [gold-bearing] soils to fall back upon, or to aid us in our forward march, and yet Singapore had a hidden treasure which has happily developed itself in industry and intelligence! For it is to those sources, aided by a liberal system of Government, that Singapore is indebted for the proud position she now holds, and so long as that system shall be maintained, so long will Singapore continue to flourish and the Government be respected by every nation and people with whom we may hold intercourse.[163]

The Inhabitants of Singapore

The streets of Singapore were an incredible mixture of races and costumes. Almost all were recent immigrants. Wallace noted that "the Chinese never depart the least from their national dress, which indeed it is impossible to improve for a tropical climate, either as regards comfort or appearance. The loosely hanging trousers and neat white half-shirt-half-jacket is exactly what a dress should be in this latitude."[164] Chinese labourers wore "only a short pair of breeches, reaching from the hips to half way down the thighs, and thus almost naked they carry heavy loads of Gambier leaves and pepper".[165] Chinese men had the front of their head shaved and grew a long pony tail or queue, often reaching down to their ankles, although contemporary drawings and photographs show most men on the streets wound them up on their heads.

The queue was said to show obedience to the Emperor. But it was not always a convenient accoutrement. About 2 am on 26 March 1856 (while Wallace slumbered a few miles away), a small boat with five Chinese men was rowing from Damar Island (then a small islet off the west coast, now part of Jurong

Harbour) to Kampung Glam on the east coast. When they were near the eastern entrance to New Harbour, and perhaps only 500 paces from shore, a long sampan manned by six Malays approached. In the darkness, the Malays bombarded the startled Chinese men with stones and then thrust spears at them. The owner of the boat, Kee Ah Hio, who was standing and rowing in the traditional Chinese fashion, was struck by a spear in the abdomen. The four Chinese passengers leapt into the water to swim to safety but the water was shallow enough to stand in. The robbers boarded the boat and stole about two piculs of black pepper. (A picul was the weight a man could carry on his shoulder.) One of the Chinese men reported, "When I was in the water one of the Malays pulled me up by the touchong (tail), felt about my body and robbed me of ten dollars which were in my waist belt."[166]

In another incident, a Chinese man was burgling a Chinese house in Bukit Timah village in the centre of the island when he was caught by his queue while trying to escape. The owner of the house held on despite receiving numerous stab wounds. As neighbours approached, the burglar finally cut off his own queue but he was caught and taken to the police.[167]

Wallace became familiar with the rows of streets with their colourful shops selling every manner of article imaginable. One of these shopkeepers no doubt inspired a little note Wallace jotted with the title "The Chinaman at Singapore".[168]

> The merchant is generally a fat round faced man, with an important
> & business-like look. He wears the plain clothes of the meanest coolie
> for he has perhaps been a coolie himself but he is always clean & neat,
> & his tail tipped with red silk hangs down to his heels. He can speak
> a little English. He has a clean & handsome warehouse in town & a
> good house in the country. He keeps a fine horse & gig & every even-
> ing may be seen driving out bareheaded to enjoy the cool breeze. He
> is rich, he has several small shops, he lends money on good security,
> he makes hard bargains & gets richer every day.

The wealthiest Chinese merchant was Hoo Ah Kay, known as Whampoa after his birthplace, a suburb of Canton. Like so many other Chinese immigrants, he came to Singapore as a penniless young man. He eventually made a large fortune from provisioning ships, merchandising and so forth. He became one of Singapore's leading figures and the first Asian member of Singapore's Legislative Council.[169] At his large house, he even had a menagerie including at one time an orangutan.[170]

China did not allow the emigration of women. By Wallace's time, Chinese men outnumbered woman fifteen to one in Singapore. This mostly male, uneducated, migrant population led to a rather wild frontier atmosphere: prostitution, gambling and opium smoking were widespread. Perhaps as many as 15,000 Chinese men were opium smokers.[171] The gambling was described by a visiting American naval captain:

> Gaming was going on in every shop, under the colonnades, in the bazaars, and at the corner of almost every street, a variety of games were playing. Of several of these I had no knowledge; some were performed with cards, and others with dice....Those who have not seen the Chinese play, have never witnessed the spirit of gambling at its height; their whole soul is staked with their money, however small it may be in amount, and they appeared to go as earnestly to work as if it had been for the safety of their lives and fortunes.[172]

The Malays lived mostly in a region in the northeast of the harbour called Kampung Glam in traditional wooden houses built on poles. Wallace described the Malay costume as a "national jacket & sarong with loose drawers". In his notebook, he recorded: "[Skin] Colour reddish brown of various shades. Hair black straight, on body & beard scanty or none. Stature low or medium, form robust, [women's] breasts very much developed."[173] The men carried the traditional long knife or *kris* in a sheath on the waist. The Malays had been Muslims since the process of Islamisation in Southeast Asia began in the early 14th to 16th centuries. Malays were mostly employed as boatmen, fishermen, woodcutters and carpenters. A few worked in the fields nearby. They dominated the boating trade that ferried people and cargo to and from the ships in harbour.

The Bugis, a Malay race from the island of Celebes (Sulawesi), were famed seafarers and sometimes notorious pirates. They also wore the *kris*. Wallace recalled, "When strolling along the Campong Glam in Singapore, I have thought how wild and ferocious the Bugis sailors looked, and how little [I] should like to trust myself among them."[174] He changed his mind after spending time in Celebes.

The Indian peoples of early Singapore, both Hindu and Muslim, were never called "Indians", but usually "Klings", a Malay term originally for southern Hindus.[175] They were also a mostly male population. They came to Singapore to earn money and then return home. Wallace described them as "handsome, dark-skinned" men who worked mostly as "petty merchants and shopkeepers. The grooms and washermen are all Bengalees, and there is a small but highly respectable class of Parsee merchants."[176] He thought these merchants bargained harder than any others in Singapore: "[They] always ask double what they will take, and with whom it is most amusing to bargain."[177] They were said to wear "white muslin garments, and a great scarlet, or white, turbin".[178]

The European residents "dress entirely in white, with pith hats, to shield them from the sun".[179] The pith hat or pith helmet is part of the classic, now almost mythical, attire of the colonial European, from the early 19th to the mid-20th centuries. But the origins of pith helmets have been forgotten. Originating from India in the 18th century, the first reference in the *Oxford English Dictionary* is only 1858. But references to "the cork or pith helmet worn by the troops" date back at least to 1835.[180]

The pith helmet was originally made of, as its name suggests, pith — the soft centre of a plant. The pith came from an Indian marsh plant (*Aeschynomene aspera*) with a white spongy interior lighter than cork. Indian fisherman used it as floats for their nets.

The hats were often half an inch thick but still very lightweight. They were covered in a light-coloured cloth. The reasons behind the design were said to be scientific. The Chinese somehow could shave the crowns of their heads and survive exposure to the tropical sun, but a European was sure to suffer sunstroke — or death! *Chambers's journal* noted in 1861 that "the pith hat is an invaluable protection against sunstroke[181]." In 1862, the Medical and Physical Society of Bombay advised that during hot times "a good pith helmet [be] constantly worn[182]". The modern imitations of pith helmets tend to be made of moulded plastic or cardboard — the moulded outlines of folded cloth still imitating the cloth wrappings that covered the original helmets — as a skeuomorph. So modern imitations can give no sense of what it must have been like to wear a very thick but extremely lightweight sunshield on one's head. A particularly high-tech version was advertised in the 1860s.

ELLWOOD'S PATENT AIR-CHAMBER HELMET.

ELLWOOD'S PATENT AIR-CHAMBER HATS AND HELMETS having been in use in India and elsewhere for several years, and being pronounced by all who have worn them to be THE COOLEST AND MOST SUITABLE HEAD-DRESS FOR HOT CLIMATES, and the only light and effectual protector of the head from the rays of the Sun…these Hats and Helmets are composed of two parts:—the outer part forming an Air-Chamber round the inner one; the non-conducting properties of the air in this chamber have the effect of *intercepting the rays of the sun*, and prevent them from passing to the head of the wearer. As the air in the chamber becomes warm or rarefied, it passes off by an aperture at the upper part of the outer crown, and is replaced by cool air entering through a series of perforations on the under side of the brim, thus causing *a rapid circulation* which has the effect of KEEPING THE HEAD PERFECTLY COOL, and renders the Hat comfortable to the wearer.…

CAUTION.

The great superiority of these Hats and Helmets over all others, for tropical climates, has induced some persons to offer to the public useless imitations, bearing some outward resemblance to them, but totally devoid of their good qualities; the public is therefore respectfully cautioned to observe that NONE ARE GENUINE UNLESS THEY BEAR THE NAME OF THE PATENTEES ON THE LINING.[183]

If only mad dogs and Englishmen went out in the midday sun (as Noel Coward put it in his famous 1932 song), Englishmen did so confident that their scientific headwear would protect them.

Finding Wallace

Unfortunately, next to nothing has been published about Wallace's time in Singapore, mostly because he gave a very thin account of it. This, he recalled, was "due to my having trusted chiefly to some private letters and a note-book, which were lost".[184] Eight notebooks survive that seem to have been used by Wallace in the Eastern Archipelago. These have not previously received the scholarly attention and meticulous examination of Darwin's *Beagle* and transmutation notebooks.[185] They are not even very consistently or accurately named.

Compared to Darwin's expensive field notebooks, mostly leather bound with heavy, high-quality paper, Wallace's notebooks are decidedly shabby with cheap paper and thin cover material often worn off.[186] They are falling to pieces. It is a telling reminder of the differences, as well as the similarities, between the two men.

The table on p. 74 correlates my short names for the notebooks with other details and the names used by earlier writers.

Wallace wrote to his mother, far away in London, "Some of the English merchants here have splendid country houses. I dined with one to whom I brought an introduction. His house was most elegant, and full of magnificent Chinese and Japanese furniture." Letters of introduction were a means of admitting previously unknown but more-or-less social equals into one's acquaintance. It was not proper to write to or call on a gentleman unannounced.[187] Etiquette books of the day advised that letters of introduction should be brief and refer to the occupation of the person presented, in this case "naturalist". When meeting, a slight bow was customary rather than shaking hands.

Wallace benefited from this kind of social arrangement throughout his travels in the East. It belies the myth that he was "working class". If he had been, such letters of introduction, and the essential context and assistance that followed, would have been inaccessible to him, and he could never have achieved what he did. Similarly, Wallace was referred to in print as "Wallace Esq". The term "esquire" once referred to a man of considerable landed property, in social rank just below a knight. But by the 1850s, it was used "as a term of courtesy, to every one who holds a respectable position in society", but was not synonymous with or as high as the term gentleman.[188]

Wallace never mentioned the name of the English merchant he dined with. It was likely George Garden Nicol or John Jarvie, the Singapore partners of Hamilton, Gray and Co. Wallace had their names at the top of his address list in *Notebook 4*. Both had splendid country houses. Nicol, who was not English but Scottish, owned the 150-acre Sri Menanti Estate in Claymore and Tanglin about two miles out of town. Wallace noted the regional monsoon seasons at the front of *Notebook 1* as "Information from Mr Nicoll". Jarvie lived on a nutmeg plantation near Bukit Timah.

Wallace's notebooks

Short names	Dates	Text on covers	McKinney 1972	Brooks 1984
Journal 1	6.1856–3.1857	"Jo<urnal> | Baly .. Mac<assar>….| …Aru 1856"	"Journal, 1856–57"	Field Journal
Journal 2	3.1857–3.1858	"18<57..58> *Journal* | Part. 2."	"Journal, 1857–58"	Field Journal
Journal 3	3.1858–8.1859	"*Journal* | *Part.* 3. 1858..59."	"Journal, 1858–59"	Field Journal
Journal 4	10.1859–5.1861	"Journal. Pt. 4"	"Journal, 1859–61"	Field Journal
Notebook 1	1854–1861		"1854 notebook"	Registry of Consignments
Notebook 2/3	5.1855–11.1860	"Insects | *Insect Notes.* 3." "Birds | Birds [and] Mammals. | 3."	"Bird and Mammal Register, 1855–60"	Species Registry
Notebook 4	1855–1860	"Notes. *Vertebrata.*" "Notes. Insects 4."	"Species notebook"	"Species Notebook" "Daily Register of Insect Collections"
Notebook 5	10.1858–1865	"Register. 1858 | Birds." "Register. 1858 | Inse<cts>"	"Bird and Insect Register, 1858–62"	Species Registry

Nevertheless, there is little evidence that Wallace interacted much with European society in Singapore during his first visit. Long-time resident J. T. Thomson recalled that the elite society was tight knit, probably not something the gauche Wallace could have entered easily. As Wallace told Silk, "I have not your talent at making acquaintances, and find Singapore very dull. I have not found a single companion. I long for you to walk about with and observe the queer things in the streets of Singapore."[189] *The Calcutta Review* described European society in Singapore in 1861.

> Trade is the principal object with most Europeans, and they are either partners in mercantile firms, clerks in commercial offices, hotel-keepers, or the floating population consisting of travellers, ship-captains, and others. We do not include here the Civil and Military Officers or the Soldiers of the garrison. If society is exclusive any where in the East, it is so in Singapore. Money-making has not many humanising tendencies. The richer adventurers exclude the poorer from social intercourse with them, and the poorer, as they increase in wealth, seek the society of those who once kept them at a distance and exclude those who take their place. The evil effects of such a system is great in a place like Singapore, where society is so contracted, literature unheeded, and amusements few. While the wealthy have their pleasant villas, their wives, and the few recreations they are in a position to command; the young assistants and clerks unite with ship-captains and other pleasure hunters to crowd the drinking shops, politely termed 'bowling alleys' and 'billiard rooms,' or frequent the brothels so numerous in all quarters of the town, or have Malay mistresses at home. This laxity of morals has however greatly diminished during the last few years, and will, we may hope, continue to do so.[190]

For the first week, Wallace and Allen stayed in a hotel. Which hotel was not recorded, only that Wallace thought it expensive. There were then three hotels in Singapore: the London Hotel on the open promenade lawn or Esplanade overlooking the harbour (now called the Padang), the Adelphi Hotel on High Street and the Auckland Hotel, Bonham Street. Wallace and Allen probably stayed at the London Hotel which was the principal hotel. As the advertisement in *The Straits Times* detailed in English, French and Portuguese, the hotel was "under the patronage of the Peninsular and Oriental Company. Commanding a splendid view of the harbour…Passengers…will find every comfort and moderate charges…Hot and cold baths at all hours. Four excellent billiard tables, and a good skittle ground."[191]

The London Hotel was kept by a crafty and enterprising Belgian named Gaston Dutronquoy. He came to Singapore from France in 1839 as a painter.[192] Six months later, he opened the London Hotel on Commercial Square. In 1841, he moved it to the recently vacated house of architect George Coleman (later 3 Coleman Street). In the same year, Dutronquoy offered carriage rentals from the hotel. In 1843, he advertised to provision ships with food and drink.[193] He sometimes used the dining room of the hotel as the "Theatre Royal" where local amateurs enacted comedies. A few talented gentlemen played the parts of ladies, so well in fact that on at least one occasion one of them was mistakenly chatted up afterwards by a visiting gentleman. In 1845, Dutronquoy offered the first photographic (daguerreotype) studio in Singapore.[194] A French traveller in the late 1840s described Dutronquoy as "a *brave Belge*, who could be French, English, or Dutch, precisely as might be desirable—as cosmopolitan in manners as the Dutch-Malay was in religion".[195] In 1851, Dutronquoy moved the hotel to a more upmarket building, the former home of English merchant James Guthrie on the Esplanade, the long open green where the Singapore cricket team played.

It was a little piece of England a long way from home. The hotel offered one wing for families and another for "invalids". Dutronquoy also began to sell ice and flavoured iced drinks for dinner parties, advertised under the heading "ICES! ICES! ICES!"[196] He even offered an "iced can or rather wig" to cool the craniums of those suffering from heat-induced "brain fever"![197] Presumably for those who neglected the protection of the pith helmet.

An unimpressed American traveller described the London Hotel in 1855 as:

> That huge pile of ugly looking buildings, covering a good sized farm…kept in a manner that would disgrace a landlord in the back-woods of Kansas, where your food looks uninviting, and is brought to you at the long, well ventilated and *puncah* cooling dining hall… where your boots and books get mildewed, and your brown leather trunk resembles the skin of a Maltese cat, it has become so mouldy— where your cocoa nut oil lamp, manufactured out of a tumbler of water, on the surface of which a little piece of pith kept floating on the oil, by means of bits of cork, answers for wicking, burns dimly, and you cannot get a candle…where the labyrinth of passages, show cases and rooms, require a man to have a compass, if he does not wish to lose his reckoning[198]

Dutronquoy mysteriously disappeared while gold mining in Johore in 1857. It was rumoured that he was murdered.[199]

From his hotel room, Wallace could probably hear the din of the "bowling alley" at the back of the hotel — also run by the enterprising Dutronquoy. A traveller in 1852 remarked, "This pandemonium is lit up every night, and is filled with the towns-people and others, who play at bowls and drink brandy-and-water until a late hour of the night....very profitable to the proprietor; but it is a great nuisance to the inmates of the hotel."[200]

Wallace lost no time in beginning his investigations. He wrote to the English popular science journal *The Zoologist*:

> I examined the suburbs, and soon came to the conclusion that it was impossible to do anything there in the way of insects, for the virgin forests have been entirely cleared away for four or five miles round (scarcely a tree being left), and plantations of nutmeg and Oreca palm have been formed. These are intersected by straight and dusty roads; and waste places are covered with a vegetation of shrubby Melastonias, which do not seem attractive to insects.[201]

He predicted that if the forest clearing continued in Singapore, the result would be that "countless tribes of interesting insects [would] become extinct".[202]

As no catalogue of species was made until many decades later, it is impossible to know for sure how many have disappeared. Biologist Richard Corlett estimates that between thirty and fifty percent of species have disappeared since Wallace's time.[203] The observation that human occupation vastly reduced the local biodiversity became a frequent refrain from Wallace in the ensuing years of his travels throughout the archipelago.

The London Hotel may have played an unsung role in promoting Wallace's collecting career. We have never had a clue how Wallace came into contact with the Paris Foreign Mission Society in Singapore. It may well have been through the French-speaking Dutronquoy. About a week after Wallace's arrival, he "got permission to stay with a French Roman Catholic missionary who lives about eight miles out of the town and close to the jungle" in a district called Bukit Timah ("tin hill" in Malay), after Singapore's highest hill (530 feet) nearby.[204]

On 26 April 1854, Wallace and Allen travelled eight miles along the dusty road out of the town to St. Joseph's in the Bukit Timah district. The mission was founded in 1846 by the Rev. Anatole Mauduit in the then wild interior of the island. Born in Normandy in 1817, Mauduit arrived in Singapore in 1844. The forest was being cleared and pepper and gambier plantations were spreading, first on the western outskirts of the town and proceeding ever deeper into the island's interior. Gambier was used in dyeing and tanning hides.

The road to Bukit Timah was completed just three years before. St. Joseph's was built on a low lying hill to cater to the hundreds of Chinese plantation labourers in the area. For his efforts amongst Chinese peasants on the other side of the world, Mauduit was paid the miserable sum of about £30 per annum.[205] After some successful fundraising, the original thatched wooden church near Kranji was replaced with a smart, neoclassical church with Palladian portico supported by six Doric columns in 1853. Wallace thought it was "a very pretty church".

Wallace later praised the proselytising efforts of the French Catholics in comparison to the inactivity of the Church of England. "My friend at Bukittima was truly a father to his flock. He preached to them in Chinese every Sunday, and had evenings for discussion and conversation on religion during the week. He had a school to teach their children. His house was open to them day and night."[206] No photographs of Mauduit are known to survive. He died in Singapore in 1858 and was buried in his church. The grave was later moved to the nearby cemetery where his tombstone can still be seen today, tilted to an odd angle by the roots and vines that have colonised the cemetery.

Ida Pfeiffer visited Singapore in 1847 and 1851. The fact that a lady travelling alone could be a tourist in Sarawak and visit Dyak head-hunters and the Moluccas makes Wallace's adventures seem rather less heroic. Perhaps it was for this reason that he never mentioned her in any of his travel writings. But his letters show he was well aware of her itinerary, which he ended up partially following. Pfeiffer provided an excellent description of the pepper and gambier plantations near Bukit Timah:

> The pepper-tree is a tall, bush-like plant, that, when trained and supported with props, will attain a height varying from fifteen to eighteen feet. The pepper grows in small, grape-like bunches, which are first red, then green, and lastly nearly black. The plant begins to bear in the second year.
>
> The greatest height attained by the gambir plant is eight feet. The leaves alone are used in trade: they are first stripped off the stalk, and then boiled down in large coppers. The thick juice is placed in white wooden vessels, and dried in the sun; it is then cut into slips three inches long, and packed up. Gambir is an article that is very useful in dyeing, and hence is very frequently exported to Europe. Pepper plantations are always to be found near a plantation of the gambir plant,

St. Joseph's Church as seen from Bukit Timah Road, 1877 (replaced by the current building in the 1960s). Courtesy of National Museum of Singapore, National Heritage Board.

as the former are always manured with the boiled leaves of the
latter....a common labourer receives three dollars a month[207]

However, these crops eventually depleted the soil and the plantations later died out.

Wallace later wrote, "The mission-house at Bukit-tima was surrounded by several of these wood-topped hills, which were much frequented by woodcutters and sawyers, and offered me an excellent collecting ground for insects."[208] The fact that the hilltops were still forested at all was thanks to the governor who believed that deforestation of the hilltops would adversely affect the climate of the island.[209] Wallace wrote to his mother a few days later. "The missionary speaks English, Malay and Chinese, as well as French, and is a very pleasant man. He has...about 300 Chinese converts....Charles gets on pretty well in health, and catches a few insects; but he is very untidy, as you may imagine by his clothes being all torn to pieces by the time we arrived here. He will no doubt improve and will soon be useful."[210]

So after travelling halfway around the world, Wallace had ended up sweating on this remarkable little island. It was a bizarre hybrid of East and West rather as he was of a hybrid of Victorian social types. It was time to get down to work. From this base camp, he and Allen set out into their first forays to collect specimens in the Eastern Archipelago. Following the roads and tracks of the plantations and woodcutters was essential as the tall lush forests were impenetrably thick and crisscrossed with vines and creepers, many with vicious thorns. A dirt road led to the top of Bukit Timah that had once been used to drive out of Singapore to enjoy the views from the summit. By 1854, it had fallen into disrepair and the jungle was starting to reclaim it.

The diversity of plant species was dazzling. Every tree and branch supported further species of ferns and creeping vines, which themselves were encrusted with smaller growths. The chorus of insect and bird calls varied from place to place in the forest like an ever-changing musical accompaniment. Loudest of all was the deafening siren of the cicadas — the never-ending music of the forest. On his first day, Wallace captured eleven species of long-horned beetles called longicorns with enormously long sweeping antennae.[211] It was a good start.

> In about two months I obtained no less than 700 species of beetles, a large proportion of which were quite new, and among them were 130 distinct kinds of the elegant Longicorns (Cerambycidæ), so much esteemed by collectors. Almost all these were collected in one patch of jungle, not more than a square mile in extent, and in all my subsequent travels in the East I rarely if ever met with so productive a spot. This exceeding productiveness was due in part no doubt to some favourable conditions in the soil, climate, and vegetation, and to the season being very bright and sunny, with sufficient showers to keep everything fresh. But it was also in a great measure dependent, I feel sure, on the labours of the Chinese wood-cutters. They had been at work here for several years, and during all that time had furnished a continual supply of dry and dead and decaying leaves and bark, together with abundance of wood and sawdust, for the nourishment of insects and their larvæ. This had led to the assemblage of a great variety of species in a limited space, and I was the first naturalist who had come to reap the harvest they had prepared. In the same place, and during my walks in other directions, I obtained a fair collection of butterflies and of other orders of insects, so that on the whole I was quite satisfied with these my first attempts to gain a knowledge of the Natural History of the Malay Archipelago.[212]

In a 28 May letter to his mother, Wallace described a typical day at Bukit Timah.

Get up at half-past five. Bath and coffee. Sit down to arrange and put away my insects of the day before, and set them safe out to dry. Charles mending nets, filling pincushions, and getting ready for the day. Breakfast at eight. Out to the jungle at nine. We have to walk up a steep hill to get to it, and always arrive dripping with perspiration. Then we wander about till two or three, generally returning with about 50 or 60 beetles, some very rare and beautiful. Bathe, change clothes, and sit down to kill and pin insects. Charles ditto with flies, bugs and wasps; I do not trust him yet with beetles. Dinner at four. Then to work again till six. Coffee. Read. If very numerous, work at insects till eight or nine. Then to bed.[213]

Wallace was enraptured with the spectacular insects. He was "delighted" with the "hosts of elegantly varied Longicorns".[214] Some he thought of as "my Singapore friends,—beautiful longicorns of the genera Astathes, Glenea and Clytus, the elegant Anthribidæ [fungus weevils], the pretty little Pericallus and Colliuris".[215] He saw the graceful flying lemurs gliding from tree to tree although as something other than "good sport". The Oriental Magpie Robin had "a very beautiful and varied note; it is the commonest bird in Singapore…it feeds much

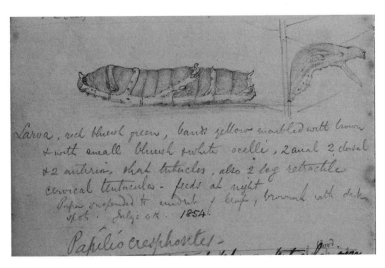

On 4 July 1854, Wallace drew a pencil sketch of a butterfly caterpillar and pupa that he kept at home to observe. Linnean Society Library MS 140b (now in a bundle of Wallace drawings from the Amazon).

on the ground, and its rich black and white plumage makes it a pleasing object".[216]

Amidst this cornucopia of exotic forms, he made one of his earliest theoretical observations in Asia, one which marked his path towards a theory of species, because he recognised "[the] Euplœas [a genus of milkweed butterflies] here quite takes the place of the Heliconidæ of the Amazons, and exactly resemble them in their habits".[217]

Bukit Timah is now a nature reserve which still possesses 840 species of flowering plants and over 10,000 species of animals (including insects) — more than all of North America or Europe. One species of monkey, the greyish brown long-tailed macaque, is still abundant and very tame. The shy black Raffles' banded langur (banded leaf monkey) is now critically endangered. There is a splendid Wallace Education Centre. One half is devoted to the Wallace Environmental Learning Lab (WELL) and there is a "Wallace trail" through the forest nearby where he once collected.

When in town, Wallace used the Singapore Library. The beginning of *Notebook 1* is filled with early reading notes. Wallace's interest in the origins of species, apart from the 1845 letters, now becomes evident for the first time. Indeed, as far as the surviving evidence reveals, all of Wallace's evolutionary theorising occurred in the Eastern Archipelago.

For example, he took notes on the geologist Joseph Jukes' *A sketch of the physical structure of Australia* (1850). Wallace noted the apparent age of Australia, and that the "fossils agree with present fauna & flora" that is that the same unique types (marsupials) which live there now were present in its ancient past. Given a history of the world then universally accepted, in which parts of continents rose and sank beneath the sea, islands were created where the species remained stranded but hence related on nearby islands: "Thus *species* of the different colonies differ; [but their] *genera* the same."[218]

Wallace also took notes on an article on the relationship between the Edentata (anteaters, armadillos and sloths) and reptiles.

> Edentata allied by internal structure to birds and reptiles-but more nearly to reptiles-
> ~~Reptile~~ Birds also allied to reptiles
> ? Mammalia and birds have both branched out of *reptiles*, not from *the other*.[219]

These notes are a tantalising beginning. Wallace was interested in the relationships between groups of living animals as part of the history of life. Language of "branching out" seems to indicate actual genealogical descent. However, the language of the time could be ambiguous so we cannot be sure, because Wallace later believed in branching descent, that this is what he meant in these early notes.

The article he was reading ended with an unattributed quotation: "The true affinities of organic structures branch out irregularly in all directions."[220] This was from naturalist Hugh Edwin Strickland who proposed "branching" relationships between different groups of animals, but did not mean an evolutionary family tree. Strickland merely meant that diagrams of animal relationships should not just have straight lines, but be drawn with branches for similar groups. And before Strickland, branching was used in yet another non-evolutionary sense in Cuvier's famous "radically distinct branches in the animal kingdom".[221]

Tigers

In his published writings, Wallace described his time at Bukit Timah in harrowing terms. "In the midst of this entomological banquet there is, however, one drawback—a sword suspended by a hair over the head of the unfortunate flycatcher: it is the possibility of being eaten up by a tiger! While watching with eager eyes some lovely insect, the thought will occasionally occur that a hungry tiger may be lurking in that dense jungle immediately behind intent upon catching you. Hundreds of Chinamen are annually devoured."[222] In *The Malay Archipelago*, he reported that tigers "kill on an average a Chinaman every day".[223] He carried a heavy Colt revolver during his early excursions. Wallace collecting in the midst of tiger-infested jungles is a romantic image repeated by all modern writers. But could this be true? A death per day? "Hundreds" annually?

Reports of people killed by tigers were published in Singapore newspapers since 1831.[224] In 1832, the *Asiatic journal* reported that tigers were beginning to infest Singapore. The same journal reported in 1843 that "more than three hundred natives are every year carried off" by tigers.[225] In the same year, the *Singapore Free Press* attributed bizarre cunning (not to mention the ability to suck blood!) to a tiger:

> A Chinaman while engaged in constructing a Tiger pit at the back of
> Mr Balestier's Sugar Plantation was pounced upon by a Tiger who

after killing him and sucking the blood, walked off into the jungle leaving the body behind. We suppose the Tiger knowing the object of the Chinaman's labours took this opportunity of giving a striking manifestation of his profound disapproval of all such latent and unfair methods of taking an enemy at disadvantage instead of meeting him face to face in fair and open fight.[226]

In fact, virtually all writers on Singapore in the 1850s and 1860s claimed that tigers killed a person per day.[227] Ida Pfeiffer wrote that in 1851 "no less than the almost incredible number of four hundred persons were destroyed by [tigers]".[228] By April 1854, the *Southern literary messenger* reported, "Incredible as it may appear, it is nevertheless true, that from the immediate vicinity of Singapore about 400 persons are annually carried off by Tigers."[229] The *Illustrated magazine of art* from New York stated in the same year that "in Singapore, a man is killed by a tiger every day, on average". Clearly, Singapore had achieved "a melancholy yet a world-wide reputation" for tigers.[230]

The widespread deforestation for pepper and gambier plantations changed the environment of the island. Only after forest was cleared and labourers began living nearby did tigers begin to attack humans. Tigers were seen swimming across the straits from the peninsula. The gambier was grown tightly packed and grew five to six feet high and formed perfect cover for tigers to stalk prey. A traveller reported, "The victim dies immediately with the vertebrae of the spine broken, and is then carried off into the jungle to be devoured by the brute at leisure."[231] One English visitor recalled, "Two Chinamen cleared a space in the woods for a garden; but, being mightily afraid of tigers, one worked, while the other beat a metal drum called a gong, the noise of which they thought would scare [tigers] away. One day the working man heard the gong cease, and, looking up, he beheld man and gong both carried off by a large tiger."[232] No wonder the Singapore Library kept a copy of "Tiger Shooting; by Lieut W. Rice, Bombay Army".[233]

The Straits Times reported in January 1854, "Tigers are committing serious havoc amongst the planters' labourers employed in the jungle, to the great terror of the Chinese. The frequent carrying off of natives must arise from the number of these ferocious animals having greatly increased, or that they are more bold and desperate than heretofore."[234] There were even reports of sightings and footprints in the town. Yet, *The Straits Times* seems to contain no reports of a death by tigers in all of 1854. While actually at Bukit Timah in 1854, Wallace wrote to his mother, perhaps to reassure her (contradicting his more exciting accounts elsewhere), "I have not seen any tigers yet & do not expect to, for there are not many in this neighborhood & there has not been a man killed at the place for two years."

But the following year, the death toll increased. In August 1855, *The Straits Times* estimated that two bodies turned up per week, which would amount to 110 deaths![235] But according to the same newspaper's year-end statistics, only twenty-one deaths were reported in all of 1855.[236] This is very far indeed from 300–400.

A few writers doubted the claims of a death per day. One of these was the naturalist Cuthbert Collingwood who referred to conversations with long-term residents of Singapore who claimed that tigers were rare.[237] Others thought that some of the bodies were in fact murder victims disguised to look like tiger victims.

In making these claims, writers, including Wallace, were in fact repeating an old, if irresistible, legend. The increase in deaths to twenty-one in 1855 only reinforced the death-per-day legend. The satirical London magazine *Punch* responded to the 1855 reports:

A SCHOOL FOR TIGERS IN THE EAST.

RAPID DEPOPULATION OF SINGAPORE BY TIGERS. — Two deaths by tigers every week (says the *Singapore Free Press*) are read of in the papers just about as much a matter of course as the arrival or departure of the P. and O. Company's steamers....

If the population of Singapore is really being converted into food for tigers, and the inhabitants are departing as regularly as the steamers, it is high time that something should be done to save the remnant of the populace. Considering that the tigers have evidently got the upper hand, we think they show a sort of moderation in taking only two inhabitants per week, and there is consequently no hope of any further diminution, for it is clear that the brutes are already on what may be considered low diet. We cannot be surprised at the anxiety of the Editor of the *Singapore Free Press*, who may any day be selected as a moiety of the weekly allowance of the somewhat abstemious tigers, who appear to be practising the negative virtue of moderation and regular living....

The Singapore journalist expresses his fear that the "evil will go on increasing,"—or in other words, that the population will go on diminishing—and we fully sympathise with his editorial fears; for even should he be so lucky as to escape till after every other inhabitant is disposed of, it would be but a sorry consolation to feel oneself constituting the last mouthful at a feast of tigers.[238]

Rewards (between $50 and $150) were offered by the government from time to time for the killing or capture of a tiger. John Cameron, the editor of *The Straits Times* who repeated the death-per-day story, recorded in 1865:

> There is an American here, an old backwoodsman, who has for many years devoted himself to the destruction of these animals; he is known as Carol, the tiger hunter; but he has had but poor sport of it in Singapore, having only upon two occasions succeeded in obtaining the reward—though I believe he has killed many tigers in Johore. He is of eccentric habits, but is kindly treated by the Chinese planters throughout the island and by the Malays in Johore, and seems content with the hunter's life.[239]

Carol the tiger hunter was mentioned a few times in the Singapore newspapers in 1863.[240] Elsewhere, he was described as a "French Canadian, named Carrol [sic], who left his country during the disturbances in 1838. He used to live in the jungle almost altogether, and he made tiger hunting a business for the sake of the rewards, which were considerable at one time, about 1860, as the Chamber of Commerce gave a reward as well as the Government, and the body was also worth money. Carrol died in the General Hospital. He was an elderly man; a very fine rifle shot, and was known because he always wore a gold ring halfway up a long greyish beard, like a necktie ring."[241]

The last person killed by a tiger in Singapore was in 1890. Today, tourists are often told that the last tiger in Singapore was shot in the Long Bar of Raffles Hotel, but this was an escaped "native show" animal in 1902.[242] It is widely accepted that the last wild tiger in Singapore was shot in October 1930. A famous photograph of the event is frequently reprinted and is on display boards at the Wallace trail and the visitor centre at the Bukit Timah Nature Reserve.[243] Yet five years later, *The Straits Times* reported that tigers had been living near Mandai Road for over two years.[244]

The Chinese Riots

While Wallace was staying at Bukit Timah, "one morning 600 Chinese passed our house in straggling single file, armed, in the most impromptu manner, with guns, matchlocks, pikes, swords, huge three-pronged fishing-spears, knives, hatchets, and long sharpened stakes of hard wood. They were going to buy rice, they said, but they were stopped on the road by a party of about a dozen Malay police, five of them shot, and the rest turned back. The disturbance lasted a week, and even now men are still occasionally killed, nobody knows why."[245] This observation has never been explained.

Wallace was witnessing one of the most destructive episodes in Singapore's history — the Chinese Riots of 1854. The conflict was between the immigrant communities of Hokkien and Teochew Chinese from different regions of China (the southern region of Fujian province and the Chaoshan region of eastern Guangdong province, respectively).[246] Part of the conflict between the two groups was control over the pepper and gambier plantations in the very area where Wallace was collecting.

The riots began on Friday morning, 5 May 1854 during a quarrel between a Hokkien shopkeeper and a Teochew customer over the price of some rice.[247] This soon erupted into large-scale and very violent riots between the Hokkiens and Teochews, which then spread over the whole island. Pitched battles took place in the town and across the countryside. Houses were burnt, shops looted and opponents killed and mutilated.

On the morning of Tuesday 9 May, a large body of Teochews marching down Bukit Timah Road was stopped by police.[248] This may have been the group recorded by Wallace. The Chinese in the interior were said to be all Teochews. As the supplies of rice in the countryside, the only food available, were either captured or stopped, groups of Teochews marched together into town to procure rice. Hence, the group Wallace saw were Teochews and indeed on their way to buy rice as he was informed. It was an exceptionally hot day. Edward Rohde, a twenty-five-year-old German farm manager died from "a stroke of the sun in the jungle at Tangling". Given such fatalities, it is easier to appreciate how protection from the sun, such as pith helmets, was taken so seriously.

On 10 May, troops were sent around the island by the government steamer the *Hooghly* to come in behind the rioters and surprise them. At the same time, forces marching out from the town came from the other direction and the large-scale riots finally ended. A few more skirmishes and murders occurred over the following days. The rioting, looting, violent and often bloody clashes lasted for ten days. According to *Allen's Indian Mail*, 220 people were wounded and 400 were killed by rioters, but 850 rioters were themselves killed; two Westerners were wounded by accident, 512 prisoners were taken, 280 houses were burnt down, fifty-three gardens destroyed and fifty-three shops ransacked.[249] It was probably the bloodiest episode in Singapore's history until the Second World War.

Yet, despite the dramatic events shuddering the island and the smoke of burning houses encircling Bukit Timah, Wallace was consumed with collecting insects in the forest. This underscores just how peripheral Wallace was at the time, an unknown European busy with an unusual occupation. It also demonstrates that social context does not necessarily impact on the development of science. This unrest could be compared to that stirred up by the freethinking radicals who challenged Christianity and the dons of Cambridge University while Darwin was a student there. It is an intriguing story happening in the same time and place, but we have no evidence that Darwin cared or took any notice.

Wallace spent a week on the tiny nearby island of Pulau Ubin (Pulau is Malay for "island") quarried for its granite.[250] He did not reap a good harvest, nor did he record specimens from Pulau Ubin separately from Singapore. This may have been when he observed "a wild pig swimming across the arm of the sea that separates Singapore from the Peninsula of Malacca".[251]

At the end of May, Wallace took his first consignment of 1,087 insects to Hamilton, Gray and Co. on the south side of the Singapore River to be shipped by care of "Mr. J. Deal Jun. Custom House Agent High Street Southampton *with great care*" to Stevens in London.[252] Wallace's notebook records that he sent the consignment on 28 May. No ships are listed in *The Straits Times* as departing for London on that day; however, the brig *Seaton* was listed as departing for London on 27 and also 31 May.[253] There may have been a delay or false start. His consignments consisted of items for sale and others that were "private" for his personal collection. He noted on 8 July that he had collected 4,380 insects in Singapore in nine weeks.[254] Wallace later wrote, "Even the best collections I have been able to make can only be looked upon as samples of the productions of these luxuriant regions."[255]

Malacca, from *Eastern Islands or Malay Archipelago*. Engraved by J. & C. Walker, 1836.

Chapter 4

MALACCA AND BORNEO

After sailing from London on 15 March 1854, the barque *Eliza Thornton* arrived in Singapore on 12 July. Amongst her cargo were boxes of equipment and books Wallace had shipped to himself via the Cape route.[256] Wallace and his family sent boxes back and forth via the Cape and normally used the faster and more expensive overland route only for letters.[257] In search of richer fields for collecting, Wallace and Allen sailed north to Malacca probably the next day on 13 July in the sixty-eight-tonne schooner, *Kim Soon Hin* "with about fifty Chinese, Hindoos and Portuguese passengers". Wallace and Allen were disappointed to learn that passengers on such vessels in the East were expected to bring their own food. They had to make do with "rice and curry".[258]

Malacca was an "old and picturesque town…crowded along the banks of the small river, and consists of narrow streets of shops and dwelling-houses, occupied by the descendants of the Portuguese, and by Chinamen. In the suburbs are the houses of the English officials and of a few Portuguese merchants, embedded in groves of palms and fruit-trees, whose varied and beautiful foliage furnishes a pleasing relief to the eye, as well as most grateful shade. The old fort, the large Government House, and the ruins of a cathedral, attest the former wealth and importance of this place, which was once as much the centre of Eastern trade as Singapore is now."[259]

Again, Wallace's French missionary friends provided the contacts. Wallace and Allen stayed two days with an unidentified French missionary. However, Wallace recorded the barometer reading of "Mr Favrex" at the front of *Notebook* 1 on 16 July 1854. This was Father Pierre Favre who was assigned to the Chinese Catholics at Malacca.[260]

Durian

In Malacca, Wallace first tried durian, "a fruit about which very little is known in England". He was not impressed. The smell was so "offensive" he could barely bring himself to taste it. The powerful smell of durian has always shocked Westerners, and it still does. It is the only food actually banned on public transportation in several countries! Another English traveller described his encounter with durian in 1852, "No sooner had we divided the shell that holds that delicious pulp, whose exquisite flavour, as we are told, no human art could equal, than our olfactory nerves were assailed with such an effluvia, as well nigh scared us from our propriety.…[as] if it was reared upon a dunghill, or…the fowl and offensive atmosphere of a common sewer."[261] Like Wallace, J. T. Thomson was first introduced to durian in Malacca. Thomson described his encounter with durian in vivid, almost eye-watering terms:

> I look and see a roughlooking substance full of yellow yokes or seeds borne forward to the table. But what is this odour? I looked about me furtively, and my friend smiled. I took a momentary glance at the lady, and she laughed outright. The fruit is placed on the table. Shades of Cloacina! [the Roman goddess of the sewers!]

what is this? I give a piteous glance to my host: he laughs immoderately. I look at the contents of the fruit dish, and learn that the atrociously foetid odours come from it....I would have held my nose did good breeding allow it, but I resigned myself to my fate, and looked on. My host proceeded to open up the disgusting entrails of the horrid-looking vegetable, and they send forth an odour of rotten eggs stirred up with decayed onions....that my polished friends should eat such an abomination as this, was beyond my conception.[262]

Unlike the more adventurous Wallace, Thomson could not bring himself to taste the offending fruit. Wallace recalled that later "in Borneo I found a ripe fruit on the ground, and, eating it out of doors, I at once became a confirmed Durian eater".

Wallace attributed his conversion to trying durian out of doors. In fact, there are hundreds of varieties of durian and they differ greatly in smell and taste. These differences are obvious in the large range of prices in the markets of Singapore today. A fine high-quality durian at the stall near my house can cost $90. Some varieties are simply too overpowering for an uninitiated Westerner to attempt. Other varieties are more sweet and subtle and offer an excellent way to cross the Rubicon of durian's odiferous barrier. Once converted, Wallace described durian as the "king of fruits" and as "a food of the most exquisite flavour it is unsurpassed". In *The Malay Archipelago,* he expanded on an earlier letter description:

This pulp is the eatable part, and its consistence and flavour are indescribable. A rich butter-like custard highly flavoured with almonds gives the best general idea of it, but intermingled with it come wafts of flavour that call to mind cream-cheese, onion-sauce, brown sherry, and other incongruities. Then there is a rich glutinous smoothness in the pulp which nothing else possesses, but which adds to its delicacy. It is neither acid, nor sweet, nor juicy, yet one feels the want of none of these qualities, for it is perfect as it is. It produces no nausea or other bad effect, and the more you eat of it the less you feel inclined to stop. In fact to eat Durians is a new sensation, worth a voyage to the East to experience.[263]

Wallace also made a more serious point with the durian fruit, using it to quite literally knock natural theology on the head.

Poets and moralists, judging from our English trees and fruits, have thought that there existed an inverse proportion between the size of the one and the other, so that their fall should be harmless to man. Two of the most formidable fruits known, however, the Brazil Nut (Bertholletia) and the Durian, grow on lofty trees, from which they both fall as soon as they are ripe, and often wound or kill those who seek to obtain them. From this we may learn two things:—first, not to draw conclusions from a very partial view of Nature; and secondly, that trees and fruits and all the varied productions of the animal and vegetable kingdoms, have not been created solely for the use and convenience of man.[264]

This was exactly the sort of anti-natural theology Wallace had been exposed to in the mechanics' institutes, here applied to new observations.

After two days in Malacca, during which Wallace met the Resident Councillor and other Europeans, Wallace and Allen travelled thirteen miles inland to stay with some Chinese Catholic converts at a mining settlement called Gading. Wallace hired two Portuguese men, "one as a cook, the other to shoot and skin birds".[265] Wallace was disgusted by a bucket used as a toilet in a corner of the house where he stayed — the contents collected for fertiliser. There were even buckets kept in front of Chinese houses to collect public excrement for this purpose! Pig manure, which smells almost as bad, was also collected. Wallace next wrote to his mother:

> My guns are both very good, but I find powder and shot in Singapore cheaper than in London, so I need not have troubled myself to take any. So far both I and Charles have enjoyed excellent health. He can now shoot pretty well, and is so fond of it that I can hardly get him to do anything else. He will soon be very useful, if I can cure him of his incorrigible carelessness. At present I cannot trust him to do the smallest thing without watching that he does it properly, so that I might generally as well do it myself.[266]

Wallace shot many beautiful and brightly coloured birds and collected insects in the forest for his collection.

After two weeks, Allen caught a fever so they returned to Malacca where Wallace too succumbed to illness. A fever was then sweeping through Malacca and a few days before had struck down the Resident Councillor, Captain Ilay Ferrier.[267] The government doctor made Wallace take prodigious quantities of quinine daily for a week and he recovered. "I see now how to treat the fever, and shall commence at once when the symptoms again appear. I never took half enough quinine in America to cure me."[268] The doctor was never named but was possibly

Theodore Cantor, a Danish physician and surgeon of the Bengal Medical Service. Cantor was one of the most skilled and accomplished naturalists in the region. He collected and published widely on insects, molluscs, fish, reptiles, birds and mammals. He also sold specimens through Stevens in London. In other words, Wallace may have found there was competition in Malacca.

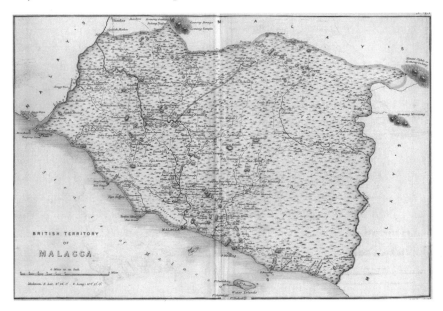

Map of Malacca. Mount Ophir is in the upper right corner. *An atlas of the Southern part of India*. Madras: Pharaoh and Co., 1854.

Wallace's next excursion was to Ayer-panas ("hot spring") about four miles to the southeast where there was a government bungalow offered to Wallace by the recently deceased Captain Ferrier. Wallace was accompanied by the twenty-year-old son of a dealer in natural history objects at Malacca named George Rappa, Jr., "who wished for change of air and exercise, and whose acquaintance with the Malays and their language was of much use to me".[269]

> Here I first saw the huge bats commonly called "flying foxes," whose wings often expand five feet. They came in the evening to the fruit-trees near the house, looking more like aerial machines than any living creatures. It was truly an extraordinary sight to behold these great-winged animals for the first time, so totally different are they from anything we can behold in Europe. They are much esteemed for food by all the inhabitants of Malacca, and we soon had an

opportunity of tasting one, but it was too tough for me to pronounce an unprejudiced opinion on its merits as an article of food.[270]

The Mountain of Gold

The most conspicuous landmark visible from Malacca is Mount Ophir (Gunung Ledang). Westerners had used the name "Mount Ophir" since the end of the 18th century. Ophir was a region in ancient mythology whose streets were made of gold from a nearby mountain. Since Mount Ophir was said to be full of gold, many mines were opened there.

Wallace, accompanied by Rappa, hired five or six Malays to act as porters.[270] As they marched through the jungle, lithe little leeches attacked from all sides. "The little creatures are as tough as leather; nothing will kill them but cutting them in pieces."[272] When his drinking water ran out, Wallace even tried the liquid from pitcher plants which, although full of insects, he found "very palatable though rather warm".[273] He seems not to have speculated on the function of the plants' liquid reservoirs. One writer in 1852 still thought that the reason pitcher plants trapped insects was because "the Creator" designed them to keep down the number of insects.[274]

The party marched thirty miles to camp for a week at the foot of the mountain. Wallace captured hundreds of new and rare insects. He heard that elephants, rhinoceroses and tigers lived in the area but they did not catch sight of any. One day, they ascended the summit of Mount Ophir. Although no geologist, he noted: "by means of careful observation with Adie Sympiesometer ascertained to be 3,920 feet above the sea. It is an isolated mountain, & in fact there appears to be no connected chain in this part of the peninsula."[275]

Darwin Turns to Species

At the same time, half a world away in Kent, Darwin finally finished his barnacles. On the very day that he packed up the last borrowed specimens, he noted in his journal: "Sept 9 Began sorting notes for Species theory".[276] At last Darwin turned full time to his species theory. "From September 1854" he wrote in his autobiography, "I devoted all my time to arranging my huge pile of notes, to observing, and experimenting, in relation to the transmutation of species."[277] To pursue his research, he experimented with the natural dispersal of seeds in artificial seawater and took up pigeon breeding. He wrote to his cousin Fox in March 1855, referring to his species work in the most casual way, "I forget whether I ever told you what the object of my present work is—it is to view all facts that I can master (eheu, eheu, how ignorant I find I am) in Nat. History, (as on geograph. distribution, palaeontology, classification Hybridism, domestic animals & plants &c &c &c) to see how far they favour or are opposed to the notion that wild species are mutable or immutable: I mean with my utmost power to give all arguments & facts on both sides. I have a *number* of people helping me in every way, & giving me most valuable assistance; but I often doubt whether the subject will not quite overpower me."[278] This was, again and again, Darwin's explicit concern, that the subject would be too much for him. He never said he was afraid.

Singapore, 25 September–17 October 1854

Wallace wrote to Norton Shaw, Secretary of the Royal Geographical Society, "At Malacca I collected insects extensively, also land shells, birds &c. & obtained an acquaintance with the inhabitants, the scenery, & the animals & vegetable productions of this portion of the East, which will be of great value in my exploration of the less known country I have now reached."[279] It was a sort of practice expedition from the safety of British Singapore to the other British territory.

At the end of September, Wallace and Allen returned to Singapore, presumably on the barque *John Bibby*.[280] Singapore was quiet and back to business as usual after the turmoil of the riots in May. They apparently stayed again with the Rev. Mauduit at Bukit Timah. Perhaps, having met the naturalists Cantor and Rappa in Malacca, Wallace realised that to find novel specimens he would have to travel farther east or north. At first, he planned to go to Cambodia with one of the French missionaries. They told him it would make excellent collecting ground. But the missionary's trip to Cambodia was several months away. Wallace needed to collect and generate revenue continuously.

Sir James Brooke had arrived in Singapore on 3 September from Sarawak on the brig HMS *Lily* to attend a Commission examining his conduct in Sarawak. The local newspapers were full of discussion. The Commission officially closed on 20 November and the commissioners departed Singapore for Calcutta aboard the steamer *Shanghai*.[281] That December, a book in support of Sir James was published in Singapore detailing the case.[282] No charges were brought against Sir James and he was effectively exonerated of the accusations.

Although he was distracted by the ongoing Commission, Sir James "most kindly offered [Wallace] every assistance in exploring the territories under his rule". He supplied Wallace with a letter to his nephew and heir apparent Captain John Brooke "to make me at home till he arrives, which may be a month, perhaps".[283]

Wallace spent most of this stay in Singapore "packing up, arranging & cataloguing all my collection: about 6000 specimens of insects, birds, quadrupeds & shells". Wallace was, after all, a commercial collector. As he replied rather sharply to Silk, "Sir W. Hooker's remarks are encouraging, but I cannot afford to collect plants. I have to work for a living, and plants would not pay unless I collect nothing else, which I cannot do, being too much interested in zoology."[284]

Wallace recorded in *Notebook 1*: "2nd small Box overland, sent Oct. 16th. 1854".[285] Only the P&O ships carried the "overland" route and none of their ships sailed on the 16th. The P&O steamer *Cadiz* departed on the 19th.[286] He eventually earned £20 from the consignment, almost equal to his annual salary at the school in Leicester.[287] He sent his third consignment to Stevens on the 534-tonne sailing ship *Royal Alice* which left Singapore on 24 October. Wallace recorded the 17th in his notebook.[288] The constant discrepancies between his dates and those of ships sailing presumably means that Wallace recorded when he deposited his consignment with Hamilton and Co. or when he believed the ship would sail.

Wallace was becoming rather tired of his young assistant, as he complained to his mother, "If it were not for the expense, I would send Charles home. I think I could not have chanced upon a more untidy or careless boy. After 5 months I have still to tell him to put things away after he has been using them as the first week. He is very strong & able to do any thing, but can be trusted to do nothing out of my sight."[289] It must have been at the same time rather satisfying to bemoan the help to his mother who once had domestic servants of her own but experienced such a socio-economic downslide after her marriage. Her son Alfred was getting back where the Wallaces belonged.

Wallace's mother had sent a care package, but most of the food was spoiled by the long transit in a sailing ship around the world. "The pudding & triffle cake were masses of mould & insects, quite useless. The covers of the jars were all eaten through by ants and small insects. The currant jam was mostly spoilt, sour. The gooseberry remained very good."[290]

He was preparing for his future travels. He forwarded his letter from the Dutch government to Batavia (Jakarta) the capital of the Dutch East Indies. He received a civil reply informing him that "I should meet with no obstructions in visiting any of their eastern possessions".[291] It was time to proceed deeper into the Eastern Archipelago.

Singapore to Sarawak. "Mr. Wallace's route" map MA1.[292]

Borneo, 29 October 1854–10 February 1856

On 17 October 1854, Wallace and Allen sailed on the little 154-tonne brig *Weraff* for the territory of Sarawak on the great island of Borneo 500 miles to the east of Singapore.[293] The *Weraff* was registered at Singapore since 1853 and regularly made journeys between Singapore and Sarawak as well as Penang and even Calcutta. Wallace and Allen arrived at Sarawak on 29 October. Wallace erroneously stated 1 November in *The Malay Archipelago*.[294] Wallace would spend longer on Borneo than any other island in the archipelago.

Wallace left no journal account of his voyage to Sarawak or his first five months there. Presumably the lost notebook covered this time. His first surviving narrative account begins at the Si Munjan coal works to the east in March 1855. Ida Pfeiffer had also travelled from Singapore to Sarawak in 1851 on the barque *Trident*.[295] It took the *Trident* twelve days to reach the mouth of the Sarawak River, then, with another half a day waiting for the tide to rise high enough to enter the mouth of the river, it could take a further three days for the twenty-five miles up the river to the capital. This explains why Wallace's journey took from the 17th to the 29th. The riverbanks along the whole journey upriver were clothed in virgin mangrove forest until a short distance from the town. The former danger of pirates in their praus with low draft had made the area far from the town too dangerous.

Sarawak is a region of northwest Borneo circumscribed from the rest of the great island by ranges of hills that form the central mountain range. It had about 300 miles of coastline. Sarawak had been governed, somewhat loosely, under the Sultan of Brunei to the north. In 1841, Brooke, in the area with his schooner *Royalist*, provided assistance to restore order there during a local power struggle. Like Raffles, Brooke offered to remove a problem of the local ruler and as reward received possession of territory. In return, Brooke was made the Rajah of Sarawak, effectively founding it as a kingdom under a line of "white rajahs". Brooke's heirs continued to rule until the Japanese invasion in 1941. Thereafter, it briefly became a British colony. It remains distinct today as a state of Malaysia, whereas the rest of Borneo (apart from Sabah and Brunei) belongs to Indonesia which consists of the former Dutch East Indies.

The population of Sarawak was about 27,000, with the capital Kuching with 12,000. Ethnographically, Sarawak still had a large proportion of tribal peoples such as the Iban and Dyaks in addition to Malays. Following Brooke's establishment, there was a surge in Chinese immigration. Brooke made Sarawak a free port like Singapore. Trade dramatically rose and flourished. There was rice cultivation, opium farming, sago, gutta-percha and gold, diamond and antimony mining. Pfeiffer described the capital in 1851:

> The town of Sarawak has neither streets nor squares; but consists of a throng of huts, crowded together without any order or symmetry. They are constructed out of the nipa palm, and stand on piles eight or ten feet high…The population of Sarawak consists of Malays and Chinese; for the few Dyaks you see form no families; they are mostly either in service, or they have come here on business. The Chinese and Malays inhabit separate quarters of the town, and the former depart in nothing from the habits of life and costume of their native country…The country round Sarawak is very pretty, and rendered prettier by a few European houses that are scattered about on the hills around, where are also a small fort, a neat church and mission house, and a court of justice. All these edifices—Rajah Brooke's residence not excepted—are built of wood.[296]

She described how the Malays were Sunni Muslims but did not practise in the same way as those she had seen in the Middle East. For example, the women were not veiled, and indeed barely clothed in a loose sarong, and were free to move about as they pleased and even divorce was said to be easy and common.

Kuching had a small European court surrounding Sir James. Several were recruited during his visit to Britain in 1847. His sister's eldest son, Captain John Brooke, was heir apparent, although later disinherited in favour of his brother Charles Brooke in 1865. A former navy midshipman and grandson of the seventh Earl of Elgin, Charles T. C. Grant (1831–1891), was private secretary and, it is rumoured, Brooke's lover. Spenser St. John (1825–1910) was Brooke's secretary and from 1851 Acting Commissioner and Consul General, Britain's diplomatic representative in Borneo, after the title was resigned by Sir James over the accusations that led to the Borneo Commission. St. John later wrote an important book about his life in the East and two biographies of Sir James.

Also recruited in 1847 was Francis Thomas McDougall (1817–1886), the first Bishop of Labuan and Sarawak from 1849 to 1868 and his wife Harriette (1817–1886). They conducted a mission house, church and school and oversaw the hospital. A Danish merchant, Ludvig Verner Helms, worked on the incipient mining concerns that would become the Borneo Company Ltd., in 1856.[297] Helms had a great interest in natural history and befriended Wallace.[298]

Wallace brought his letter of introduction to Captain Brooke and so was allowed to stay in the house of Sir James. The house, his second in Sarawak, was built in 1844 on the left bank of the river. St. John described it as oblong and situated

> on a rising knoll between two running streams, with the broad river flowing below. It was a pretty spot. A four-roomed, lofty house, surrounded by broad verandas; in front his well-stocked library, a splendid hall or dining-room, with a couple of bedrooms behind them. When I knew it, a special wing had been added for Mr Brooke's own use, and the rest was given up to his followers. Around the house was the thick foliage of fruit-trees, with lawns and paths bordered by jasmine plants and the Sundal Malam, that only gives out its fragrant perfume during the night. Pigeon-houses, kitchens, and servants' rooms were partly hidden by trees, and here and there were planted and tended with uncommon care some rose plants, Mr Brooke's favourite flower[299]

James Brooke's second bungalow. J.A. St. John, *Views in the Eastern Archipelago: Borneo, Sarawak, Labuan*. London, 1847.

Harriette McDougall described the house in 1848: "amidst gardens, and fruit-trees, stands the Rajah's house, and several other pretty Bungalows, belonging to English gentlemen…The roses and jessamines, which grew luxuriantly under the verandahs, perfumed the air, and the flights of cooing blue and white pigeons, which had their dove-cot near the house, gave us a gentle greeting."[300] St. John noted the "noble library which was once the pride of Sarawak" consisted of over 2,000 volumes of "the best historians and essayists, all the poets, voyages and travels, books of reference, and a whole library of theology—books on every side of the question; and I well remember a sneaking parson from Singapore, who came on a visit, examining the library, and when he found the works of Priestley and Channing along-side of those of Horsley and Pye Smith, going away and privately denouncing 'the Rajah as an infidel and an atheist, or, worse still, a Unitarian.'"[301] This house and its library were destroyed by fire in February 1857 in an attempted coup by a Chinese faction.

Sir James and St. John were still in Singapore for the Borneo Commission when Wallace and Allen arrived. St. John returned around the beginning of December and Sir James, who left Singapore on HMS *Rapid* on 6 December, returned in mid-December.[302] St. John recalled, "Now commenced a really quiet life. Sir James was free from the anxieties caused by the coming of the Commission, and devoted himself to the happiness of the people....This was perhaps the happiest time he ever spent. The country was progressing...he could live in the capital or in his country cottage as he felt inclined, and he returned to a course of chess and pleasant reading."[303]

Wallace had been collecting up and down the river as best as could despite the frequent rains of the wet season. In late November, he wrote a brief article on the insects of Malacca for the *Zoologist*.[304] He then spent a warm tropical Christmas with Sir James and his circle in Sarawak. Wallace wrote to his mother, perhaps a little pleased with his proximity to such a famous man. "I have now seen a good deal of Sir James, and the more I see of him the more I admire him." He was "a gentleman and a nobleman in the noblest sense of both words".[305] Clearly, Wallace found more society and inclusion in Sarawak than he had in Singapore where most Westerners were merchants or administrators and were less inclusive and less interested in natural history. Wallace and Sir James became good friends. They shared an interest in chess and discussed many social and natural history topics.

St. John's recollections of Wallace at this time are often quoted. It should be remembered, however, that they are retrospective and written in 1879 after (and clearly in light of) the publication of Darwin's *Origin of species* and after Wallace had become a prominent name in British science.

> We had at this time in Sarawak the famous naturalist, traveller, and philosopher, Mr Alfred Wallace, who was then elaborating in his mind the theory which was simultaneously worked out by Darwin— the theory of the origin of species; and if he could not convince us that our ugly neighbours, the orang-outangs, were our ancestors, he pleased, delighted, and instructed us by his clever and inexhaustible flow of talk—really good talk. The Rajah was pleased to have so clever a man with him, as it excited his mind, and brought out his brilliant ideas.[306]

Clearly, Wallace made a lasting impression, even if the details of St. John's recollection are not to be trusted. Wallace never hinted that humans were descended from orangutans.[307] Bishop McDougall recalled that one evening "Captain Brooke's insect treasures were produced, for a visit from Mr. Wallace the naturalist had given rise to a rage for collecting".[308] On 10 February 1855, Wallace sent two consignments to Stevens which left Sarawak on the *Weraff*. The 500 land and water shells alone netted £15.[309]

Jungle view, Sarawak. J.A. St. John, *Views in the Eastern Archipelago: Borneo, Sarawak, Labuan.*
London, 1847.

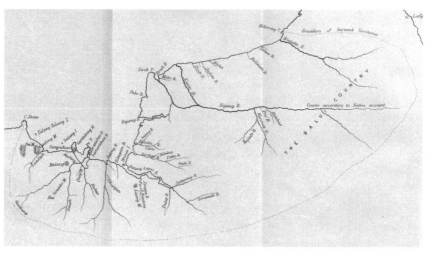

Extract from map of Sarawak. S. St. John, *Life in the forests of the Far East,* 1862.

Chapter 5

TESTING THE WATERS

The wet season in Sarawak lasts from December to March. Wallace made trips along the river from the "picturesque limestone Mountains and Chinese gold-fields of Bow and Bedé" down to the estuary, collecting wherever he could. But the wet weather meant a poor harvest.[310] In February, he later recalled he stayed in a "little house at the mouth of the Sarawak river, at the foot of the Santubong mountain".[311] This was likely a house owned by Sir James, which Harriette McDougall described as "a little Dyak house built on high poles…It was an inconvenient little place, into which you climbed up a steep ladder—only one room, in fact, with a verandah…the beauty of that shore made the house a secondary consideration. A small Malay village nestled in cocoa-nut palms at the foot of Santubong; in front lay a smooth stretch of sand, and a belt of casuarina-trees always whispering, without any apparent wind to move their slender spines."[312] St. John recalled, "The gems of the scene are the little emerald isles that are scattered over the surface of the bay, presenting their pretty beaches of glistering sand, or their lovely foliage, drooping to kiss the rippling waves."[313]

Wallace recollected that he was alone in the house with only "one Malay boy as cook". This is possibly the first reference to Ali. "When I was at Sarawak in 1855 I engaged a Malay boy named Ali as a personal servant, and also to help me to learn the Malay language by the necessity of constant communication with him. He was attentive and clean, and could cook very well."[314] He was also a good boatman. Ali was about fourteen or fifteen years old. The term "boy" meant servant in the colonial east.[315] Wallace did not mention where Allen might have been.

Outside the sky was as grey as an English autumn though the thick torrents of rain and balmy humidity were unmistakably tropical; "during the evenings and wet days I had nothing to do but to look over my books".[316] His reading during the past months clearly reflected his growing interest in the history of life. He read some of his copy of Lyell's *Principles of geology*. Lyell summarised the current state of knowledge for the successive appearance and disappearance of species over geological time. Lyell was unusual amongst geologists of the day in not accepting the progressive nature of the fossil record. But his interpretation, just like his dismissal of Lamarck's transmutation, was often not as influential for his readers as the evidence surveyed. Probably in Singapore, or perhaps in Sir James' library at Sarawak, Wallace took notes on the Swiss zoologist François Jules Pictet de la Rive's *Traité élémentaire de paléontologie* with its Cuvierian reflections on the character of palaeontology. Pictet's summary of the fossil record made a deep impression on Wallace.[317]

Wallace had already read the second edition of Darwin's *Journal of researches* with its remarks on fossil succession of the same mammal types in South America. "This wonderful relationship in the same continent between the dead and the living will, I do not doubt, hereafter throw more light on the appearance of organic beings on our earth, and their disappearance from it, than any other class of facts". The same book had highly suggestive remarks about the species of the Galápagos: "Most of the organic productions are aboriginal creations, found nowhere else" yet "all show a marked relationship with those of America".[318] In the first edition, Darwin had noted "the confirmation of the law that existing animals have a close relation in form with extinct species....The law of the succession of types, although subject to some remarkable exceptions, must possess the highest interest to every philosophical naturalist, and was first clearly observed in regard to Australia, where fossil remains of a large and extinct species of Kangaroo and other marsupial animals were discovered buried in a cave."[319]

Also carried halfway around the world by steamships was a recent article on the same subject by the naturalist Edward Forbes, lately president of the Geological Society of London and newly appointed professor of natural history at the University of Edinburgh.[320] Forbes was a philosophical naturalist of great ability. He surveyed the current state of knowledge of the distribution of fossil species and genera across geological time. He denied that the fossil evidence lent any credibility to theories of evolution because "the replacement [of fossil species] is not necessarily that of a lower group in the scale of organisation by a higher".[321]

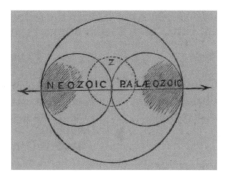

Forbes' polarity theory. *Notices of the Proceedings of the Royal Institution of Great Britain* 1: 428–433, 1854, p. 432.

What great pattern did the evidence reveal? Forbes believed that a greater abundance of fossils at the beginning of the record, interrupted by a dearth and then followed by a great burst of new species presented a pattern that was evidence of a divine plan of "polarity". It was an hour-glass shape for creation.

In his rain-soaked hut overlooking the estuary, Wallace read Forbes and was "annoyed to see such an ideal absurdity put forth when such a simple hypothesis will explain all the facts".[322] It was not just polarity that annoyed Wallace but also Forbes' dismissal that the progression in the fossil record was evidence for an evolutionary view. Wallace had long been convinced by the progressive theory of *Vestiges* that species change.

He got out his pen and ink and began writing an article to contradict Forbes — and to summarise his own views. It would become one of Wallace's most remembered publications. It was sweeping in scope and ambitious in its conclusions. The article was a combination of second-hand knowledge from *Vestiges*, Knight, Darwin, Lyell and Pictet and his own observations of species distributions since he now had an outstanding and "vivid impression of the fundamental differences between the Eastern and Western tropics".[323]

The paper showed that Wallace was no "mere collector", but a profound thinker on some of the most fundamental scientific questions of the time, or indeed of any time. His style was carefully crafted to represent himself as a respectable and accredited scientific participant in elite discussion on these questions. The historian Jonathan Hodge observed that the paper "not only took its title from Lyell quite silently, it began defining its very objectives in three opening paragraphs that insisted on the 'light thrown' on geographical distribution by 'geological investigations'; the three paragraphs being nothing more nor less than an encapsulation of Lyell's teaching, but with no reference to the *Principles*, which, Wallace clearly assumed, would be instantly recognised as this source".[324] Such things are lost on a modern reader who approaches the essay with modern biology in mind. Or worse, readers unfamiliar with Wallace's sources are in danger of attributing all of the details it contains to Wallace.

This highlights a major difficulty in understanding historical documents. Without an understanding of their original context, it is all too easy to interpret them as saying something quite different from what they originally meant. Take for example the phrase "daylight robbery". This phrase survives well today because it seems to mean a robbery so blatant that it occurs in broad daylight or in plain view. Fair enough. But apparently, the phrase can be traced back to the English window tax that saw many houses brick up their windows to avoid tax and thereby losing daylight. Regardless of how true the daylight robbery story may be, it highlights how different meanings are possible when something is interpreted outside its original context. The out-of-context meaning might seem to be perfectly obvious, but can nevertheless be quite wrong.

The Sarawak Law Essay

Wallace began by pointing out that the "geographical distribution of animals and plants" in the world presents two principal puzzles: why they are placed where they are and how to classify them? Geology shows the earth has changed over a very, very long period of time. It also shows that the current arrangement of life on Earth is only "the last stage of a long and uninterrupted series of changes". The oldest species have vanished and new ones have gradually appeared, and the major kinds of living things had appeared in a gradual and distinctly non-random order. "Mollusca and Radiata existed before Vertebrata, and the progression from Fishes to Reptiles and Mammalia, and also from the lower mammals to the higher, is indisputable."[325] Reminiscent of his Pictet notes, he pointed out that "no group or species has come into existence twice". Furthermore, as Darwin and Jukes had shown, "the natural sequence of the species by affinity [similarity] is also geographical". And, as the taxonomic catalogues of the British Museum and his own experience demonstrated, "no species or genus occurs in two very distant localities without being also found in intermediate places". Similar types of organisms are found near each other.

As later species closely resembled earlier extinct species in the same region, Wallace described the latter as "formed" or "created" according to the earlier "anti-type" or models for them. Often, however, "two or more species have been independently formed on the plan of a common antitype", hence "the series of affinities [similarities]...can only be represented by a forked or many-branched line".[326]

Branching

Several conspiracy theorists, especially Brackman, Brooks and Davies, have argued that Wallace's mention of "branching" here "planted the seeds of the principle of divergence" later to appear in Darwin's 1858 paper and *Origin of species*.[327] But Wallace referred to an 1840 article by Strickland who opposed idealised arrangements of species such as symmetrical plans or circles. Strickland proposed instead that species should be arranged according to their degrees of similarity or "series of affinities". If there were only two similarities for a species, on either side, then one should arrange them along a straight line. For example, if there was a large species, a very small one, and a third that was medium-sized, one should arrange them in a line according to size. But as there was sometimes more than one kind of similarity, Strickland proposed that "the natural system may, perhaps, be most truly compared to an irregularly branching tree, or rather to an assemblage of detached trees and shrubs of various sizes and modes of growth".[328]

Wallace put it like this: "the analogy of a branching tree, [is] the best mode of representing the natural arrangement of species". But Wallace moved a step

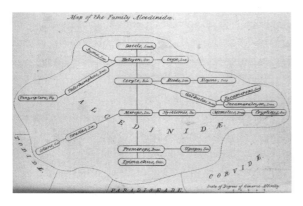

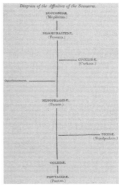

"Map" of the river kingfishers from Strickland, 1840.

Wallace's 1856 "diagram" of the Fissirostres. S28.

beyond Strickland by suggesting that a tree was not just an organisational feature but a historical one because it also represented the "successive creation" of species.[329]

If one were to apply the standards of the Wallace–Darwin conspiracy theorists, Wallace would be classed as an egregious plagiarist! The similarity between this article and Lyell, Strickland and so forth is far more pronounced than Wallace's writings are to Darwin's. But Wallace was no plagiarist by the standards of his own time, nor ours. He was simply influenced by the writings of his predecessors. But this should demonstrate just how absurd the accusations against Darwin really are.

The higher law

There is another law in Wallace's essay that has been overlooked. The essay also addressed "rudimentary organs" or what are now called vestigial structures such as "the complete series of jointed finger-bones in the paddle of the Manatus and whale".[330] Wallace pointedly declared, "To every thoughtful naturalist the question must arise, What are these for? What have they to do with the great laws of creation? Do they not teach us something of the system of Nature?" Modern readers naturally assume that Wallace meant vestigial organs are evidence of evolutionary leftovers from earlier ancestors. But this is incorrect.

Wallace explained that these structures are pre-made or pre-adapted. They occurred without reference to the need of the organism or the environment around it. He gave a few examples. The wings of the penguin are not a remnant of flying bird ancestors, but something that may lead to flight wings in a subsequent creation based on the penguin model or antitype. Similarly in *Notebook* 4, Wallace described the air spaces in ostrich bones as evidence for the same phenomenon.[331]

This is reminiscent of *Vestiges* where evolution up the scale of being was pre-built into organisms. Occasionally, these inbuilt future steps were evoked by "a higher law".[332] Like *Vestiges*, Wallace even suggested that in the future a "race of intelligent beings" might follow humans. Wallace was hinting at a higher natural law, one that sometimes superseded like-begets-like.

Wallace later discussed this higher law in his 1856 paper "On the habits of the Orang-utan". There, he stated quite defiantly that "many animals are provided with organs and appendages which serve no material or physical purpose". Such features simply appeared but they were evidence of a higher natural law.

Wallace remained vague as to what exactly this law was. At least two features are clear. The law explained the "beauty, and the harmony of the organic world". Apparently, beauty was just part of nature. Second, structures will appear that will be intermediate between the species that acquires them and species that would be formed later. The law seemed to be an inherent tendency for elaborations to emerge in living things that would allow them to change and evolve. But Wallace kept the evolution part private for the moment.[333] For now, it was enough to insist that novel structures were destined to appear in organisms, as we might say today, apparently randomly.

Bold conclusion

Wallace pointed out that Forbes' polarity theory assumed that our knowledge of extinct life was complete. Wallace countered that the current picture of past life was more likely to be very incomplete. Darwin had stressed in *Journal of researches* how haphazard and incomplete the fossil record must necessarily be. From all of these observations and generalisations, Wallace boldly concluded, "*Every species has come into existence coincident both in space and time with a pre-existing closely allied species.*" In another passage, he nicely rephrased his point, "The great law which has regulated the peopling of the earth…is that every change shall be gradual; that no new creature shall be formed widely differing from anything before existing; that in this, as in everything else in Nature, there shall be gradation and harmony." It was probably the best summary of the state of scientific knowledge of life on Earth written in the first half of the 19th century.

Wallace signed the paper "Sarawak, Borneo, Feb. 1855." It may have been sent to London with his consignment of specimens to Stevens on 10 February on the *Weraff*. He recorded the next consignment of 4,750 insects sent on the "Dido, left Sarawak, March 3rd." In fact, the schooner *Dido* left Sarawak on 6 March and arrived in Singapore on the 10th.[334] The Sarawak law paper was printed in the *Annals and magazine of natural history* in August for their September issue.[335]

What was the Sarawak law?

But today, almost every commentator on this famous essay fundamentally misreads it as Wallace's first public declaration of evolution.[336] It is said to outline the modern theory of evolution minus only natural selection. One writer even called it "the most significant advance in evolutionary theory since the time of Lamarck".[337] Seen in this way, it is not surprising why so many people are puzzled why Darwin did not get it. Why couldn't Darwin see how close Wallace was?

But a careful historical reading of the paper can clear this up. Wallace nowhere indicated that new species were descended from earlier ones.[338] Instead, he employed exclusively vague language of "creation" whereby "species might have been formed", but he never stated how. This is more than just not having a "mechanism" like natural selection; far more importantly, it leaves out genealogical descent, which is the fundamental core of the theories Darwin and Wallace would eventually announce together.

Instead, the paper presented a theory of *succession*. New species have appeared, by means unspecified, after similar ones existed in the same locale. Succession was a hot new idea, and it was halfway between orthodox views of static creation and evolutionary explanations. Darwin once mentioned his views of succession to the geologist Adam Sedgwick, to the latter's disbelief.[339] Wallace's later views clearly led him to see the Sarawak paper in a different light. As an old man, he described the paper as his first publication on evolution. In a sense this is true because it discusses the history of life on Earth and the way the evidence is summarised clearly suggests evolution as a likely explanation, but the transformation of species is never mentioned. But while still in the East, he referred to the paper as "On Law of Succession of Species".[340] In a 1903 recollection, Wallace conceded that he left out descent: "I left it to be inferred".[341]

The word "creation" did not necessarily mean divine special creation, but could also mean naturally produced by unknown laws. As William John Hamilton, president of the Geological Society, said in 1856, "In a geological point of view the term 'Creation' signifies the fact of origination of a particular form of animal or vegetable life, without implying anything as to the precise mode of such origination."[342]

Wallace's ambiguity about how species were "created" according to "antitypes" meant that some contemporaries read the paper as suggesting evolution and others did not. The zoologist Edward Blyth in Calcutta seemed to think it indicated evolution and wrote to Darwin to mention it. Lyell too was shaken by Wallace's lucid summary. Lyell opened his own species notebooks in November 1855 and wrote Wallace's name at the top of the first page of text.[343]

Other readers, such as Darwin, concluded that Wallace meant a series of supernatural creations in coincident times and places.[344] Such a view would have been hardly revolutionary or even original. Darwin noted in the margin of his copy of *Annals* "nothing very new", "uses my simile of tree" and added "it seems all creation with him…It is all creation…I sh^d state that put generation [i.e. reproduction from parent to offspring] for creation & I quite agree".[345] Hence, when Darwin first mentioned the article to Wallace, he stated, "Though agreeing with you on your conclusion's in that paper, I believe I go much further than you."[346] Only later in *Origin of species* (1859) did Darwin note, "I now know from correspondence, that this coincidence [Wallace] attributes to generation with modification."[347] Such were the mixed consequences of leaving descent to be "inferred".

Sir James Brooke also read Wallace's article, but was not sure whether species grew one into another or whether there had been a series of successive creations giving the same appearance. But Sir James was happy to accept an evolutionary interpretation since, for him, it was perfectly consistent with a divinely created universe. After all, Sir James wrote to Wallace, imagining a creator who constantly needed to create new species to fill gaps in his own system sounded like an argument against a creator!

But this letter reveals something new about Wallace. His preceding letter to Sir James does not survive. But from Sir James' reply, we learn that Wallace had said at the time that his Sarawak law paper was intended to test the waters, or as Sir James put it, "to feel the pulse of scientific men in regard to this hypothesis". Wallace had stressed, in Sir James' words, "the bigotry & intolerance at which views or facts apparently adverse to received systems & doctrines are received".[348] This probably explains why Wallace's paper was written in such ambiguous language. Wallace had brilliantly and clearly generalised the current state of scientific knowledge about the history and current distribution of life on Earth in a way that was entirely consistent with an evolutionary explanation. He was easing closer to declaring his belief in evolution.

Reactions to the paper were decidedly mixed and deeply disappointing for Wallace. "Soon after this article appeared, Mr. Stevens wrote me that he had heard several naturalists express regret that I was 'theorizing', when what we had to do was to collect more facts."[349] In his anniversary address to the Geological Society of London, Hamilton said of the Sarawak paper, "Mr. Wallace is a naturalist of no ordinary calibre.…The question is one of great importance, and deserving the careful investigation of every geologist; but I think it may be doubted whether this assumed law can be maintained as a universal generalization."[350]

In May 1857, Darwin wrote to Wallace in the first letter of their surviving correspondence.

> By your letter & even still more by your paper in Annals, a year or more ago, I can plainly see that we have thought much alike & to a certain extent have come to similar conclusions. In regard to the Paper in Annals, I agree to the truth of almost every word of your paper; & I daresay that you will agree with me that it is very rare to find oneself agreeing pretty closely with any theoretical paper; for it is lamentable how each man draws his own different conclusions from the very same fact.—
>
> This summer will make the 20th year (!) since I opened my first-note-book, on the question how & in what way do species & varieties differ from each other. — I am now preparing my work for publication, but I find the subject so very large, that though I have written many chapters, I do not suppose I shall go to press for two years.[351]

Most modern authors interpret this as a half-veiled warning to keep off Darwin's patch. I read it differently. Not only is such a move, in my opinion, totally uncharacteristic of Darwin, it assumes that Darwin thought Wallace was anywhere close. But Darwin saw Wallace as just another creationist, if a particularly clever one. There was no need to warn off a creationist from his evolutionary patch. After all, Darwin had come to the same conclusions on succession way back in 1837.

Furthermore, the off-my-patch interpretation overlooks the fact that by the mid-1850s, Darwin was referring more and more often to his upcoming species book to correspondents.[352] So the comments to Wallace were characteristic of Darwin's letters at this time. Biographers Desmond and Moore try to have it both ways. According to them, Darwin referred to his evolution theory as "species & varieties" to conceal that he really meant evolution.[353] If so, why use language not indicative of evolution to warn Wallace away from evolution?

In November, Bates wrote to Wallace about the paper, "I was startled at first to see you already ripe for the enunciation of the theory." He went on to nudge Wallace. "The theory I quite assent to, and, you know, was conceived by me also."[354] Bates was reminding Wallace that he felt a co-ownership of "the theory". If so, these lines were Wallace's first priority dispute. But what did

Wallace and Bates regard as "the theory"? If species changing over time was from *Vestiges* and succession from standard works on geology like Pictet, Lyell and Darwin, what was left? Possibly common descent, since that was not derived from common sources and Bates may have known how to read it in between the lines, just as modern readers do. Another possibility is the theory embedded in the essay to explain vestigial organs. We could call it Wallace's "higher law theory".

Si Munjon, Borneo, 14 March–27 November 1855

After finishing his Sarawak law paper and sending consignments to Stevens, Wallace resolved to leave for further collecting in a less visited area. There was plenty of discussion of a new coal works to the east up the mighty Sadong River in an area which had lately been ceded to Sir James by the Sultan of Brunei, hence Wallace could still be under British protection. Ludvig Helms recalled:

> The clearing of ancient forests at these mines offered a naturalist great
> opportunities, and I gave Wallace an introduction to our engineer in
> charge there....Many notes must at that time have passed between us,
> for I took much interest in his work. We had put up a temporary hut
> for him at the mines, and on my occasional visits there I saw him and
> his young assistant, Charles Allen, at work, admired his beautiful
> collections, and gave my help in forwarding them.[355]

The new mine was aimed at producing a reliable and profitable source of local coal. The coal used by the P&O and Royal Navy ships in Singapore and throughout the region was laboriously, and expensively, brought all the way from Europe in sailing ships. So the incentive to find a local source was considerable. St. John later remarked, "This attempt to open a coal-mine in Borneo was a costly failure — rumour said £20,000 — and all because, to save a few hundreds, the opinion had been taken of a practical miner who could have developed a real seam, instead of obtaining a report from a scientific engineer, who, by boring or other means, would have discovered if a workable seam existed."[356]

Wallace left no record of how he and Allen travelled from Kuching to the Sadong River. It was certainly in a ship that was going there anyway. An English miner later sailed to Si Munjon in the *Water Lily* in five days.[357] Wallace and Allen sailed up the Sadong River for about twenty miles before a narrow tributary, the

Si Munjon, poured in from the southeast. They reached the landing place for the mine works on 12 March according to his *Journal*, but in *Notebook 4* and *The Malay Archipelago*, he stated the 14th.

The area was very swampy and mostly impassable. A road made of logs was laid over the mud from the landing place leading to a hill about a mile and a half inland where the mines were being opened. At the base of the hill, there were several rough thatched houses. About 100 Chinese workmen and Dyak labourers were working under the direction of an English mining engineer named Robert Coulson. He had worked for the Eastern Archipelago Company in Labuan in 1851 and later settled in Singapore.[358] Helms provided Wallace with a letter of introduction to Coulson who was absent when the ship arrived. The weather was wet as Wallace and his servants spent the day carrying "things to Hill". This probably barely dampened Wallace's anticipation. He was in the most remote spot he had ever reached. The next day, the sunshine was back and so was Coulson. He invited Wallace to stay in his bamboo house until a new one could be built.

Insect captures

While the coal works extracted wealth from the ground, Wallace set about his specimen mining. "The country all round us is dead level and a perfect swamp, the soil being a vegetable mud, quite soft, and two or three feet deep, or perhaps much more. In such a jungle it is impossible to walk; a temporary path has, however, been made from the river…Along this path is very good collecting-ground, but many fine insects are daily lost, and butterflies can hardly be captured at all, from the impossibility of stepping out of the path, and the necessity of caution in one's movements to preserve balance and prevent slipping, not at all compatible with the capture of active tropical insects."[359] Nevertheless, Wallace thought it "the best locality for beetles I found during my twelve years tropical collecting". On one day, he collected seventy-six different kinds. By the end of the following month, he had gathered over 1,000 species.

He discovered a new butterfly, "the magnificent *Ornithoptera Brookeana* [see colour insert], perhaps the most elegant butterfly in the world". He named it after Sir James. The males' velvety black wings are decorated with dramatic series of metallic green triangles.

Wallace described a typical day collecting at Si Munjon. It probably reflected many of Wallace's days in the East. From the time of waking until breakfast, he labelled and noted the "captures of the previous day", all the while keeping watch for ants that might attack his collections. It was necessary to keep the legs of his work table standing in coconut shells full of water, like little moats to prevent ants from reaching his castle. Animal skins were laid out to dry and any insects that came during the night were bottled. At about 10 am, he set out for the day.

> My equipment is, a rug-net, large collecting-box hung by a strap over my shoulder, a pair of pliers for Hymenoptera [wasps and bees], two bottles with spirits, one large and wide-mouthed for average Coleoptera [beetles], &c., the other very small for minute and active insects, which are often lost by attempting to drop them into a large mouthed bottle. These bottles are carried in pockets in my hunting-shirt, and are attached by strings round my neck; the corks are each secured to the bottle by a short string.[360]

A "rug-net" is a misprint. The only instance recorded by the *Oxford English Dictionary* of a rug-net for insects is in fact this quotation from Wallace! Victorian naturalists used many kinds of nets for collecting insects, including a water net similar to that used by anglers to land fish but lined with gauze, a butterfly net with a pole from six to thirty feet long, a large clap net like a sheet on long sticks and a bag-net or ring-net. Wallace clearly meant one of these ubiquitous terms. A bag-net was any type of net on a pole, with a round or triangular mouth. A net with a round loop of wire was called a ring-net. The net was of book-muslin, the finest and most transparent grade of dress cotton, less likely to be torn than gauze. A ring-net, used to capture butterflies or moths on the wing, was narrower than a Y-shaped sweeping net and may have proved more workable for Wallace in dense forests where vines and thorns snag hanging cloth. Stevens recommended a ring-net in his instructions for tropical collecting: "about a foot and a half in diameter and two feet deep… attached to a stick three to six feet long".[361] The "pair of pliers" were not the sort used for repairing bicycles (which would crush wasps and bees!), but forceps or tweezers about six inches long. They were used for taking insects out of holes, dung, etc.

Near some dead trees, Wallace hunted for glossy iridescent jewel beetles. He saw a bright golden one. He tried to sneak up on it, "but before I can reach him, whizz!—he is off, and flies humming round my head. After one or two circuits he settles again in a place rendered impassable by sticks and bushes, and when he leaves it, it is to fly off to some remote spot in the jungle." It was a lucky escape for the beetle.

After picking his way through the swamp along the log road, Wallace spotted "a pretty little long-necked Apoderus", a type of beetle with its head perched on the end of a bizarrely long neck. Pop! it went into a bottle of alcohol. Farther on, Wallace swept his net over a dry log in the sun. One or two weevils were caught, and popped into the bottle to swirl about in Wallace's pocket for the remainder of the day amidst an ever-growing collection of drowned fellow insects.

Wallace appreciated the attractiveness of each new insect with the eye of a connoisseur. After a long day of looking in different types of habitat, he returned to the hut. Allen had also returned with a "fair" collection of insects. They had collected "94 beetles, 51 different species, 23 of which are new to my collection: I have 5 new Longicorns, 2 new Buprestidæ, and 5 new Curculionidæ. I have been out five hours, and consider this a very good day's work."

But collecting a beetle or shooting a bird was only the beginning. In fact, collecting could almost be called specimen manufacture. It took a lot of equipment and expertise. Stevens published an eight-page pamphlet entitled *Directions for collecting and preserving specimens of natural history in tropical climates*.[362] Stevens never visited the tropics so his directions no doubt reflected the knowledge of Wallace and other experienced collectors.

> The collector should be furnished with knives, scissors, scalpels, pliers, nets, a large assortment of pins of various sizes, needles, a hammer, small hatchet, packing-cases and small, including cork boxes for lepidoptera and other insects, and a great number of pill boxes in nests), cotton and paper, and also with a folding-net, hoop-net, water-net, forceps, digger, glass phials, &c, for collecting insects: he must also have a good supply of prussic acid and arsenical soap.

Insects, depending on their size, fragility, etc., were usually captured in a net or by hand and then put in a bottle of alcohol or other poison to kill them. Otherwise, they would eat or damage each other in the bottle. Some insects were pinned right away and put in a large box carried on a strap over the shoulder. Once back at base at the end of the day, the insects were further prepared. They were soaked for fifteen minutes in warm water and then dried on paper. Then they were pinned on cork or similar material. Stevens noted, "Beetles should always be pinned through the right elytrum, or wing case, so that the pin may come out between the first and second pair of legs; but all other insects may be pierced vertically through the thorax." One had to take care that the legs or antennae were not broken off "for then their value is greatly diminished". The insects then had to be left to dry and finally packed in boxes with "a small quantity of powdered arsenic or camphor, to prevent the attacks of small insects".

Small mammals and birds were usually shot. Collectors like Wallace used double-barrelled percussion guns. Percussion guns were popular since the 1820s as they were less susceptible to wet and damp conditions than their predecessor, the flintlock. They had to be loaded through the muzzle for each shot. Hence, a shooter had to carry a powder flask, a bag of copper percussion caps, a cleaning rod, linen wadding and different sizes of lead shot. The size of the shot chosen depended on the bird. Shooting a bird with shot too large would destroy its feathers or even blast the bird into a cloud of feathers. Birds once shot were carried in a bag until returning to base where they were skinned similarly to mammals. The animal was skinned with a knife, only the "skull and the bones of the legs and feet are to be left". The skin was then rubbed with arsenical soap and left to dry, thereafter stuffed with cotton or a local substitute and the belly sewn up with thread. The head and legs were preserved and the tail bone left in, otherwise the tail feathers would fall out.[363] Notes needed to be taken to record the colour of the eyes, beak and legs as these would change or were lost during drying. One was then left with an outstretched and lightweight bird that could last indefinitely.

Labels

Many specimens had to be labelled. Round ones could be easily made in the field with a standard bit of kit that came with Wallace's gun. A steel wadding punch, about five inches tall, was placed on a few layers of linen and hit smartly with a hammer. The ring-shaped cutting base would neatly cut a series of circles. The same could be done with paper to make an infinite supply of circular labels about half an inch across. Wallace usually made his own labels and then wrote an abbreviation on them in ink to indicate the collecting location: "SING" for Singapore, "Mak" for Macassar and so forth.[364] He had small rectangular labels pre-printed for birds and mammals: "Collected by A. R. Wallace 185_" with blank lines to add a species name and locality by hand.

The sight of a white man writing on little circles and attaching these to dead insects for hours utterly bewildered local people on Kaióa in October 1858:

> Of course when I sat down to work the house was surrounded with men women & children lost in amazement at my extraordinary & inexplicable operations, & when I proceeded to write the name of the place on small circular tickets & attach one to each insect, [the local people] could not repress signs of their astonishment. Had they been a little more enlightened in the ways of white men they would have looked upon me as a fool or a madman, as it was they evidently accepted my operations as worthy of all respect though as utterly beyond their comprehension as would be a steam engine or an Electric telegraph.[365]

The practice of collecting

The collecting practices of Darwin and Wallace were not identical.[366] These differences derived in part from the different reasons they collected and where their collections were destined. Both scrambled about the countryside enthusiastically with guns or nets bagging all the specimens they could locate through their experience and cunning. Both gradually gave up shooting during their voyages, leaving this more to assistants. But the differences are perhaps as numerous as the similarities.

Darwin collected information about a specimen for the use of a specialist in Britain. He typically recorded the date, locality, behaviour, colours, local name and other information about a specimen. So, for example, he recorded the habits of a South American bird he collected in his *Falkland notebook* on 31 May 1833, "Furnarius walks" (as opposed to hopping).[367] Then the specimen was given a 140-word description in his *Zoology notes*.[368] After the voyage, a scientific description of the bird was published.[369]

Darwin aimed to identify the kinds of living things present in the places he visited and of course to find novelties. As naturalist to the expedition, it was expected that his specimens would be added to national collections. Historians often blame Darwin for "neglecting" to label his Galápagos finches by island because this caused him some confusion later.[370] But Wallace routinely lumped together collections from nearby islands such as Singapore with Pulau Ubin.[371]

Wallace's specimens were gathered for sale and for his private collection. This meant he gathered from the "collectable" types of species almost exclusively and in vast numbers with minimal information. His instructions for other collectors suggested listing only the number of each order/family of insects per day. He employed many assistants to increase the size of his collections. Run like a business, Wallace and his men did not collect on Sundays. This was mass collecting for sale. Hence, historian D. B. Baker's overly harsh assessment that "so far as the advancement of scientific knowledge of the fauna of the Malay Archipelago is concerned, Wallace's efforts were largely wasted".[372]

"Remarkable beetles found at Simunjon, Borneo." MA1:58.

Output by Wallace and Darwin were described and published mostly by other naturalists. The *Wallace Online* project at the National University of Singapore has identified more than 130 publications which described Wallace specimens from the Eastern Archipelago. This is about the same number of publications on Darwin's *Beagle* specimens identified by the *Darwin Online* project.[373] Darwin tried to place his collections by group, such as mammals or birds, with specialists and sent his specimens to national collections, whereas most of Wallace's collections were dispersed amongst the cabinets of individual collectors.

What did field collectors and those who purchased these specimens actually do with them? It is remarkable that people like Henry Bates spent their entire careers collecting beetles but did not delve into their habits, ecology or internal structures. They did not dissect them. They instead studied external appearances — thanks to the wonderful structure of beetles with their hard exoskeletons, which means they look the same when dead and their colours tend not to fade.

All of this activity resulted in rows of beetles perched on pins like bizarre miniature statues or worse, those convicted of treason whose corpses were erected on pikes. The number of beetles were so great and their arrangement and relationships to each other were by themselves such vastly complex problems that this was enough to provide a lifetime of study.

Was Wallace a naturalist or a "mere collector"? A writer in *The Edinburgh Review* in 1812 made this distinction: "The cultivators of Natural History, like the objects they consider, admit of classification into genera and species, which hold very different stations in philosophical science. We must place in the lowest rank, the mere collector of specimens" contrasted by the "superior classes of naturalists".[374] In a review of some of Darwin's favourite Cambridge reading, J. F. Stephens *Illustrations of British entomology*, probably by the entomologist J. B. Burton, we read, "But the mere collector is not and cannot be justly considered as a naturalist."[375] In fact, there were no fixed definitions. Wallace was a collector but he clearly saw himself as a naturalist, and his publications reflect this. This created an unspoken tension that Wallace strove to correct in his frequent communications in scientific periodicals.

Bats and cats

On the evening of 21 April 1855, a massive flock of giant fruit bats passed overhead for three hours with their slow, prehistoric-looking shapes silhouetted against the orange stain of sunset. Wallace estimated there were 30,000 bats.[376] At some time, he acquired the rarest specimen he ever collected. It was a large rust-coloured wild cat. Now called the Borneo Bay Cat, it is still the rarest cat in the world. For many years, Wallace's was the only known specimen. Even today only twelve specimens are known![377]

While in his bamboo house, Wallace read the Scottish science writer Mary Somerville and took notes on the comparative abundance of species in different parts of the globe, though often disagreeing.[378] The journalist William Knighton had written of the Eastern Archipelago: "Nowhere is vegetation more rich & luxuriant, nowhere is woman more delicately moulded, more finely"; Wallace retorted dismissively in his notebook: "first line true, rest absurd; -the women are *absolutely ugly.*"[379]

I am often asked whether Darwin or Wallace got up to any hanky panky during their travels. Historians normally pass over this issue in silence. It is a fair question however. The short answer is, no. We have no evidence at all that they did. But then little evidence is likely to have been left behind anyway. For example, no one dreamed that Darwin's cousin W. D. Fox was up to no good. But his recently published diaries, with comments about a girlfriend or servant girl written in Greek code, show that he was.[380] Wallace's friend Richard Spruce apparently lived with a "moca of the mountains" in Brazil.[381] As for Wallace, far away from home in lands full of nubile topless girls and with many opportunities to experiment, such as the "brothels so numerous in all quarters of [Singapore]", it is impossible to say. But Wallace was no rake.

His periods of reading often corresponded with that other indoor activity — writing. On 25 May, he wrote a letter about his collecting. "The only striking features in the animal world [here] are the hornbills, which are very abundant, and take the place of the toucans of Brazil, though I believe they have no real affinity with them, and the immense flights of fruit-eating bats, which frequently pass over us."[382]

The men of the forest

Only a week after his arrival at Si Munjon, Wallace was rewarded with the other great incentive that had drawn him there, the only great ape living outside Africa — the orangutan. In Malay, "orang" is man and "utan" is forest, thus according to Wallace's *Malay-English dictionary*, "man of the woods or forest".[383] He preferred the local Dyak term "mias". It was shorter and easily pronounced. An entry in his notebook records his excitement when one was first sighted.

> Monday March 19th. This was a white day for me. I saw for the first time the Orang utan or 'Mias' of the Dyaks in its native forests. I was out after insects not more than a quarter of a mile from the house when I heard a rustling in a tree near & looking up saw a large red haired animal moving slowly along hanging from the branches by its arms. It passed in this manner from tree to tree till it disappeared in the jungle which was so swampy that I could not follow it. On a tree near I found its nest or seat formed of sticks & boughs supported on a forked branch.[384]

The orangutan was also the first non-human great ape seen by Darwin, though he saw his in the London Zoological Gardens in 1838.

Orangutans were first brought to Europe at the end of the 18th century for royal zoos, first in the Netherlands and later Paris. Between 1816 and 1830, seven individuals are known to have been taken to the West.[385] But they did not live long in captivity. The cold temperatures and unfamiliar diet were no doubt much to blame.

Wallace hoped to answer one of the unsolved mysteries about these extraordinary creatures: were there one, two or even three species? From the dead and living specimens so far brought back to Europe it was hard to tell. It was well known that the animals lived only on Sumatra and Borneo. One problem was that some males had large fleshy flanges on the sides of their faces, while others did not. Were these different species?

The same naïveté that allowed Wallace to believe so readily in phrenomesmerism was still in him. He recorded the extraordinary report of the chief of the Balow Dyaks that "no animal dare attack [a mias] but the Alligator & the boa constrictor. It always kills the alligator by main strength pulling open its jaws & ripping up its throat, standing upon it. If the boa attacks a mias he seizes and bites it in two."[386]

Around 23 May, Wallace returned home in the afternoon from collecting insects and was preparing to bathe when Charlie came rushing in out of breath. "'Get the gun sir, -be quick, -such a large mias, -oh!' — 'Where' said I, 'Close by' -he can't get away' So the gun was got out & one barrel being ready loaded with ball I started off calling upon two Dyaks who happened to be in the house at the time to accompany me & ordering Charles to bring all the ammunition after me as quick as possible."[387]

They set off into the forest after this "monster". For a long time they could not find it until Wallace heard a rustling high in the trees overhead. Still they could not see anything until one of the Dyaks spotted it and called out. Wallace "saw the huge red hairy body & a huge black head looking down surprised at the disturbance. I immediately fired, & he made off rapidly towards the road moving with very little noise for so large an animal." Wallace reloaded and the party pursued as best they could amongst the dense forest of boulders and hanging vines. Wallace fired two more rounds.

Like clockwork, his well-practised hands reloaded both barrels — pouring in powder, a lead ball and wadding tamped down with a whalebone cleaning rod. Two little copper percussion caps were placed on the nipples at the back and with hammers pulled back, the gun was ready to fire again. He fired two more rounds high into the towering tangle of leaves and shadows. "Once while loading I had a splendid view of him walking along a large limb of a tree in a semi erect posture, & showing him to be an animal of the largest size. At the path he got on to one of the loftiest trees in the jungle. We here saw one leg hanging down broken by a ball & there was no doubt he had several other wounds."[388] Wallace continued to reload and fire until the orangutan moved off. Again Wallace had a good sight and fired. Riddled with bullets, the orangutan finally settled in a clump of branches to rest, and die.

A Dyak began to climb the tree, but the orangutan mustered the last of its strength and moved to an adjoining tree to hide in dense foliage. A Chinese mine worker with an axe appeared and the smaller tree was cut. But the tree was so entwined with creepers and vines that it did not fall to the ground but merely leaned at an angle, supported by the living web of the forest. Wallace's team pulled and yanked at the vines to shake the tree and dislodge the orangutan; "after a few minutes when we least expected it down he came with a crash like the fall of giant; & he was a giant! his head & body being as large as a man's." It was a large male with face ridges. Wallace took these to be signs of a distinct variety or species. "His legs and arms were tied together & two men carried him home on a pole. His outstretched arms measured 7 feet 3 inches from finger to finger. His wounds had been fearful. Both legs were broken, one hip joint shattered to pieces; the root of the spine completely shattered & two bullets flattened in his neck & jaws. Yet he was still alive when he fell."[389] The skin of one was worth £50 to the British Museum.

A long series of orangutan sightings and shootings follow in Wallace's notebook which make gruesome reading today. While a modern reader may wince at Wallace blasting now endangered orangutans, there is little point in condemning historical figures for shooting animals that were then common. Virtually every animal was shot on sight in the 19th and earlier centuries with little or no compunction. Coulson and Sir James also shot orangutans. And more importantly, condemning historical figures for not sharing modern values completely blinds us to understanding the past — it was after all very different from the present. For example, in the Amazon, Wallace had marvelled at the first monkey he saw shot. "The poor little animal was not quite dead, and its cries, its innocent-looking countenance, and delicate little hands were quite childlike." But no matter how anthropomorphic the little monkey might be, it was still meat!

"Having often heard how good monkey was, I took it home, and had it cut up and fried for breakfast."[390] Yet others, such as Spenser St. John, wrote of orangutans in a more sympathetic and gentle manner. He could never bring himself to shoot one.[391]

Historian Janet Browne has written about the great transformation in descriptions of gorillas from the mid-19th to the late 20th centuries.[392] Early accounts such as those by the American adventurer Paul du Chaillu portrayed gorillas as ferocious monsters, the embodiment of wild, untamed savagery. By the 1970s, when millions of television viewers saw David Attenborough nestled in the grass with a family of wild mountain gorillas gently sitting about and playing with his shoe laces, the gorilla's image had come a remarkable full circle to gentle giant, wrongfully maligned and killed by its more violent cousin, mankind.

The orangutan has undergone similar transformation in its public image in the West. The gruesome killings with a razor in Edgar Allen Poe's story "The murders in the Rue Morgue" (1841) were committed by an orangutan. Often referred to as the first fictional detective story (at least in English), Poe's detective, Dupin, was the model for Sherlock Holmes. Indeed, the first Sherlock Holmes story is clearly heavily drawn from Poe as is the "Dancing men" from Poe's "The Gold-Bug".

In "The murders in the Rue Morgue", Dupin, informed only by newspaper reports of the crimes, solves the mystery through his powers of deduction alone. Dupin learns about orangutans from reading Cuvier. The orangutan seems to

A female orangutan. MA1:64.[393]

symbolise brute strength and violence which is conquered by the ratiocination of Dupin. But today, the orangutan is seen as a gentle forest grazer, also sadly endangered.

Wallace's baby

On 16 May, some Dyaks spotted another orangutan in the trees for Wallace. It took three shots to kill her. They found her small baby, about a foot long, under the tree in the mud nearby. It was still alive. Wallace carried the baby home while his Dyaks carried the corpse of the mother back on a pole. The frightened baby curled its fingers tightly into Wallace's beard. He had "great difficulty in getting free". Carrying anthropomorphism to a surprising degree, Wallace tended the infant like a human baby, constructing a bottle to feed it rice water and a crib and a soft mat to lie on. Wallace was fascinated to observe the habits of the tiny creature, so human-like and yet at the same time so different with its flowing red body hair and round protruding snout. Wallace bought a long-tailed macaque to keep it company while he was out collecting.

Wallace wrote a humorous letter home and to *Chambers's journal* about "the addition to my household of an orphan baby", continuing the joke about feeding and tending the crying baby as if it were human until at last revealing that it was in fact an orangutan.[394] After about three months, he recorded telegraphically in his notebook: "Taken ill gave it castor oil. Got better & want of proper food. Ill again fever, dropsy. Loss of appetite torpidity — Died."[395] He later wrote:

> I much regretted the loss of my little pet, which I had at one time looked forward to bringing up to years of maturity, and taking home to England. For several months it had afforded me daily amusement by its curious ways and the inimitably ludicrous expression of its little countenance. Its weight was three pounds nine ounces, its height fourteen inches, and the spread of its arms twenty-three inches. I preserved its skin and skeleton, and in doing so found that when it fell from the tree it must have broken an arm and a leg.[396]

In all, Wallace shot about fifteen and procured five other orangutans around Si Munjon. He wrote to Bates that these specimens were "proving, I think, satisfactorily the disputed point of the existence of two species".[397] In June 1856, Wallace's conclusions were confidently announced in *Annals and magazine of natural history*: "two species of Orang have been ascertained to exist in Borneo".

When Sir James read this article in Singapore in October 1857, he wrote to caution Wallace, "On the whole before coming to any positive conclusion on the number of species it will be advisable to wait for more facts and a larger field of investigation. Your collection was made in a limited space of country and should not therefore be held conclusive as settling the general question."[398] Such a caution could well have been Wallace's own words and are reminiscent of his own strictures on Forbes and natural theologians.

Today, two species are recognised, one in Borneo (*Pongo pygmaeus*) and one in Sumatra (*Pongo abelii*). However, the Borneo species is sub-divided into three sub-species. Wallace saw only one of these. Even by the standards of the time, Wallace, who collected in only one location, was rather hasty if he hoped to settle the question of the number of orangutan species in Borneo. Spenser St. John, who lived many years in Borneo, noted that "my friend, Mr. Wallace...unfortunately sought them in the Sadong river, where only the smaller species exists".[399]

The Dyaks had names for three or four types which Wallace took to be species. The case was made more difficult because male orangutans only develop large cheek pads when they are sexually mature. So Wallace found females with no cheek pads, males with no cheek pads and males with cheek pads.[400]

At the end of June, Wallace wrote to his sister Fanny. "Madame Pfeiffer was at Sarawak about a year or two ago and lived in Rajah Brookes house while there. Capt Brooke says she was a very nice old lady something like the picture of Mrs. Harris in 'Punch'.[401] The insects she got in Borneo were not very good, those from Celebes & the Moluccas were the rare ones for which Mr Stevens got so much money for her. I expect she will set up regular collector now, as it will pay all her expenses & enable her to travel where she likes. I have told Mr Stevens to recommend Madagascar to her."

In fact, Pfeiffer travelled to Madagascar in May 1857. Whether this was related to Wallace's suggestion is unknown. It is certainly a striking coincidence. Unfortunately, she became involved in a disastrous attempted coup against Queen Ranavalona I. Pfeiffer and a few other Westerners were to be executed but saved through the intervention of Crown Prince Rakoto. Pfeiffer was expelled in July 1857, while Wallace was in Macassar. Pfeiffer had become ill in Madagascar and, after her return to Vienna, died in October 1858. She and Wallace never met.

Fanny had found another boy to send out to work as an assistant. But Wallace was already exasperated with Charlie Allen.

> Charles has now been with me more than a year, and every day some such conversation as this ensues: "Charles, look at these butterflies that you set out yesterday." "Yes, sir." 'Look at that one—is it set out evenly?" "No, sir." "Put it right then, and all the others that want it." In five minutes he brings me the box to look at. "Have you put them all right?" "Yes, sir." "There's one with the wings uneven, there's another with the body on one side, then another with the pin crooked. Put them all right this time." It most frequently happens that they have to go back a third time. Then all is right. If he puts up a bird, the head is on one side, there is a great lump of cotton on one side of the neck like a hen, the feet are twisted soles uppermost, or something else. In everything it is the same, what ought to be straight is always put crooked. This after twelve months' constant practice and constant teaching! And not the slightest sign of improvement. I believe he never will improve. Day after day I have to look over everything he does and tell him of the same faults. Another with a similar incapacity would drive me mad. He never, too, by any chance, puts anything away after him. When done with, everything is thrown on the floor. Every other day an hour is lost looking for knife, scissors, pliers, hammer, pins, or something he has mislaid.[402]

To feed his small establishment, Wallace planted onions and pumpkins and acquired chickens and three pigs. He paid a Chinese boy to tend them. The boy assisted Allen in skinning orangutans. Asians appear in Wallace's writings as native peoples do in Darwin's travel writings, a knowledgeable surrounding presence that is never really in the foreground. Wallace's informants were usually left unnamed as the "Chinese workman" at Si Munjon who brought a flying frog or the boys in Aru who brought shells. Wallace probably did not know their names.

The flying frog was a totally unknown and unimagined type of creature. The workman who brought it to Wallace said he had seen it glide down from a tree at an angle. Perhaps it was escaping from a tree snake. Wallace examined the frog and found "the toes very long and fully webbed to their very extremity, so that when expanded they offered a surface much larger than the body. The fore legs were also bordered by a membrane, and the body was capable of considerable inflation."[403] He drew a sketch of the frog. The specimen does not survive and seems not to have made it to England. Wallace may not have known how to preserve amphibians since they were not part of the commercial specimen trade and he never collected them. It is now known as Wallace's Flying Frog.

Flying frog (now *Rhacophorus nigropalmatus*). MA1:60.

Housebound again

At the end of June 1855, Wallace "had the misfortune to slip among some fallen trees, and hurt my ankle, and, not being careful enough at first, it became a severe inflamed ulcer, which would not heal, and kept me a prisoner in the house the whole of July and part of August".[404] The weather was very hot and dry.[405] Confined once again, as at Sarawak, Wallace took to his books with his infant orangutan across the room for company. After his orangutan notebook entries are a series of undated reading notes which apparently date from July 1855. These include some of the most important notes on species theory Wallace ever wrote. Both his notes and choice of reading reveal his annoyance with divine design arguments.

In the article on the Coconut family in the *English cyclopaedia* was the remark: "the soft scar…is to allow of a passage through the shell of the nut for the germinating embryo, which, without this wise contrivance, would be unable to pierce the hard case".[406] Making notes in pencil under his own heading "Proofs of Design", Wallace responded with a devastatingly perceptive critique that would have won nods of approval back in the mechanics' institutes in England.

Is not this absurd? To impute to the supreme Being a degree of intelligence only equal to that of the stupidest human beings. What should we think, if as a proof of the superior wisdom of some philosopher, it was pointed out that in building a house he had made a door to it, or in contriving a box had furnished it with a lid! Yet this is the kind and degree of design imputed to the Deity as a proof of his infinite wisdom. Could the lowest savage have a more degrading idea of his God.[407]

On the next page, Wallace noted the article on Birds where the number of neck bones was explained as created according to the needs of different kinds of birds. Wallace mocked, "The writer seems to have been behind the scenes at the creation & to have been well acquainted with the motives of the creator."[408]

Wallace next made notes on Lyell's *Principles of geology*. On a following page, using the same pencil, Wallace made a start on his own account of the way the living world changes under the heading "Note for Organic law of change".[409] We must assume uniformity in nature as Lyell showed. Geological change throughout Earth's history is due to secondary laws. For Wallace, secondary laws echoed with the meanings from Combe and *Vestiges*. The organic changes that have taken place during Earth's history are probably due to secondary laws too. These changes must have been slow and gradual. Wallace summarised, "We cannot help believing the present condition of the Earth & its inhabitants to be the natural result of its immediately preceding state modified by causes which have always been & still continue in action."[410]

After barely 200 words of speculation, Wallace returned to noting Lyell. What caught Wallace's attention time and again were Lyell's objections to evolution. "Lyell says the Didelphys of the Oolite is fatal to the theory of progressive development." *Didelphis bucklandi* was a sensational early mammal find at the beginning of the century in rocks thought to be older than the creation of mammals. Wallace countered:

> Not so if low organized mammalia branched out of low reptiles, fishes. All that is required for the progression is that some reptiles should appear before Mammalia & birds or even that they should appear together. In the same manner reptiles should not appear before fishes but it matters not how soon after them. As a general rule let Naturalists determine that one class of animals is higher organized than another, & all that the development theory requires is that some specimens of the lower organized group should appear earlier than any of the group of higher organization.[411]

Wallace's language here of "branched out" strongly suggests he was now thinking in terms of common genealogical descent — a step beyond his ambiguous use of Strickland's "irregularly branching tree" in the Sarawak law paper. The order in which different kinds of living things had appeared was undeniably progressive for Wallace, and indeed almost everyone else at the time except Lyell.

Wallace switched to Lyell's Volume 2, Chapter 1 "Changes of the organic world now in progress". Wallace responded:

> Lyell, says that varieties of some species may differ more than other species do from each other without shaking our confidence in the reality of species. But why should we have that confidence? Is it not a mere preposesson or prejudice like that in favour of the ~~immutability~~ stability of the earth which he has so ably argued against? In fact what positive evidence have we that species only vary within certain limits?[412]

A few pages later Wallace noted:

> In a few lines Lyell passes over the varieties of the Dog & says there is *no transmutation*. Is not the change of one original animal to two such different animals as the Greyhound & the bulldog a transmutation? Is there more essential difference between the ass the giraffe & the zebra than between these two varieties of dogs. Do the carrion crow & the rook differ more essentially in specific characters than the Polish hen & the Dorking fowl. And is there any other reason why these are not distinct species, than that we believe them to have been derived from a common stock, in the one case which we do not believe it in the other.[413]

The problem of assigning the origin of a species to creation but the origin of a variety to descent would later be taken up in a short article in 1857. His notes on Lyell and species origins continue for a further eleven pages, but these will be discussed when we come to Ternate, where Wallace used them in writing his famous essay.

Wallace and head-hunters

In November 1855, Wallace sent Allen and the collections by ship back to Kuching. On the 27th, Wallace left Si Munjon to travel upriver to the sources of the Sadong River and then down the Sarawak Valley and so back to Kuching overland. He travelled by boat with "a Malay lad named Bujon, who knew the language of the Sádong Dyaks, with whom he had traded".[414]

They stopped first at a village called Gúdong for supplies. This area was so rarely visited by Westerners that the sight of the long-shanked, bearded and bespectacled Wallace sent one Dyak girl about twelve screaming in fright. She jumped into the river and swam away to escape from the monster! At the next village, Jahi, they had to change to a smaller and lighter boat because of the heavier current in the river. A few hours onwards, they passed all signs of cultivation and human habitation. The following morning, they reached another small village, Empugnan, at the base of a large mountain. "Early in the afternoon we arrived at Tabokan, the first village of the Hill Dyaks" (or Senankan Dyaks).

Westerners of the time referred to two types of Dyaks: land Dyaks and sea Dyaks. The latter were often said to be pirates. Modern anthropologists recognise over 200 ethnic groups, each with their own Austronesian dialect, and trace their ancestry, along with those of most other indigenous peoples of Southeast Asia, from a large series of migrations from Asia about 3,000 years ago.

Dyaks lived in communal bamboo long houses on stilts — a tradition that still endures in some remote parts of Borneo. The Dyaks were most notorious amongst Westerners for their tradition of head-hunting. Rajah Brooke had outlawed the practice, but Dyak homes were still proudly decorated with the suspended shrunken heads of fallen enemies. Wallace met a newly arrived

Interior of a Dyak village. Wallace, *Australasia*, p. 360.

English missionary, "When he saw how we lived in open houses & open doors at night surrounded by Chinese & Dyaks he said 'People in England wouldn't believe this.' He said 'I met a Dyak on the path with a long knife & I expected to have my head cut off.'" Wallace's expedition has consequently been seen as a brave adventure amongst head-hunters. "The old men here relate with pride how many heads they have taken in their youth, and though they all acknowledge the goodness of the present Rajah's government, yet they think that if they could still take a few heads they would have better harvests."

Yet Ida Pfeiffer, the fifty-six-year-old Austrian widow, travelled alone amongst the Dyaks. Her adventures in Sarawak seem more harrowing than Wallace's. She reflected, "I shuddered, but I could not help asking myself whether, after all, we Europeans are not really just as bad or worse than these despised savages? Is not every page of our history filled with horrid deeds of treachery and murder?" She added, "I should like to have passed a longer time among the free Dyaks, as I found them, without exception, honest, good-natured, and modest in their behavior. I should be inclined to place them, in these respects, above any of the races I have ever known." Wallace agreed, "They are a very kind, simple and hospitable people....They are more communicative and lively than the American Indians, and it is therefore more agreeable to live with them."[415]

Wallace and Bujon proceeded upriver with Dyak boatmen who punted a long narrow boat with bamboo poles. In the afternoon, they arrived at a village called Borotói. Wallace was again an exotic object of wonder and amazement. "On entering the house to which I was invited, a crowd of sixty or seventy men, women, and children gathered round me, and I sat for half an hour like some strange animal submitted for the first time to the gaze of an inquiring public."[416] The next day, Wallace stayed at a Dyak village called Budw where he "slept very comfortably with half a dozen smoke-dried human skulls suspended over my head". A few days later, they reached Sarawak.

During or perhaps on reflection after his visits with the Dyaks, Wallace was reminded of the population theory of Malthus. "I was much struck by the apparent absence of those causes which are generally supposed to check the increase of population, although there were plain indications of stationary or but slowly increasing numbers." After his return to Singapore, he seems to have noted while reading Boswell's *Life of Johnson*, "Dr. Johnson said 'Marriage not natural to man. In savage state, man & his wife have dissentions & part. When a man sees another woman that pleases him better he will leave the first." Ten copies of

Boswell were in the Singapore Library, making it one of the most popular biographies in the collection, second only to the memoirs of Raffles by his widow.[417] But Wallace disagreed with the opinion of the irascible Dr. Johnson, stating, "Incorrect - The indians of the Amazon, the Dyaks of Borneo never leave their wives, nor the Papuans, the woman always submits."[418]

Sarawak, c. 6 December 1855–10 February 1856

After his return to Sarawak, Wallace and Allen were reunited. From 13–20 December 1855, Wallace went to Bukit Serambu, about twenty miles upriver from Kuching. There, climbing the "very steep pyramidal mountain" covered with luxuriant forest "on a little platform near the summit", Sir James had a "rude wooden lodge" called Peninjauh ("lookout") for relaxation in the cool mountain air.[419] "A cool spring under an overhanging rock just below the cottage furnished us with refreshing baths and delicious drinking water." The whitewashed wooden cottage turned out to be an excellent place to capture nocturnal moths; "during the whole of my eight years' wanderings in the East I never found another spot where these insects were at all plentiful".[420]

On the 20th, they returned to Kuching to spend Christmas with Sir James and his circle. Wallace wrote, "All the Europeans both in the town and from the out-stations enjoyed the hospitality of the Rajah, who possessed in a pre-eminent degree the art of making every one around him comfortable and happy."[421] On 31 December, Wallace and Allen returned to Peninjauh for further collecting. Here, Wallace passed his thirty-third birthday. For the first time, we learn that Wallace took Ali with them. Wallace and Allen collected snail shells, butterflies and moths as well as ferns and orchids.

At the same time that Wallace was at Peninjauh, Charles Darwin was seeking pigeon and poultry skins from Westerners living or working all over the world. Earlier that year, he had begun to keep fancy pigeon breeds himself and soon launched an ever-expanding network of enquiries for different national and international breeds from pigeon fanciers to gather evidence for variation and the species theory.[422] Of the almost thirty letters he dispatched was one for "R. Wallace" sent via Stevens.[423] It would be the first written contact between Darwin and Wallace.

A draft of Darwin's letter survives which gives some idea of his queries. But even more tantalising are the three surviving letters to other men on Darwin's list.

"View from near the Rajah's cottage". S. St. John, *Life in the forests of the Far East*. 1:166, 1863.

In all three of these, Darwin mentioned, "I have for many years been working on the perplexed subject of the origin of varieties & species, & for this purpose I am endeavouring to study the effects of domestication, & am collecting the skins of all the smaller domesticated birds & quadrupeds from all parts of the world."[424] Wallace must have been quite intrigued when he received this letter.

Wallace and his team returned to Kuching on 19 January 1856.[425] It was time to return to Singapore. Allen wanted to stay in Kuching with Bishop McDougall rather than continue as the constantly corrected assistant to the freethinking Wallace. Wallace later wrote to Fanny, "Charles has left me. He has staid with the Bishop at Sarawak who wants teachers & is going to try to educate him for me. I offered to take him on with me paying him a fair price for all the insects &c. he collected, but he preferred to stay. I hardly know whether to be glad or sorry he has left. It saves me a great deal of trouble & annoyance & I feel it quite a relief to be without him. On the other hand it is a considerable loss for me, as he had just begun to be valuable in collecting."[426]

Sir James later wrote to Wallace, "Charles alias Martin alias Allen was miserable at the mission — the constraints were more than he could bear, which might have been foreseen had his previous life been considered before putting him into theological harness. He came to government employ though I had nothing for him to do, but I dare say he will get on in the employ of the Company."[427]

At some time, Wallace captured a tiger beetle "on sand in sunshine, between high & low water mark active".[428] Tiger beetles are fast running predatory beetles that often inhabit open ground. This beetle, "singularly agreeing in colour with the white sand of Sarawak", was remarkable.[429] The phenomena of such matching colours intrigued Wallace. It would be two years before he would explain it. These tiger beetles would become the unsung inspiration for Wallace's evolutionary breakthrough.

Sarawak proved to be one of Wallace's most successful collecting locations. He and his team procured more than 25,000 insects as well as birds, shells, plants and mammals including the orangutans and Bay Cat.[430] Wallace had also pushed far ahead with his evolutionary speculations and published a theoretical article on the succession of species that would be remembered to this day as the Sarawak law paper.

"Malay houses at the anchorage, Sarawak". Collingwood, 1868, p. 201.

"A Dyak or head-hunter of Borneo." Bickmore, 1869, p. 206.

Chapter 6

CROSSING THE LINE

According to *The Malay Archipelago,* Wallace left Sarawak on 25 January 1856. There is no record of such a ship. However, on 10 February, the barque *Santubong* left Sarawak and arrived in Singapore on the 17th.[431] Wallace's dates are often wrong. This fact is important for the vexed questions of where he conceived of natural selection and when he sent the famous letter to Darwin two years later.

Allen remained in Sarawak, but Wallace brought his "Malay boy" Ali with him. Nowadays, Ali is always referred to as Wallace's able collecting assistant. This Ali was, but only later in the voyage. At first, Ali was a cook and servant. That is why when Allen left Wallace wrote to Fanny, "I must now try and teach a China boy to collect and pin insects."[432] Only at the end of their stay in Lombok do references to Ali show that he had begun to skin birds.

Wallace and Ali stayed in town rather than with the missionaries at Bukit Timah. Wallace wrote to his brother-in-law, the early London photographer Thomas Sims (1826–1910):

> I quite enjoy being a few days at Singapore now. The scene is at once so familiar and strange. The half-naked Chinese coolies, the neat shopkeepers, the clean, fat, old, long-tailed merchants, all as busy and full of business as any Londoners. Then the handsome Klings, who always ask double what they take, and with whom it is most amusing to bargain. The crowd of boatmen at the ferry, a dozen begging and disputing for a farthing fare, the Americans, the Malays, and the Portuguese make up a scene doubly interesting to me now

that I know something about them and can talk to them in the general language of the place. The streets of Singapore on a fine day are as crowded and busy as Tottenham Court Road, and from the variety of nations and occupations far more interesting. I am more convinced than ever that no one can appreciate a new country in a short visit. After two years in the country I only now begin to understand Singapore and to marvel at the life and bustle, the varied occupations, and strange population, on a spot which so short a time ago was an uninhabited jungle. A volume may be written on Singapore without exhausting its singularities. "The Roving Englishman's" is the pen that should do it.[433]

"The Roving Englishman" was the pen name of travel writer George Sala who wrote for Dickens' magazine *Household words*, available in the Singapore Library.

Staying in the town gave Wallace the opportunity to use the Library again. He made notes about the construction, by Chinese labourers, of a "Singapore lighthouse". Wallace and Ali sailed by it on their return from Sarawak.

> At Singapore lighthouse.
> Stones of 660 lbs. carried by 4 chinamen up an inclination of 15° to height of 20 feet = 165 per man.
> largest stone carried by 4 men 990 lbs = 247 lbs per man.
> 4 men raised 3918 lbs 20 feet in 4 hours = 326 lbs raised 1 foot per minute for each man & working 9 hours a day.
>
> ——
>
> A European can do nearly double, but these chinamen were working by *day* & not by *contract work*.
> J.G. Thomson. Government Surveyor *at Singapore*.[434]

A pamphlet by Thomson was printed in Singapore in 1852.[435] Thomson, who had the harrowing encounter with durian in Malacca, had been invalided home in 1853 after two years of exposure working on the lighthouse. It was built on a rocky islet twenty-four miles east of Singapore Island in the Strait of Malacca. It was known for centuries as Pedra Branca, Portuguese for "white rock", a reference to its original whitish appearance from accumulated bird guano. It was a nesting site of the black-naped tern. The Portuguese were the first Western sailors to frequent the straits in the middle of the 16th century. The lighthouse was named after the Scottish hydrographer of the East India Company who surveyed the waters of the area. It is still in use today. Starting in 1979, the possession of

Pedra Blanca became an international dispute between Malaysia and Singapore that dragged on for almost thirty years. The case finally went before the International Court of Justice in the Hague which ruled in May 2008 that Pedra Branca is under Singapore's sovereignty.[436]

In addition to Dickens' *Household words,* Wallace read and made notes on scientific publications relevant to his interest in evolution. From the *Transactions of the Geological Society of London,* he made notes on the anatomist Richard Owen which supported Wallace's expectation that fossil reptiles had a "more generalized structure as compared with the more specialized structures of existing species".[437] In Lindley's *Introduction to botany* (1832), Wallace noted in Chapter 2 "Irregular metamorphosis", "New forms, miscalled species, are always starting up in every Botanic Garden. In the garden of Berlin Link states that *Zizyphora dasyantha* after many years changed to another form which might be called *Z. intermedia.*"[438] Progressive historical development and a plethora of new varieties always emerging — these two themes would be integral to Wallace's evolutionary theory.

Wallace may by now have received Darwin's first letter. It was around this time that Wallace transcribed a passage from Darwin's *Journal of researches* (1845) on the tucu-tucu, a burrowing rodent whose tiny eyes were often damaged. Darwin remarked, "No doubt Lamarck would have said that the tucu-tucu is now passing into the state of the Proteus & Asphalax." It was one of the most suggestive evolutionary passages in the book. Wallace concluded with his own speculations about animals living in dark environments. Those with large eyes, he supposed, were obliged to be nocturnal because the structure of their large eyes was unsuited for daylight. Their eyes were not an adaptation to darkness.[439] Such passages show just how anti-adaptationist Wallace was before 1858.

On the following page, Wallace analysed an article on varieties by the naturalist Edward Blyth from an old issue of the *Magazine of natural history.*[440] Wallace also noted a curious remark from the recent work of a French missionary traveller, Évariste Régis Huc, also in the Singapore Library. A type of rice that sprouted unusually early would grow, uniquely, north of the Great Wall of China. It was developed into a useful variety. Wallace noted that if birds carried this rice away to another country with no rice, the useful variety would be considered "a species peculiar to that country".[441] The remark shows Wallace's interest in how to define species. It's not really about the origin of species, nor selection. Although it seems to suggest that, this passage is about problems with determining what species are. Just because a species is found locally did not mean it had arisen locally. Wallace was convinced that the history of a species coming from elsewhere was more reasonable than attributing it to a local creation.

Writing

Wallace spent part of these months in Singapore writing. He penned an article for *Annals* entitled "Attempts at a natural arrangement of birds". Conspiracy theorist Roy Davies waxed lyrical in his belief that this paper was evolutionary and that Darwin stole ideas from it. In fact, there is no hint of evolution or descent, just grouping birds by affinities according to Strickland's system and noting that existing gaps were once filled by extinct groups. Noting that missing groups filled gaps makes sense to us today in terms of branching descent. But the exact language used by Wallace was that of the static arrangements of "affinities" of Strickland. There were no such implications of evolution for contemporary readers, just calls for more examination to determine how species should be correctly grouped together according to their similarities.[442]

Anti-Adaptation

In an article on orangutans, Wallace noted that although they possess large canine teeth, these were not needed and served no purpose. He used this example to illustrate one of his general beliefs about structure versus habits.

> We conceive it to be a most erroneous, a most contracted view of the organic world, to believe that every part of an animal or of a plant exists solely for some material and physical use to the individual...The separate species of which the organic world consists being parts of a whole, we must suppose some dependence of each upon all; some general design which has determined the details, quite independently of individual necessities. We look upon the anomalies, the eccentricities, the exaggerated or diminished development of certain parts, as indications of a general system of nature, by a careful study of which we may learn much that is at present hidden from us; and we believe that the constant practice of imputing, right or wrong, some use to the individual, of every part of its structure, and even of inculcating the doctrine that every

modification exists solely for some such use, is an error fatal to our complete appreciation of all the variety, the beauty, and the harmony of the organic world.[443]

Then, quoting from "Plurality of worlds" (1853) by Cambridge professor William Whewell, "Do not all these examples, to which we might add countless others, prove that beauty and regularity are universal features of the work of Creation in all its parts, great and small?"[444] This was another reference to Wallace's

"Orang utan attacked by Dyaks". Frontispiece to MA1.

higher law. It underlaid the variations that appeared in nature, and they were being misunderstood as adaptive. Reading Wallace's article, one could be forgiven for thinking that he would never propose a theory for adaptation.

Curiously, Wallace's insistence that the orangutan never used its teeth in defence is contradicted by his own account of a Dyak who attacked an orangutan which in defence "got hold of the man's arm, which he seized in his mouth, making his teeth meet in the flesh above the elbow, which he tore and lacerated in a dreadful manner".[445] This incident was depicted by the German natural history illustrator Josef Wolf for the frontispiece to Wallace's *The Malay Archipelago*.

Back to Bukit Timah, March–May 1856

At the end of March, Wallace returned to stay at the French mission at Bukit Timah. In Europe, the Crimean War was ending with a treaty signed on 30 March 1856. Because of the speed of steam navigation, the Crimean War was the first to be reported in British newspapers day-by-day by war correspondents. In his 20 February letter to Fanny, Wallace mentioned his own opinions of the war, based on newspapers which now quickly reached the East via the overland route. He thought the war was "noble" and "necessary" because it would "ensure the future peace of Europe" against the caricature of Russia prevalent in the newspapers as a militaristic and despotic eastern power bent on domination.

But Wallace's mind was occupied most with the birds and insects of Singapore. He spent seven weeks at Bukit Timah "going daily into the jungle". Many hours were spent with his head craned back looking up into the forest canopy to locate a bird he heard calling somewhere in the foliage. It is difficult to spot a bird if it does not move. From the shade of the forest floor, the sky visible through the underside of the canopy looks like a kaleidoscope with the light filtering through given a greenish tinge by the leaves. The overlapping leaves appear as dark spots with patches of blue sky or white cloud visible in the gaps. He found "many pretty new things showing that Singapore is far from exhausted yet, & will furnish hosts of novelties to a resident collector".[446] Wallace recorded in his notebook:

> *March. 1856. Singapore. Bee Eater (Merops)*
> This like a swallow but slower, very graceful circles round & settles on sticks & twigs & posts.

Seizes insects on the wing & rests to swallow them cleans its bill against the perch. Chirps or twitters during flight.

At Singapore & Malacca migratory - appears in November, leaves in March — April.—[447]

This was apparently the Blue-throated Bee Eater (*Merops viridis*) [see colour insert], one of the most colourful and striking birds in Singapore with deep chestnut-coloured head, nape and upper back, its lower back and tail glistening pale blue, pale green wings and a pastel green breast. On 4 April, he made notes on a long-tailed macaque.[448] He also saw the strange "Galeopithecus, or flying lemur" gliding between forest trees.[449]

The tigers were still at large. On the morning of 4 May, an "agricultural labourer" was killed by a tiger in the northwest of the island, a few miles away.[450] Six days later, a very large male tiger fell into a twenty-five-foot pit a few yards from the summit of Bukit Timah.

In *The Malay Archipelago,* Wallace described the tiger pits as "carefully covered over with sticks and leaves, and so well concealed, that in several cases I had a narrow escape from falling into them. They are shaped like an iron furnace, wider at the bottom than the top, and are perhaps fifteen or twenty feet deep, so that it would be almost impossible for a person unassisted to get out of one. Formerly a sharp stake was stuck erect in the bottom; but after an unfortunate traveller had been killed by falling on one, its use was forbidden."[451]

News of the capture reached the town that evening. At 9 am the following morning, Police Inspector Arthur Pennefather, together with a Jemadar, police peons and a party of European gentlemen went to see the spectacle. The tiger roared with rage as the crowds of Chinese labourers, Indian convicts and Westerners crowded around the opening of the pit to catch sight of the enormous cat, probably a man-eater. Ropes and strong wicker baskets were brought to the spot to attempt to cage the animal alive. A carnival-like atmosphere emerged as the workmen and convicts attempted to lasso their prickly prey with a suspended rope. One observer drolly noticed that "though truth was generally represented to be at the bottom of a well, he had no desire to elucidate the fact under existing circumstances".[452]

Wallace was living only a short walk away and it is hard to imagine that he would not have heard of the commotion and he might have joined the throng. It took until 4 pm to loop the rope around the tiger's neck and one leg, to avoid suffocation, and elevate it enough to put into baskets which were then securely wrapped with additional rope. Chinese coolies were engaged to carry the tiger down the hill where it was transferred to a cage. It was kept on display at the Bukit Timah police station for two or three days where Wallace would have seen it every day. The tiger was then brought to the police station in town near the Esplanade. *The Straits Times* recommended visiting the curious sight. The "natives" who captured the animal were not further identified by the newspapers, but they received not only the government reward of $50 but also planned to sell the tiger for transport to Europe. A fantastic sum of $500–$600 was confidently expected.[453] Wallace should have considered bagging some bigger prey.

On the same day the tiger was captured, Wallace finished a letter full of insect news to Bates. Two days later, he wrote another to Stevens, "I have been making a small collection of crustacea from the market. The small ones I can succeed with pretty well, but those of larger size will rot & fall to pieces notwithstanding all my care to dry them. Will the B.M. buy fish from here or from Celebes &c. There is a young man here who would make a collection. How did Madame Pfeiffer preserve hers. She must have had a good many to fetch £25." It is odd that Wallace did not know how to preserve crustaceans and fish in spirits, a common practice for decades.

Wallace asked Stevens to assist in selling two boxes of books to help his friend George Rappa, Jr. "He is the son of the collector who lived many years at Malacca, but has quarrelled with his father & is very badly off." Wallace's assistance may have helped because in 1859, Rappa became a partner of Philip Robinson, founder of Robinson and Co., the oldest department store in Singapore.

Wallace also met the British botanist Thomas Lobb (1817–1894) who was on his third collecting trip in the East. "He appears a first rate collector & has had great experience in the East."[454] Lobb had been to Moulmein (Burma) and was setting off for Labuan, in Borneo. Lobb also lived from specimens, collecting plants for the Veitch Nursery in England, well known for its abilities to raise exotic plants for British collectors and museums. They supplied Darwin with orchids. On his previous trip to India, Lobb had met the botanist Joseph Dalton Hooker who would later play an unexpected role in Wallace's life.

Darwin's Big Book

Back in England, during a fine spring weekend, the Lyells paid a visit to the Darwins. Lyell was shown the pigeon coops. Darwin explained their relevance to his species theory of descent from common ancestors. Lyell already knew that Darwin believed species change. Darwin now explained the details of natural selection, the focus of his current work. Lyell, who had been so impressed with Wallace's Sarawak law paper, was shaken, though still not convinced. Nevertheless, after thinking it over at home, Lyell suggested in a letter that Darwin publish a sketch of his species theory soon to avoid being forestalled, rather than waiting until the species research was completed.

After consulting with other friends, Darwin decided to put further research on hold and begin writing a smaller book. On 14 May 1856, he "began by Lyells advice writing species sketch".[455] From this point, Darwin worked almost exclusively on his projected "big book" as can be seen from the entries in his "Journal" listing the completion of chapter after chapter.[456] Darwin had in mind a scientific treatise like Lyell's classic three-volume *Principles of geology*. The "big book" would probably not have been finished until 1860 or so.[457] This is about eight years beyond Darwin's 1845 estimate of when he would publish his theory. This is exactly the amount of time spent working on barnacles and lost to ill health. His barnacle work, like all his projects, took considerably longer than he anticipated.

Just as Darwin began his species book, Wallace deposited his "10th Consignment" of insects, shells and birds at Hamilton, Gray and Co. to be shipped to London. The consignment departed on the *Dunedin* on 1 June 1856.[458] Wallace was clearing the decks to prepare for his next expedition to the Dutch port of Macassar on the great island of Celebes, where Ida Pfeiffer had collected such valuable insects.

Wallace told Stevens, "I have made preparations for collecting extensively by engaging a good man to shoot & skin birds & animals, which I think in the countries I am now going to will pay me very well."[459] This was a Portuguese man named Manuel Fernandez from Malacca. Perhaps Rappa helped with the introduction. Wallace longed to explore the East. "I look forward, in fact, with unmixed satisfaction to my visit to the rich and almost unexplored Spice Islands,—the land of the Lories, the cockatoos and the birds of paradise, the country of tortoise-shell and pearls, of beautiful shells and rare insects."[460] He was leaving British-controlled territory for the first time.

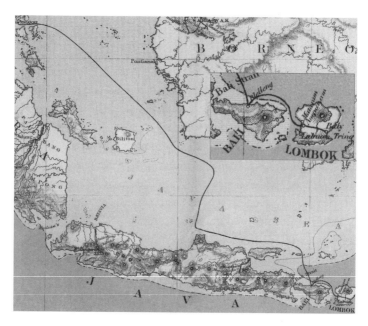

Singapore to Bali (Bali and Lombok inset). "Mr. Wallace's route" map MA1.

Bali, 13–15 June 1856

On Friday 23 May 1856, Wallace, Ali and Fernandez sailed with the barque *Kembang Djepoon* ("Japanese blossom" in Malay).[461] She was bound not for Macassar directly, but deeper into the heart of the archipelago. The ship was as diverse as the streets of Singapore. A Dutch vessel, her owner was Chinese, the crew Javanese and the captain English. The captain of a merchant vessel in the archipelago was called a "Nakoda". The word, derived from Persian, preserved a trace of some of the first international navigators to reach the archipelago centuries before. After a journey of twenty days against the monsoon winds from the East, the *Kembang Djepoon* anchored off Bileling (Buleleng) on the north side of the island of Bali.

Now famous as a tourist attraction, the small island of Bali is separated from the great island of Java by a strait only two miles wide. The centre of Bali is dominated by the steep volcanic mountain range running east to west, which gave birth to the island. The great volcanoes, covered in perennial verdure, towered as high as 3,143 metres above the village along the shore. When they were free of the clouds that often obscured them by day, Wallace thought the peaks "magnificent objects at sunrise & sunset".[462] Bali had several rajahs who had recently come under Dutch control. Dutch administrators first arrived the previous year. The Hindu religion still remained from previous centuries, but since the introduction of Islam throughout the archipelago, Bali was an isolated Hindu island in a sea of Muslim lands.

Wallace went on shore with the Nakoda to meet the Chinese merchants. They dressed in local costume. Wallace found them difficult to distinguish from Malays, suggesting to him "the close affinity of the Malayan and Mongolian races". Walking inland, Wallace was astonished by the sight of the extensive landscaping of the terraced hillsides with their complex irrigation, bursting with rice ready to harvest. "I had not yet beheld so beautiful well cultivated a country out of Europe & was both astonished & pleased with it." Wallace thought that the terrace system must have come from India with Hinduism. A true Victorian, he said nothing of the bare-breasted women of the island. As usual, he collected insects and birds. "On the sandy beach I found a dark coloured *Cicindela* [tiger beetle] which could hardly be seen upon the grey volcanic sand."[463] It was another curious case of matching colouration.

Lombok, 17 June–30 August 1856

Leaving Bali on 15 June 1856, the *Kembang Djepoon* sailed two days to the east to anchor in the smooth and sheltered bay of Ampanam (Ampenan) on the west coast of the smaller island of Lombok. Like Bali, Lombok was an idyllic emerald island set in a crystal blue sea dominated by a towering 3,600-metre volcano, Mount Rinjani, perched at its northern end. Lombok was not part of the Dutch East Indies, but since 1839 formed the independent Hindu kingdom of Mataram.

Here, Wallace, Ali and Fernandez would stay until they could catch a ship to Macassar in Celebes. Wallace noted, "There are now two races the Sassaks or indigines of Lombock who are Mohamedans & the Balinese who are Brahmins and are the dominant race having many privileges."[464] Lombok was also extensively landscaped into terraces: "the country cultivated like a garden". It exported rice, coffee, cotton and tobacco, especially to markets in China.

Wallace's party landed the boxes ashore still dry despite the heavy surf on the black sandy beach. There were only a handful of Westerners on Lombok. Wallace was met on the beach by one of them; he recorded only as "Mr Carter". This was Joseph Carter, an English merchant in the East for eighteen years "as a commander of vessels & as a merchant, [and] is well educated" and on Lombok since 1855, presumably taking over the factory of the first English merchant there, Mr. King, resident from 1832.[465] Long strings of pack horses brought rice from the terraced interior to Carter's yard.

A German-born American merchant sailor, Charles Nordhoff, whose ship was at Ampanum around this time, noted that it took two weeks to fully load his ship with rice for China.[466] Carter offered Wallace assistance and the use of his house during his stay. "His house storehouses etc. were enclosed by a high bamboo fence & were all of bamboo with a thatch of grass that being the only available building material."[467] Wallace tried to repay the favour by asking Stevens if William Saunders, the wealthy insurance broker buying so many of Wallace's insects, could use his influence to have Carter appointed the Lloyd's agent at Lombok.

Perhaps Wallace's evolutionary views again formed part of the conversation. Carter told Wallace that he had seen "in the interior of the Coti River in Borneo a man with a tail - about 4 or 5 inches long. Had examined it, but does not know if there is a tribe of the same structure."[468] Stories of men in the Eastern Archipelago with tails had been repeated in Europe for many years and even Linnaeus and Lord Monboddo believed them, causing some to believe in a race intermediate between humans and apes. Perhaps the man with the tail was just an anomaly.

Wallace devoted himself to "shooting the birds of the neighbourhood". What the inhabitants thought of the bearded foreigner blasting all the birds in the area went unrecorded. Local boys caught dragonflies which were fried in oil and eaten as a delicious snack. At least the dragonflies escaped Wallace's collecting box. Wallace caught two more species of tiger beetles on the "dark sand" of some river banks, differently coloured from the species on Bali and Sarawak.[469] Like Darwin's Galápagos mockingbirds, these tiger beetles were different on different islands. But in this case there was an obvious relationship to the environment.

The evolution of the Wallace Line

The birds were not very numerous so Wallace decided to take his party in a small outrigger to a harbour called Labuan Tring at the southern extremity of the great bay for about ten days.[470] He took a letter of introduction to a "very civil" Mr. Daud, an Amboynese Malay; "his accommodations were limited, and he could only give me part of his reception-room. This was the front part of a bamboo house (reached by a ladder of about six rounds very wide apart)." The specimen factory cranked into production again. Fernandez shot birds and Wallace skilfully swung his net after insects. Back at Mr. Daud's house with its "beautiful view over the bay", the dead were prepared for their bizarre afterlife as European collectibles.[471]

All seemed routine except for what happened next. It is one of the most important episodes in Wallace's life and one of the discoveries for which he is still justly famous — the "Wallace line". It has been called "the boldest single mark ever inscribed on the biogeographical map of the world".[472] The reason the Wallace Line is so extraordinary, even today, is because animal types as different as those on either side of his Line, are normally separated by some massive barrier like an ocean or continent, but in this case, thousands of islands are scattered all over the southern seas under the same climate. There is no barrier. Yet, on either side are dramatically unrelated animals. On one side are marsupials with pouches, on the other placental mammals with a similar separation in the families of birds. This completely breaks the normal sort of pattern seen in the world. It remains Wallace's most famous landmark, if an invisible one. The entire region is called Wallacea, also in his honour. But few seem to realise what Wallace actually did.

Some think Wallace discovered that Asian species were on the western side of the archipelago and Australian ones farther east. But this was common knowledge. One historian has even claimed that finding the line between the two faunas had "captivated naturalists for more than two centuries".[473] But no one was looking for a line. If they had it could have been quickly identified. A decade before, the Swiss naturalist Heinrich Zollinger noticed some animals were on Lombok but not Bali or farther west.[474] The problem was that typical zones for Asian and Australian forms were already known, but a sharp boundary or line between the two was not dreamt of in anyone's philosophy.

The other story is that Wallace discovered the separation between the two regions by his extensive observations. But the truth is revealed in his first mention of this subject in his *Journal* written while at Labuan Tring. Based on his knowledge of global bird distribution from the scientific literature and, apparently, the testimony of Mr. Daud, Wallace jotted, "Plenty of new birds....Australian forms appear. These do not pass further West to Baly & Java & many Javaneese birds are found in Baly but do not reach here." The notes suggest that Wallace's first hint was local information, not discovery from his own observations. He had not been farther east or made sufficient journeys back and forth to come to such a conclusion. He was only two days on Bali and never visited Australia.

One would expect local people to be perfectly familiar with the fact that animals common on Bali were not found on Lombok and vice versa. To transform this old local knowledge into a wider scientific generalisation, it took someone like Wallace who was well versed in the scientific literature on worldwide distribution, engaged in collecting over a wider region and most importantly, driven to speculate on distribution patterns. But even more appropriate for such a discovery, Wallace was a former surveyor who was particularly *au fait* with maps. Still, he was not very systematic nor had he mastered the relevant literature, but he had the talent of noticing things overlooked by others. This combined with an unfettered imagination and a knack for brilliant insights laid the groundwork.

But the idea of the Wallace Line did not emerge all at once — it evolved. It is all too easy to read "Wallace's line" into his earliest writings when in fact there is no notion of one. In this first journal entry for example, Wallace noted not a line or boundary between Asian and Australian zones, but the extent these two regions reached on Bali and Lombok. He had yet to draw the line.

Writing to Stevens from Lombok on 21 August 1856, Wallace mentioned his thoughts which would have such profound, if for some time inexplicable, consequences for biogeography.

> The birds...throw great light on the laws of geographical distribution of animals in the East. The Islands of Baly and Lombock, for instance, though of nearly the same size, of the same soil, aspect, elevation and climate, and within sight of each other, yet differ considerably in their productions, and, in fact, belong to two quite distinct zoological provinces, of which they form the extreme limits. As an instance, I may mention the cockatoos, a group of birds confined to Australia & the Moluccas, but quite unknown in Java, Borneo, Sumatra & Malacca.[475]

The science writer Simon Winchester wrote that the ornithologist Philip Lutley Sclater proposed the line before Wallace and so Sclater might have a rightful claim to the line being named after him. There are two serious problems with this. First, Wallace's letter to Stevens was published in the *Zoologist* in January 1857, six months before Sclater presented his important paper to the Linnean Society. Second, Sclater did not propose any lines at all. Instead, he organised the birds of the world into six regions, including Asian birds in "Malacca and Southern China, Philippines, Borneo, Java, Sumatra and adjacent islands…but probably not Celebes" and Australian forms to the east in New Guinea "and adjacent islands, Australia, Tasmania and Pacific Islands".[476]

In a letter to Bates in January 1858, Wallace first referred to "the boundary line" between these provinces.[477] But in March 1859, Wallace wrote to Sclater about his Linnean paper, proposing some refinements to the various zones, but made no mention of a line. Even in an 1860 paper on the zoology of the archipelago, sent via Darwin to the Linnean Society, Wallace described "the limits of the two regions", not a line.[478] After his return home, Wallace published a paper along with a map. Although this paper too did not actually mention a line, only that Bali and Lombok "belong to two quite distinct zoological provinces", Wallace provided a map of the archipelago with a bold red line showing the "Limits of the Indo & Austro-Malayan Regions". This, at last, was what Thomas Henry Huxley five years later dubbed "Wallace's line" [see colour insert].[479]

Although the Wallace Line is often mentioned, modern writers often fail to show how Wallace explained it. We now know it is the result of plate tectonics. The Australian region originated as part of South America and over millions of years has slowly drifted up alongside the islands of Indonesia, carrying the distinct Australian fauna up to close juxtaposition with the Asian zone and its fauna, hence the bizarre puzzle of two radically different kinds of living things found right next to each other and not separated by an ocean or continent.

In Wallace's time, a horizontal movement of the earth's crust was utterly unimaginable. But subsidence and uplift were well attested. Wallace believed "the W. part to be a separated portion of continental Asia, the eastern the fragmentary prolongation of a former Pacific continent".[480] Two great continents had broken up and partly sunk and this explained why such different animal types were scattered across islands so close to one another. These sunken continents could be interpolated from the areas of shallow seas connecting large parts of both sides of the Wallace Line.

Alas for Wallace, this too was an independent discovery years after someone else had made it. The old colonial hand George Windsor Earl published an important paper on this point in 1845.[481] Earl divided the archipelago into eastern and western zones based on geology and ocean depths that showed the islands were connected by shallow shelves. Wallace apparently did not read Earl until pointed out by Darwin.

The Lombok letter to Stevens contains Wallace's first mention of contact with Darwin. "The domestic duck var.[iety] is for Mr. Darwin & he would perhaps also like the jungle cock, which is often domesticated here & is doubtless one of the originals for the domestic breeds of poultry." Darwin's letter seeking exotic poultry and pigeons sent while Wallace was at Sarawak was finally answered almost a year later.[482]

Fernandez shot and stuffed birds but, as far as Wallace recorded, Ali cooked, collected wood and fetched water. Their diet was mostly "metallic green pigeons" "with other smaller pigeons & doves". Specimen collecting had at least some useful by-products. The pigeons were "excellent eating".[483] Both of his servants were frightened by rumours that the Rajah had sent an order to Labuan Tring to supply a certain number of human heads to ensure a good rice harvest.

Sitting on his specimen box inside Mr. Daud's hut and skinning the birds of the day, Wallace was amused to overhear Fernandez preaching to the local Malays and Sassaks (presumably in Malay). "Allah has been merciful today and has given us some very fine birds; we can do nothing without him." Equally, if he was unable to bring down a bird after several shots, "'Ah!' say the Malays 'its time was not come & so it was impossible for you to kill it.' A doctrine this, which no doubt quite accounts for all the facts but is nevertheless not altogether satisfactory."[484]

Wallace also overheard a conversation between a Bornean Malay and Fernandez and jotted a quick line at the front of *Notebook 4*, "In Lombock *hantus* are scarce!!" Hantus was the local word for "ghosts". Later, Wallace wrote the story up in his *Journal*.

> "One thing is strange in this country;— the scarcity of Gools." "How is" said Manuel. "Why you know in our country westward if a man dies or is killed, we cannot pass near the place at night-all sorts of noises are heard, which shows that Ghools are about. But here there are numbers of men killed & their bodies lie in the fields & by the wayside and you can walk by them at night & never hear any noises at all; which is not the case in our country as you know very well". "Certainly I do" said Manuel & so it was settled that Ghools were scarce, if not altogether absent, in Lombock.[485]

By the time Wallace finally published this story in *The Malay Archipelago*, he had himself become a spiritualist. Maybe this is why he changed "ghouls" to "ghosts".

Wallace wrote a long lament about the hardships of a "travelling collector of limited means like myself", a rare self-reference in his own terms. The surrounding country was also rather hard. "The most characteristic feature of the jungle here is its thorniness. The shrubs are thorny, the creepers are thorny. The bamboos are thorny."

> When my bird box was nearly filled, I was alarmed by an irruption of red ants, which began eating away the outer skin of my birds & threatening to do much injury. I therefore made a sudden return to Ampanam to clean & pack away what I had hitherto collected & enjoy for a few days a little European society. I then returned for another attack upon the birds.[486]

The most beautiful bird he found was a very shy ground thrush (*Pitta elegans concinna*). It was difficult to procure. He eventually found that by stalking through the dry thickets very quietly, and standing very still for half an hour, he could imitate the bird's whistle. Sometimes one would come hopping among the dry leaves. Then the silence was shattered by the smoky blast from his gun. He could then "secure my prize, and admire its soft puffy plumage and lovely colours".

On 2 August, Wallace set out with a "Mr. Ross" for an excursion into the interior of the island. Mr. Clunies Ross was "born in the Keeling Islands, and now employed by the Dutch Government to settle the affairs of a missionary who had unfortunately become bankrupt here". Clunies Ross, when a child, may have seen Darwin when the *Beagle* visited the Cocos (Keeling) Islands in April 1836.[487] They rode on horseback to Mataram, the capital of Lombok and residence of the Rajah. In *The Malay Archipelago*, Wallace devoted an entire chapter to an anecdote he heard about the Rajah devising an ingenious method to detect tax evasion by his vassals. The same story was published four years earlier by John Cameron.[488] The Rajah, never named by Wallace or Cameron, was Gusti Ngurah Ketut Karang Asem, who reigned from 1839–1870.

Wallace and his companions ended at their destination at the village of Coupang in the centre of the island. Wallace was quizzed about why he wanted dead birds, rather reminiscent of the suspicious questioning Darwin endured in South America where local people were incredulous and even suspicious that an Englishman would travel such vast distances just for dead birds. Surely they were after something else?

On 10 August, Wallace joined a party of Westerners to ride inland four miles past Mataram to Gunung Sari where the Rajah had a country seat, Hindu temple and extensive pleasure gardens. Here, Wallace earned a few more shillings in the forms of a new kingfisher and a pretty ground thrush. They returned to Ampanam where Wallace was still awaiting a ship that could take him to Macassar.

Wallace heard here again stories of Malays running amok as a form of suicide and of local punishments for infidelity which, at least on this occasion, were not just rumours, but actually took place during his stay. A married woman and her lover were caught and lashed together back to back and then thrown off a cliff into the sea "where some large crocodiles are always on the watch to devour the bodies".[489] The sensitive Wallace took a walk to avoid seeing this horrific execution.

Fernandez now left Wallace, saying he was homesick and wished to return to Singapore. Wallace suspected instead that it was "from the idea that his life was not worth many months purchase among such a blood thirsty & uncivilised people. I was now therefore again left alone to work at all the various branches of Natural History in which I feel interested."[490] This makes it clear that Ali had not been a collecting assistant. But now, for the first time, Wallace mentioned Ali taking part in collecting: "However Ali has learnt to skin a little & I must make him work exclusively at it when I get to a productive ornithological region."

Macassar, Celebes, 2 September–13 December 1856

On 30 August 1856, the 130-tonne schooner *Alma* arrived at Ampanam. Here at last was a ship heading for Macassar on the island of Celebes (Sulawesi), the sprawling four-limbed island between Borneo and the Moluccas. The voyage took three days. The *Alma* dropped anchor in the harbour of Macassar near the end of the southernmost peninsula. Wallace saw "a fine 42 gun frigate, the guard ship of the place, a small war steamer & three or four little cutters used for cruising after the pirates which infest these seas. Besides these, there were a few square rigged trading vessels & twenty or thirty native praus of various sizes."[491]

Ida Pfeiffer thought Macassar "a small, almost European-looking, town, with a fort. The Government-house is small and insignificant, and the Europeans live in poor-looking little stone houses, lying close together, along the side of a beautiful piece of meadow land, called *Hendrik's-pad.*" Once again, Wallace followed in Pfeiffer's footsteps. His mouth watered at the thought of the "fine large species described by the old naturalists, some of which have recently been

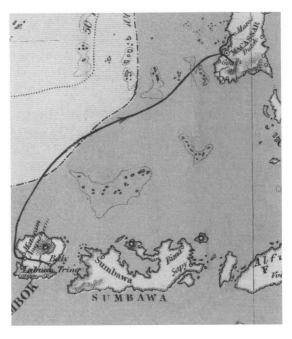

Lombok to Macassar. "Mr. Wallace's route" map MA1.

obtained by Madame Pfeiffer, [and] give promise of what systematic collection may produce".[492]

After seven months of waiting, Wallace finally arrived on Celebes. It was his first stay in a Dutch settlement. Macassar was the chief settlement of the Dutch on Celebes. Only the town and an area extending two miles outside the town was Dutch territory. Most of the island was divided amongst princely states.

The Dutch began journeying to the East Indies in 1595 to purchase spices like pepper, nutmeg and clove directly from their source and thus undercut the vast series of intermediaries between the East and Europe. These spices that are so common and cheap today were then literally worth their weight in gold. The Dutch East India Company or VOC (Vereenigde Oostindische Compagnie) was formed in 1602 — the world's first international conglomerate. Like the British East India Company, it ruled parts of the archipelago like a sovereign state until it was dissolved by the Dutch government in 1800. By Wallace's time, the Dutch government from its capital at Batavia ruled stations on Sumatra, all of Java ("the granary of the Asiatic archipelago"), a small area of Borneo, Macassar and Menado on Celebes and parts of the Moluccas.[493]

Macassar had been Dutch since the late 17th century. In 1847, it was made a free port, following the example of Singapore. It was one of the busiest ports in the whole archipelago. In 1857, imports amounted to £382,288 and exports £385,010.[494] When Wallace visited, it was inhabited by several thousand Westerners.

The long-term presence of the Dutch meant that there were far more European naturalists and collecting going on than is dreamt of in books on Wallace, partly because it is in the Dutch language. In fact, state-sponsored scientific research was established by 1820 with the Natuurkundige Commissie voor Nederlandsch Indië (Commission for the Study of the Natural Sciences of the Netherlands East Indies). Results were published in *Verhandelingen over de Natuurlijke Geschiedenis der Nederlandsche Overzeesche Bezittingen* (*Transactions on the natural history of the Netherlands overseas possessions*) between 1839 and 1847. In 1850, the Indies Commission was dissolved. After this, as historian Daniel B. Baker noted, "Investigation of the natural history of the region devolved on government officials, who were subsidized for their work in this rôle, and on interested residents, an arrangement that was to prove more efficient."[495]

There was no hotel in Macassar, so Wallace and Ali stayed at "a Society or Club house in the most fashionable part of the town". This was the Sociëteit De

Wallace slept here. The Sociëteit De Harmonie in Macassar c. 1865. Tropenmuseum.

Harmonie located on Prins Hendrik Pad, a pretty piece of meadow land near the sea. Every Dutch town in the East had its "Society" or clubhouse. Now a gutted ruin, the much altered Society building in Macassar still stands as an arts centre, perhaps the only building in the archipelago where Wallace slept that survives.

Wallace brought letters of introduction to "a Dutch gentleman, Mr. Mesman, and also to a Danish shopkeeper".[496] Wallace never gave Mesman's full name or that of the shopkeeper. The following day, Wallace paid a visit to the Governor, whose name he also never recorded. "The Governor [Colonel Cornelis Albert de Brauw] was very polite & offered me every facility for travelling about the country and prosecuting my researches. Our conversation however was but lamely carried on owing to my scanty knowledge of French between which & Malay lay our choice of a medium of communication." Wallace was accompanied by Willem Leendert Mesman, a wealthy Dutch merchant born in Macassar who could speak English. He became the model for Joseph Conrad's character Mr. Mesman in the novella *Freya of the seven isles* (1912).[497]

Ali fell ill with a fever the next day and was unable to cook for Wallace, "which put me to more than usual inconvenience as at the house where I was staying nothing could be got from the servants but meals at certain hours".[498] At the end of the week, Wallace moved into "a little bamboo house" called Mamajam lent by Mesman. "It was situated about two miles away, on a small coffee plantation and farm, and about a mile beyond Mr. M.'s own country-house. It consisted of two rooms raised about seven feet above the ground, the lower part being partly open (and serving excellently to skin birds in) and partly used as a granary for rice. There was a kitchen and other outhouses."[499] The house was surrounded by roses and jasmine, and every morning Wallace was given a few flowers for his breakfast table and had fresh milk in his coffee from Mesman's cows. Wallace engaged another man as a servant.

The house was near paddy fields and the few villages scattered about contained fruit trees that provided the only collecting in the neighbourhood. Wallace soon exhausted it. He appealed to the office of Governor de Brauw for a letter of permission to travel in the territories of the local Sultan whose land, Goa, enclosed the Dutch district of Macassar. The capital of Goa was four miles south of Macassar. Wallace was given a letter of introduction to the Sultan (Wallace mistakenly called him a rajah), Abdul Kadir Muhammad Aidid (1825–1893). Mesman was a friend of the Sultan's and spoke the Macassar language so he accompanied Wallace. When they arrived, they found the Sultan seated outside watching the construction of his new house. Ida Pfeiffer had also visited the Sultan and explained the context more fully.

> The Sultan of Goa inhabits a much handsomer house than his royal colleagues of Sidenring and Pare. It is boarded and adorned with carving; the interior presented much the same scene as that of the other royal mansions, — a superfluity of attendants and servants, a chaos of clambus, and innumerable chests and boxes, piled one above another. The sultan was just having a new house built, although his old one was still in perfect preservation. He would not inhabit it any more because his father had died in it.[500]

The Sultan granted Wallace permission to travel where he liked in the territory.

A fever was spreading through Macassar. It struck Wallace's servant who went home to his wife. Wallace then fell ill with a "strong intermittent fever every other day". In about a week, he recovered "by a liberal use of quinine". But as soon as he recovered, "Ali again became worse than ever". "His fever attacked him every day." Even illness did not excuse Ali from his duties to his master since "early in the morning he was pretty well and then managed to cook me enough for the day". In another week, Ali recovered. Wallace engaged another boy named Baderoon to cook and shoot. Since Lombok, Ali had become "a pretty good bird skinner". Wallace also "succeeded in getting hold of a little rascal of 12–14 called Baso who can speak a little Malay & whose duty is to carry my gun or insect net when I go out, & to make himself generally useful at home".

With his retinue of assistants, Wallace made many excursions inland to collect birds and insects. They started in the early morning, often taking breakfast with them to eat during a rest along the way. His "Macassar boys" would lay aside a tiny offering of rice to the deity of the spot "though nominal Mahometans the Macassar people retain many of their pagan superstitions, & are but lax in their religious observances". They quaffed the alcoholic toddy or "very refreshing" palm wine with Wallace at roadside stalls. Sometimes an old friend, the black wire-tailed drongo Wallace first saw at Singapore, fooled him with its variety of notes so that he thought a new species of bird was nearby.

On one expedition to a large patch of forest, Wallace set out with his whole team: Ali and Baderoon each carried a gun to shoot birds and Baso carried the food and storage box. Wallace was armed with his insect net and collecting bottles. He bagged some "lovely green & gold shrub beetles". The insects dropped into his bottle like shillings into his piggy bank. Hearing some gun shots, he walked in the direction of his boys to find they had shot "two specimens of the fine cuckoo *Phoenicophaus callirhynchus* so named from its beak of brilliant yellow red & black".

> It was coming back in the heat of the day that I had the greatest luck capturing three specimens of *Ornithoptera*, the largest, the most perfect and the most beautiful of butterflies. I trembled with excitement as I took it out of my net & found it to be imperfect condition. The grand colour of this superb insect was a rich shimmering bronzy black, unspotted on the upper wings but on the lowers delicately grained with white & having a marginal row of large spots of the most brilliant satiny yellow. The body was marked with shaded spots of white yellow & firey orange while the head & thorax were intense black. On the underside the lower wings were satiny white, with the marginal spots half black & half yellow. When it is considered how limited is the geographical range of these fine insects, they being entirely confined to the Indian Archepelago, the rarity of most of the species, the esteem in which they are held by Entomologists & collectors & the number of years that generally elapses between the discovery of each new species, it may be conceived with what intense interest I gazed upon my prize.[501]

Wallace wrote to Samuel Stevens on 27 September, "A friend here, seeing I had my mattress on the floor of a bamboo-house, which is open beneath, told me it was very dangerous, as there were many bad people about, who might come at night and push their spears up through me from below, so he kindly lent me a sofa to sleep on, which, however, I never use, as it is too hot in this country." The same day, Wallace's Bali and Lombok collection left Singapore on the *City of Bristol*, bound for London and Stevens.

On 10 October, Wallace wrote to Darwin. The letter does not survive, but from Darwin's long and generous reply of 1 May 1857, one can reconstruct some of what Wallace must have said.[502] Wallace suggested that Darwin (who requested domesticated poultry and pigeons in December 1855) keep domestic varieties himself. Darwin replied politely that he was following this advice, although he had in fact started long before contacting Wallace. Wallace also cast doubt on the evidence for the sterility of hybrids. This was a curious subject for Wallace to mention. He may have been responding to something in an earlier letter from Darwin, also now lost. Finally, as far as we can glean from Darwin's reply, Wallace suggested "climatal conditions" had little effect on species and their distributions. This was indeed Wallace's own thinking. He was convinced that it was the preceding anti-types, indeed the ancestors, of species that were the true clue, and that species were not somehow created to suit their climate, as Lyell argued. Past geological connections through sunken land bridges could explain present distributions. Otherwise, the similarity of climate in tropical South America and tropical Southeast Asia would result in the same plants and animals. But they were fundamentally distinct like the toucans of America and hornbills of Asia.

In October, Wallace moved to a new collecting area near the forest a few miles further inland at a place called Samata. Abdul Aidid, the Sultan of Goa, ordered that a house be made available to Wallace there. After much trouble and delay, Wallace moved in on 13 October 1856.[503] The thermometer reached 33°C (92°F). As in Borneo, Wallace was annoyed with how his appearance struck local people with "excessive terror".

> Wherever I go dogs bark, children scream, women run & men stare with astonishment as though I were some strange & terrible cannibal monster. Even the pack horses on the roads & paths start aside when I appear & rush into the jungle, & as to those horrid ugly brutes the buffaloes, they cannot be approached by me, not on account of my own but of other's safety. They first stick out their necks & stare at me, & then on a nearer view break loose from their halter or tethers & rush away helter skelter-as if a demon was pursuing them, without any regard to what may be in their way....If I come suddenly upon a well where women are drawing water or children bathing, a sudden flight is the certain result, which things occurring day after day are to say the least of them very unpleasant & annoying, more particularly to a person who likes not to be disliked, & who has never been accustomed to consider himself an ogre or any other monster.[504]

Wallace may not have been an ogre, but it was not just that he was a tall, pale-skinned European — but that he was dressed in strange apparel, with a long beard, spectacles on his face, large straw hat, and was carrying an outlandish array of unrecognisable items such as a net with a long handle, strings about his neck tied to collecting bottles bulging from his shirt pockets and other tools and collecting equipment clanking from his shoulder and belt. All of this must have combined to make Wallace the most bizarre human spectacle the local people, or their animals, had ever seen!

At the end of October, Wallace was in rather poor health and the collecting was slowing because of the frequent rains. He and his team returned to his house Mamajam. In his *Journal*, he wrote that he returned "about the middle of Nov'". But his collecting records in *Notebook* 4 extend only from 13–26 October. This inaccurately recalled date in the *Journal* should be remembered when we come to the old question of whether Wallace's famous eureka moment happened in Ternate or Gilolo.

It was time to pack up his collections. They were sent first to Batavia and thence, via the barque *Margaret West*, to Singapore and London. Wallace's collections reached Stevens in July and by the following September, the collections from Macassar had produced a handsome £130.[505]

Wallace wrote many letters at the beginning of December as he prepared for a six-month expedition to the remote Aru Islands 1,000 miles to the east. In one letter to John and Mary Wallace in California on 6 December, he wrote, "Excuse me writing more now as am very busy preparing for my six months sojourn among the natives of Arru (a place you never heard of) and have lots of letter to write previous to so long an absence from communication with civilized humans."[506] He also wrote to the Dutch Governor, Carel Frederik Goldman, at Amboyna (Ambon), requesting assistance and protection for the visit to the Aru Islands.

Wallace recorded in his *Journal*, "The wet season has now (Dec' 10th) regularly set in. Westerly winds & driving rain prevail for days together, the fields are all underwater and the Buffaloes & ducks are enjoying themselves to their hearts content."[507] On the same day, he wrote to his sister Fanny about his plans to visit the Aru Islands. He gave lists of supplies he was taking and a rather more frivolous, or perhaps not, piece of advice: "Has Eliza Roberts got rid of her moustache yet? Tell her in private to use tweezers. A hair a day would exterminate it in a year or two without any one's perceiving."[508]

A friend, possibly Mesman, introduced Wallace to an experienced Nakoda of a prau that sailed to the Aru Islands named Abraham van Waasbergen. He was half-Dutch and half-Javanese, married to a pretty Dutch lady, and could read and write both Dutch and Malay. Wallace misspelled his name as "Warzbergen" (there is no such name). Van Waasbergen assured Wallace that "two sorts of Birds of Paradise were abundant [at Aru], the large yellow and the small red kinds".[509] Wallace was sold. There were no more valuable specimens to collect than Birds of Paradise.

Van Waasbergen's prau was part of the annual trading fleet that rode the monsoon winds to the Aru Islands in the east, just south of New Guinea. In six months, when the monsoons blew in the other direction, the praus returned from Aru to Macassar and so Wallace would tag along not only on a native ship but on this centuries-old biyearly ocean highway. Once again, his science was piggy-backing on commerce. Aru lay beyond the trading routes of most Westerners and Wallace saw it as an "'Ultima Thule' of the East", a place beyond the borders of the known world. Darwin used the same expression to describe Tierra del Fuego.[510]

Wallace's party of three servants was now headed by Ali, who "could turn his hand to anything, and was quite attentive and trustworthy. He was a good shot, and fond of shooting, and I had taught him to skin birds very well."[511] Baderoon was also retained, despite his addiction to gambling. Wallace had given him four months' wages in advance shortly before departure "under pretence of buying a house for his mother, and clothes, for himself", but poor Baderoon had lost it all. "He had come on board with no clothes, no betel, or tobacco, or salt fish, all which necessary articles I was obliged to send Ali to buy for him." The third servant was Baso, always referred to as a "rascal". "He was to fulfil the important office of cook and housekeeper, for I could not get any regular servants to go to such a terribly remote country."[512]

At 6 am on 13 December 1856, the wind howled and the rain pelted down from a low dark sky as Wallace rowed out in a small boat to an exotic looking seventy-tonne wooden prau. It was unpainted wood and its deck sloped steeply down towards the bow. It had two A-frame wooden masts with

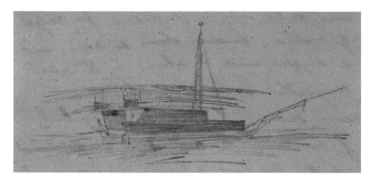

Wallace's unlabelled sketch of a prau at the end of *Journal 1*. Linnean Society of London.

enormous triangular sails made of matting. At the stern, two long oar-like rudders stuck to the side like forks. The prau had a crew of thirty Bugis sailors. There were ten Chinese and Malay men taken as part of van Waasbergen's trading operation and about half a dozen other passengers. Wallace was the wealthiest and highest status passenger.

Day after day the rain pelted down and the winds were impossible, so the vessel remained almost a week at Macassar. Wallace chose to stay on board as his cabin was dry and surprisingly comfortable. He used the time to dry his things and describe the prau in his *Journal*.

> In the widest part from the poop to midships is a little thatched house on deck. The after part of this I have quite to myself. It is a nice little berth six feet six inches long by five foot six wide & four feet high in the middle. It is entered by a low door on one side & has a little window on the other. I have it covered with my cane mats; on the further side are arranged my gun case, insect box clothes books etc. next to these are my bed & on the other side is plenty of room for my little stores for the voyage which are also hung round to the walls & roof. During the four days (most trying ones for wet) I have as yet occupied it I have been more comfortable that I have ever been on shipboard (P. & O. steamer not excepted)

The house of the Sultan of Goa. J.C. Rappard, 1883–9. Tropenmuseum.

The pond in the pleasure gardens of Gunung Sari in 1894. A. van der Kraan, 1980.

"…a large fish-pond, supplied by a little rivulet which entered it out of the mouth of a gigantic crocodile well executed in brick and stone. The edges of the pond were bricked, and in the centre rose a fantastic and picturesque pavilion ornamented with grotesque statues. The pond was well stocked with fine fish, which come every morning to be fed at the sound of a wooden gong which is hung near for the purpose. On striking it a number of fish immediately came out of the masses of weed with which the pond abounds, and followed us along the margin expecting food." Wallace, MA1:269.

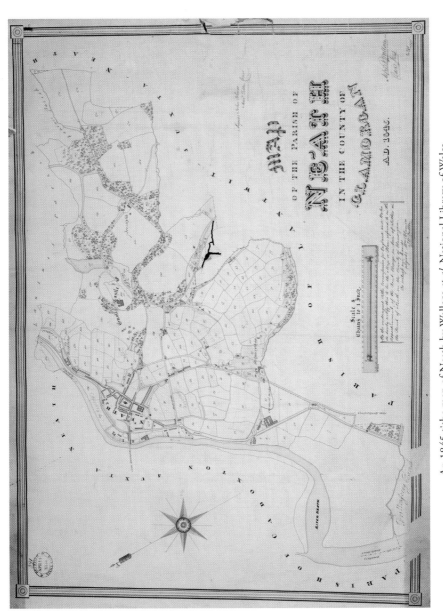

An 1845 tithe map of Neath by Wallace *et al.* National Library of Wales.

Zoological gallery of the British Museum in 1841. © Trustees of the British Museum.

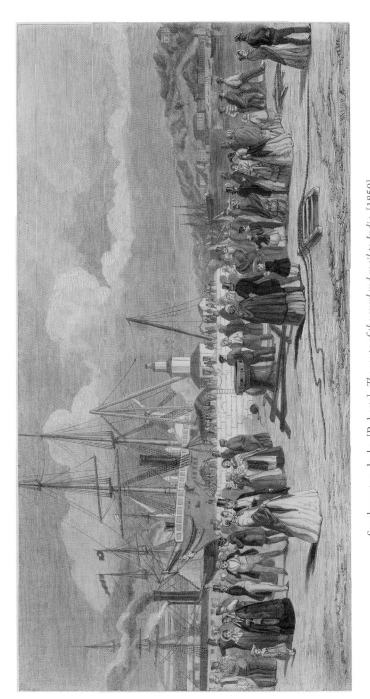

Southampton docks. [Roberts], *The route of the overland mail to India*, [1850].

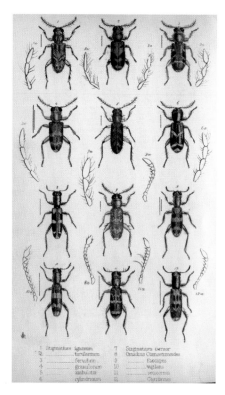

Checkered beetles from Singapore. J.O. Westwood, Descriptions of some new species of Cleridæ, collected at Singapore by Mr. Wallace. *Proceedings of the Zoological Society of London* 23:19–26, 1850, pl. XXXVIII.

Blue-throated Bee Eater. H.E. Dresser, *A monograph of the Meropidae, or family of Bee-eater.* London, 1884.

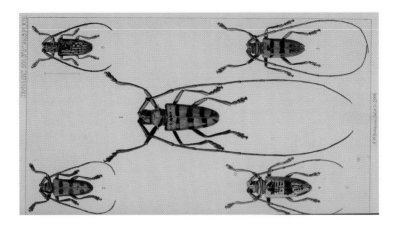

Longicorns collected by Wallace at Amboyna, Dorey and Macassar. F.P. Pascoe, Longicornia Malayana; or, a descriptive catalogue of the species of the three longicorn families *Lamiidoe, Cerambycidae and Prionidae*, collected by Mr. A.R. Wallace in the Malay Archipelago. *Transactions of the Entomological Society of London (series 3)* 3:1–712, 1864–9.

Coloured lithograph drawing by G.H. Leutemann to accompany the article 'Die Tigernoth in Singapore', [The tiger peril in Singapore] *Die Gartenlaube* (1865), p. 613. Courtesy of National Museum of Singapore, National Heritage Board.

The rarest cat in the world. The Borneo Bay Cat. J.E. Gray, Description of a new Species of Cat (*Felis badia*) from Sarawak. *Proceedings of the Zoological Society of London*, (1874), pp. 322–323.

Ornithoptera brookiana [now *Troides brookiana*]. W.C. Hewitson, *Illustrations of new species of exotic butterflies: selected chiefly from the collections of W. Wilson Saunders and William C. Hewitson*. vol. 1. London, 1855.

Ornithoptera poseidon. R.H.F. Rippon, *Icones ornithopterorum: a monograph of the Papilionine tribe Troides of Hubner, or Ornithoptera* [bird-wing butterflies] *of Boisduval*. London, 1907, pl. 7.

Wallace's Golden Birdwing Butterfly (*Ornithoptera croesus*). G.R. Gray, On a new species of the family Papilionidae from Batchian. *Proceedings of the Zoological Society of London*, 27:424–425, 1859, pl. LXVIII.

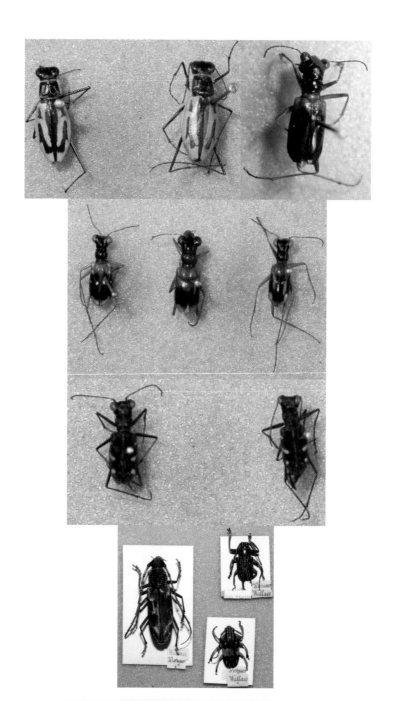

Top three rows: Tiger beetles collected by Wallace on Celebes.
Bottom row: Longicorn beetles collected by Wallace on Borneo. Courtesy of the Hunterian, University of Glasgow.

"The fine racquet-tailed kingfisher of Amboyna, Tanysiptera nais, one of the most singular and beautiful of that beautiful family." R.B. Sharpe, *A monograph of the Alcedinidae or family of Kingfishers*. London, 1868–1871.

Red-bellied Pitta. D.G. Elliot, *A monograph of the Pittidæ, or, family of Ant Thrushes*. London, 1867.

Wallace may have seen this captured tiger. Watercolour by E. Schlüter, 1858. Courtesy of National Museum of Singapore, National Heritage Board.

View of Ternate. J.C. Rappard, 1883–9. Tropenmuseum.

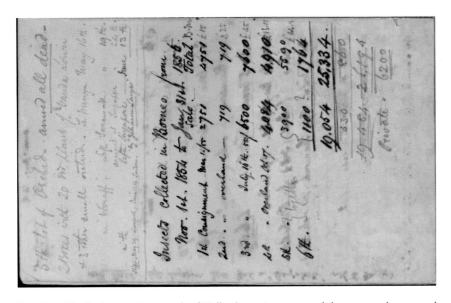

Page from *Notebook* 1 showing records of Wallace's consignments and the amounts later earned from them. Linnean Society of London.

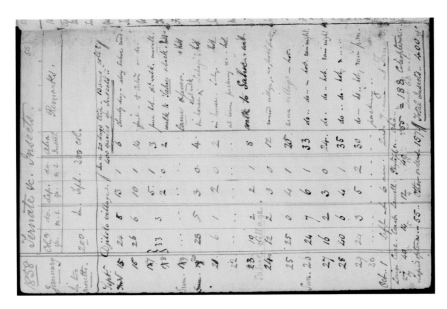

Notebook 4, p. 54b showing the almost blank collecting register for January 1858. Linnean Society of London.

Wallace's specimen of an Asian Brown Flycatcher preserved in the Raffles Museum of Bio-diversity Research, Singapore. Photograph by Tan Heok Hui. See John van Wyhe & Kees Rookmaaker, Wallace's mystery flycatcher, 2013.

PARADISEA PAPUANA.

Lesser Bird of Paradise. J. Gould, *The birds of new Guinea and the adjacent Papuan Islands*. London, 1875–1888.

PARADISEA SANGUINEA

Red Bird of Paradise. J. Gould, *The birds of New Guinea and the adjacent Papuan Islands.*
London, 1875–1888.

i Jennens lith   Stannard & Dixon.

SEMIOPTERA WALLACII.

Wallace's Standardwing. P.L. Sclater, Note on Wallace's Standard-wing, Semioptera wallacii. *Ibis*, 2(5 Jan.):26–28, 1860, pl. II.

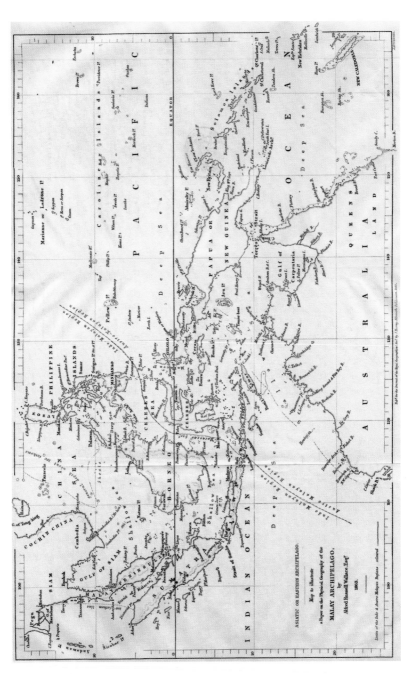

Wallace Line, as conceived in 1863. A.R. Wallace, On the physical geography of the Malay Archipelago, 1863.

Macassar to Aru. "Mr. Wallace's route" map MA1.

Chapter 7

IN SEARCH OF PARADISE

Wallace's and another prau finally sailed from Macassar on 18 December 1856, heading east on a voyage of 1,000 miles. He recorded in his *Journal* on the 22nd, "At night it is a beautiful sight to look down on our rudder, from which rushes an eddying stream of phosphoric light gemmed with whirling sparks of fire. It resembles more closely than any thing to which I can compare it, one of the large irregular, nebulous star clusters seen through a good telescope with the additional attraction of dancing motion & ever changing form."

The crew caught and fried a shark. Wallace tried a piece. It was "firm & dry but very palatable". On Christmas day, they were south of the island of Bouru. Christmas was not celebrated aboard the prau. Wallace treated himself by opening a jar of currant jelly, having an extra glass of wine and thinking of home. Wallace, the respected man on board with the largest accommodation, attended by his servants and with the freedom to eat and sleep whenever he liked, felt he had never enjoyed a voyage by sea so much. The prau itself had none of the irksome clanking engine noises or smells of coal smoke, oil or paint of a European steamer. He thought of the great steamer *Bengal* which carried him from Suez to Ceylon. Even if she was "that highest result of our civilization", he still preferred the prau.[513]

The next day, the island of Amboyna was in sight to port. Wallace's excitement grew. "The knowledge that every land about one now teems with scarlet lories & white cockatoos, & equally brilliant insects, & that hosts of unknown species await my researches, makes [me] more eager than ever for the time when

I may explore these '*terrae incognitae*' of the Naturalist."[514] On the 28th, they sailed between Ceram (Seram) and the Banda group. Ceram was obscured by cloud, but Banda's perfect volcanic cone was clearly visible with a wreath of smoke or steam encircling the summit. It was the first active volcano Wallace had ever seen but he was not moved. Perhaps his stomach was reacting to the heavy swell of the sea.

On the 30th, they passed just north of the tiny island of Teor. The bright tropical sun was reflected in every wave. The wooden praus pushed along slowly and steadily, their triangular sails filled by the winds and the green waters churned white where they cut the water. All day flying fish were seen suddenly emerging in a flash from the sea to glide extraordinary distances. "As they skim along the surface they turn on their sides so as fully to display their beautiful fins, taking a flight of about a hundred yards rising & falling in a most graceful manner. At a little distance they exactly resemble swallows."[515]

Around noon on 31 December 1856, the praus reached the northern point of Ké Island (Kai Besar).[516] Ké (pronounced "kay" according to Wallace) is a small, long and narrow island occupying 550 square kilometres. Its centre is dominated by a rugged mountain chain 3,000 feet high. Wallace painted the scene in a beautiful passage of *The Malay Archipelago*:

> The coast of Ké along which we had passed was very picturesque. Light coloured limestone rocks rose abruptly from the water to the height of several hundred feet, everywhere broken into jutting peaks and pinnacles, weather-worn into sharp points and honeycombed surfaces, and clothed throughout with a most varied and luxuriant vegeta-tion....Here and there little bays and inlets presented beaches of dazzling whiteness. The water was transparent as crystal, and tinged the rock-strewn slope which plunged steeply into its unfathomable depths with colours varying from emerald to lapis-lazuli. The sea was calm as a lake, and the glorious sun of the tropics threw a flood of golden light over all. The scene was to me inexpressibly delightful. I was in a new world, and could dream of the wonderful productions hid in those rocky forests, and in those azure abysses. But few European feet had ever trodden the shores I gazed upon its plants, and animals, and men were alike almost unknown, and I could not help speculating on what my wanderings there for a few days might bring to light.[517]

When three or four native boats arrived, Wallace had his "first view of the Papuan race in their own country". They were on board for only five minutes before Wallace was convinced that his earlier opinion, based on those he had seen occasionally before, was correct. They were certainly not of the Malay race. He was as much impressed with their behaviour and expressions as their physical appearance. He saw no evidence that the Malay race gradually blended into the Papuan as the Chinese and Malays seemed to on Bali. There was instead a discontinuous break between them. Wallace noted there were two types: aborigines and a mixed Malay race from Banda. He noted their language and dwellings. The natives lived on sago instead of rice. The economy of the northeastern corner of the island revolved around the trade with the Aru Islands as all the praus from the west stopped at Ké Island for water and to buy food, fuel oil for lamps and supplies.

New Year's Day 1857 was spent exploring the interior of the island with Ali and Baderoon. After a short walk along the coastal path, they headed uphill but eventually the jagged rocks, so characteristic of the island, were too much for the bare feet of Ali and Baderoon which were cut and bleeding. Yet the native village boys, who ran along with them to see what the odd strangers were up to, bounced along on the rocks without concern. Wallace recorded the day in his *Journal*:

> This has been a luxurious day for me as a Naturalist. I have wandered in the forests of an island, which I believe no Naturalist has trodden before me. I obtained about 50 species of insects & four birds none of which I had ever found before though I was acquainted with a few of the *Lepidoptera*. Among the beetles was a magnificent *curculis* blue & black banded & several pretty insects of a small size. A magnificent yellow and black papilis and the curious *Hamadryas*, the sole representative in the East of the S. American *Heliconidae* also rewarded my excursion.[518]

Wallace found the collecting rewarding, but the lack of trails into the interior meant that he was unable to penetrate far into the forest. The rest of their stay was spent along the coast. He paid the natives "most fragrant tobacco" to bring him insects. One of these was "the most beautiful [jewel beetle] I have seen".[519] The beetle was a glittering iridescent blue and green which seemed to change colour as he turned it in his fingers. Flying overhead "my eyes were feasted for the first time with the sight of the splendid scarlet lories on the wing as well as of that most regal butterfly *Anithoptera Priamus* with the large & very different female, but flying so high as to be out of any reach. One of these splendid insects was brought in a bamboo with lots of beetles of course torn & scratched to pieces."[520]

One day a man saw him collecting. "He stood very quiet till I had pinned and put [an insect] away in my collecting box, when he could contain himself no longer, but bent almost double, and enjoyed a hearty roar of laughter....their excitement on very ordinary occasions, are altogether removed from the general taciturnity and reserve of the Malay."[521]

At 4 pm on Tuesday 6 January, van Waasbergen's prau raised anchor and sailed east for Aru giving Wallace a few hours to watch the long rugged coast of Ké stretching away to the south as far as the eye could see, a dark green silhouette sinking into the midst of the boundless sea.

Aru Islands, 8 January–2 July 1857

About 9 pm on 8 January 1857, Wallace's thirty-fourth birthday, the praus reached Dobbo on the small low-lying island of Wamar nestled in an arm of the larger island of Wokam to the east. The Aru Islands (the name means "casuarina-trees") lie between Western New Guinea and Australia. In his essay "On the Aru Islands", Wallace speculated on the origins of these islands. "There are three distinct modes by which islands may have been formed, or have arrived at their present condition,—elevation, subsidence, and separation from a continent or larger island."[522] He concluded that, given the similarity of their species, Aru must have been connected to New Guinea by a land bridge that had since subsided. He was unaware of Earl's earlier proposal of a connection to New Guinea based on shallow seas in-between.

The islands are now understood as fragments of continental shelf and not volcanoes like Ké. The main island is so low that shallow channels cut all the way across it in three places. The Dutch had visited Aru in the early 17th century and had even built a fort on Wokam that commanded the channel between Wokam and Wamar as part of their spice empire. It was abandoned in 1808 and by 1857 it lay in ruins. However, the stone church was still in use as the islanders had been largely converted to Christianity. Aru was still nominally under Dutch control although there was no permanent Dutch presence. Every year a schooner brought a commissioner for about a month.

Dobbo was a small trading settlement on the north end of the island just a mile across the channel from the old Dutch fort. It was inhabited by Malays and a few Chinese merchants with just three rows of wooden houses with steep pointed roofs of palm thatch. A few tall and slender coconut trees dotted the

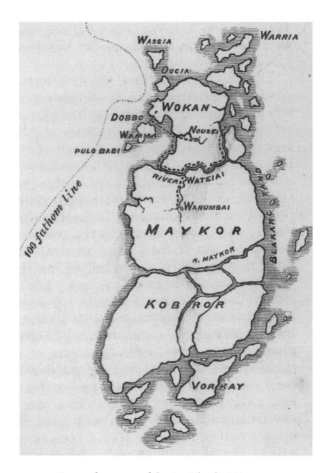

Extract from map of the Aru Islands. MA2:219.

town. There were barely a dozen people there when Wallace arrived, but over the next month, as more and more praus came, the population rose to about 500. Wallace estimated there were only forty to fifty women. Van Waasbergen had here a "Macassar girl of 15 or 16" as mistress, which was looked on as the norm.[523] Whether his pretty Dutch wife was aware of her was not stated!

Eventually, there were about fifteen large trading praus from Macassar, 200 smaller boats from Goram and Ceram and a few boats from Ké. Wallace estimated that that the exports from Aru were worth about £15,000. Aru offered mother-of-pearl, tortoise shell, pearls, sea cucumbers, edible birds' nests and timber. Imports included 45,000 half-gallon bottles of arrack, a liquor made from fermented coconut sap. There were also sago-cakes, tobacco, cotton, English and Chinese crockery as well as cutlery, muskets, gunpowder and elephant ivory.

"Dobbo, in the trading season." MA2:266. Drawn by Thomas Baines based on an original sketch by Wallace now in the Natural History Museum.

Wallace was soon settled in a thatched house near van Waasbergen. Tired of sitting on his collection box, Wallace had brought a wicker chair to make his indoor work more comfortable. The following morning, he set off after breakfast with Baderoon and a native boy as a guide. Wallace was delighted to capture thirty species of butterflies, more than he had ever captured in one day since the Amazon.

> The next two days were so wet and windy that there was no going out; but on the succeeding one the sun shone brightly, and I had the good fortune to capture one of the most magnificent insects the world contains, the great bird-winged butterfly, Ornithoptera Poseidon. I trembled with excitement as I saw it coming majestically towards me, and could hardly believe I had really succeeded in my stroke till I had taken it out of the net and was gazing, lost in admiration, at the

velvet black and brilliant green of its wings, seven inches across, its bolder body, and crimson breast. It is true I had seen similar insects in cabinets at home, but it is quite another thing to capture such oneself- to feel it struggling between one's fingers, and to gaze upon its fresh and living beauty, a bright gem shining out amid the silent gloom of a dark and tangled forest. The village of Dobbo held that evening at least one contented man.[524] [see colour insert]

The rains continued for most of the next two weeks, providing few opportunities for collecting. Wallace was disappointed to learn that the Aru Birds of Paradise were out of their colourful breeding plumage until September or October. The trading praus returned to Macassar with the reversed trade winds in July.

Hermit crabs swarmed everywhere from the white sandy beach to the forest of coconut trees; "that nasty-looking Chinese delicacy the sea-slug" was sometimes washed up on shore. Every day in the trading town men went from house to house selling fish and seafood. Wallace was disgusted with sea cucumbers which he thought looked like "sausages which have been rolled in mud and then thrown up the chimney"![525] In the evenings, the widest street was busy with cockfighting, impromptu football matches and wild binge drinking that sometimes literally brought the house down around the revellers. The women took care to remove the knives and weapons when the men started to quarrel.

On 15 February, Wallace recorded a tiny insect battle in the jungle.

> Saw a Clytus [longhorn beetle] attacked by Ants. Each leg & antenna was seized by one & he was held stretched out while dozens swarmed over him intending I suppose to eat him alive. It was not till I had driven away his persecutors, who were extremely loth to let go their prey, that I discovered he was alive, so I kindly inserted him into my bottle to be honoured by the gaze of European eyes.[526]

Observing the native islanders, even though he thought they were of mixed race, Wallace was convinced that Papuans were "negros". They had "a nearly black skin and woolly or frizzly hair. They are taller than the Malays, and more slenderly made; have a flatter forehead, more projecting brows, larger and thicker nose, with the apex rather bent down, and thick lips". He "studied their physical and moral peculiarities", language reminiscent of Combe's *Constitution of man*.

His letter was later read at the Royal Geographical Society on 22 February 1858. One of those listening was the old colonial administrator John Crawfurd, the author of Wallace's Malay dictionary. The punctilious Crawfurd objected to papers on the archipelago so often that he was nicknamed the Society's "objector-general". Crawfurd objected, "Mr. Wallace concluded they were negroes, similar to the negroes of New Guinea; but [Crawfurd] had seen them as more nearly resembling the inhabitants of the north of Australia."[527]

Wallace believed he saw evidence of Portuguese admixture in the natives and thought he heard remnants of the Portuguese language. Although some Portuguese words had diffused throughout the Moluccas as part of the lingua franca since the 16th century, there is no evidence that the Portuguese ever visited Aru.[528] Wallace had a tendency to see and hear Portuguese everywhere he went.

In mid-February, he visited Wokam Island. In his *Journal* and *The Malay Archipelago* he gave 5 February, but this is contradicted by his collecting register which shows 16 February.[529] He noticed that the fauna was different. On 18 February, Wallace received a reply from Carel Frederik Goldman, the Dutch Governor of the Moluccas, via a brig he called "Antanilla". The ship was the 275-tonne *Antilla*. Goldman politely agreed to help and sent instructions to the island's chiefs that Wallace was to be supplied with boats, men and protection. Wallace now felt able to move to the main island.

However the next day, a small prau arrived with the alarming news that pirates were in the area for the first time in several years. The prau had been attacked and one of her crew wounded but a fair wind allowed them to escape. The pirates were said to be from Maguindanao, the southernmost large island of the Philippines. They operated from several small praus. Praus from Dobbo were sent out in search of the pirates. On the evening of the 23rd, one of van Waasbergen's small praus returned. She had been attacked in broad daylight by pirates firing muskets six days before. Van Waasbergen's crew escaped in their boat and hid in the forest while their prau was plundered. On 1 March, a small prau arrived from the east coast of Aru with the news that one small pirate prau had been taken.

Wallace wrote a letter to update Stevens and left it to be taken by the *Antilla* back to Macassar in April. Wallace had purchased a boat and found two men to join Ali and Baderoon to go to the main island. They departed on 13 March. By noon they entered a small river and penetrated the thickly forested island moving upstream. Two hours later, they reached a small settlement with "a house or rather small shed of the most miserable description". The guides proposed that Wallace stay here. He paid to stay in part of the house and his Malay boys slept in the boat. But "sand flies were very abundant at night biting over every part of the body & producing a more permanent irritation than mosquitoes. My feet & ancles particularly suffered & were completely covered with little red swollen specks which itched horribly." Again the weather was rainy and the collecting of birds and insects not impressive.

In the early morning of 28 March, they moved on. Wallace wanted to explore the shallow Watelai Channel (Sungai Manumbai) which bisected the mainland of Aru. The wind was against them and after rowing hard until midday, they turned up a small river to a little village of sea cucumber fishermen. No doubt Wallace turned up his nose at the disgusting dirty sausages. He rented a shed for one chopper.

In the evening, there was a sudden cry of alarm, "'Bajak! Bajak!' (Pirates!)" The villagers grabbed their spears and bows and Wallace took his gun and ran down to the beach. But it was a false alarm; it was only one of their own fishing boats returning. Clearly tensions were running high. The collecting here was very poor so on 30 March they moved on again. The required winds were finally blowing so about noon they entered the Watelai channel and saw the islands of the east coast, home to the productive fisheries.

After a couple of days trying to get about, they turned south out of the Watelai Channel into the River Wanumbai and reached a tiny village of only two houses called Wanumbai. Many years later, Wallace received a letter from the man who piloted him up the Wanumbai.[530] After a difficult bargaining exchange, with the pilot acting as intermediary, Wallace rented accommodation in one of the houses on stilts from a man named Kamis for some cotton cloth.

> Here I tried to make a bargain for house accommodation, but the owner made many excuses & difficulties & fears first of not pleasing me & then his own son, till I was just on the point of going off to seek a more hospitable place but made a last attempt by sending my pilot to reason with him. He returned with a demand of about half the cost of the house for the use of a small slice of it for a fortnight, but as it was now put on the footing of a pecuniary difficulty only I soon settled it by getting out a bale of cotton cloth of which I offered him ten yards, which seeing I was determined not to give more he after a little hesitation accepted.[531]

The German naturalist Hermann von Rosenberg travelled here eight years later and stayed with the same host, Kamis, "a very influential man amongst his countrymen".[532] Von Rosenberg liked to call the Wanumbai "Naturalist creek" in honour of the visits there of Wallace and himself.

On 6 April, Wallace recorded in his *Journal*:

> Our house which is a very good one contains about four or five families & there are generally from 6 to a dozen visitors besides. They keep up a continual row from morning to night, talking laughing shouting without intermission; not very pleasant, but I take it as a study of national character & submit. My boy Ali says "Banyak quot bitchara orang Arru" (The Arru people are very strong talkers) never having been accustomed to such eloquence either in his own or any other Malay country we have visited.

At least the house was free from sand flies and mosquitoes. But during his collecting walks in the surrounding plantations, "the day biting mosquitoes swarmed & seemed especially to delight in attacking my unfortunate legs". Near the end of April, "after a months incessant punishment" his legs "rebelled against such treatment, & broke into open insurrection, throwing out numerous inflamed ulcers, which are very painful & prevent walking".

But despite being housebound, the collecting went on as Ali and Baderoon scoured the forests for birds and animals. They brought him three marsupials. Finally one day, Baderoon brought home a Great Paradise bird. Wallace learnt from his boys and the natives about the birds' habits.

Wallace criticised the self-trained amateur René Primevere Lesson (1794–1849), the first European naturalist to observe the mysterious Birds of Paradise alive during the voyage of *La Coquille* (1822–5). "He visited, I find, only the north coast (Dorey Harbour) and the islands of Waigiou." Yet Wallace would do the same himself. Wallace also doubted Lesson who "concluded that the bird is polygamous! but I have no doubt that what he took for females were mostly young males". There may have been young males about, but Wallace was wrong. Male Birds of Paradise are polygamous and display to attract females to mate. The strongly differing features of the two sexes are now explained by Darwin's theory of sexual selection.

When the bird box was nearly full, Wallace and his team returned in a day to Dobbo on 7 May.[533] He found the *Antilla* still at anchor so he was able to continue his letter to Stevens:

> Rejoice with me, for I have found what I sought; one grand hope in my visit to Arru is realized: I have got the birds of Paradise (that

"Natives of Aru shooting the great bird of paradise." Frontispiece to MA2. The vertical distance has been greatly shortened to fit the page.

announcement deserves a line of itself); one is the common species of commerce, the Paradisea apoda; all the native specimens I have seen are miserable, and cannot possibly be properly mounted; mine are magnificent. I have discovered their true attitude when displaying their plumes, which I believe is quite new information; they are then so beautiful and grand that, when mounted to represent it, they will make glorious specimens for show-cases, and I am sure will be in demand by stuffers. I shall describe them in a paper for the 'Annals.'

"Native house, Wokan, Aru Islands (Where I lived two weeks in March, 185[7])." "a photograph of a native house in the island of Wokan, which was given me by the late Professor Moseley of the *Challenger* expedition, because it so closely resembles the hut in which I lived for a fortnight, and where I obtained my first King bird of paradise, that I feel sure it must be the same, especially as I saw no other like it." ML1:356–357.

> The other species is the king bird (Paradisea regia, Linn.), the smallest of the paradisians, but a perfect gem for beauty…I believe I am the only Englishman who has ever shot and skinned (and ate) birds of Paradise, and the first European who has done so alive, and at his own risk and expense; and I deserve to reap the reward, if any reward is ever to be reaped by the exploring collector.[534]

The swollen insect bites and then another bout of fever kept Wallace housebound for six weeks. Confined indoors again, he wrote journal entries and speculated about human societies. In a long *Journal* entry, he considered the clues that biogeography threw on species origins.

> The existence of a species of Kangaroo in Aru & several in N. Guinea as well as almost all the Mammalia being Marsupials points to a connection of both countries with Australia…We thus see how beautifully the present geographical distribution of animals & plants may illustrate & throw light upon their geological history, while the origin & antiquity of any country deduced from geological observations will generally enable us to explain & account for all the anomalies of its natural productions, by the simple law, that in all the changes of the organic world, closely allied species if any have always been substituted for those which have become extinct.

Here was the Sarawak law with its coincident relationship between new and extinct species. Presumably this was because his *Journal* was intended to be read by others, whereas his notebooks contained explicitly heterodox views on species.

Wallace was annoyed with Baderoon's laziness and gave him a good scolding. Offended, the boy asked for his wages and left. Like at Macassar before sailing, once he had some money, Baderoon took to constant gambling.

> At first he had some luck & Ali told me he had got plenty of money & had been buying ornaments but he afterwards lost everything ran into debt, & has now become the temporary slave of the woman who has paid the amount for him…he will now most likely stay here, the year round & if as very probable he gets deeper into debt may remain a slave for life. He was a quick & active lad when he pleased but his idleness & incorrigible propensity to gambling made me not very sorry to part with him.

Baderoon was never mentioned again so it is uncertain if he remained a slave on Aru. Wallace could easily have bought his freedom but apparently did not.

By the end of June, the merchants were hitching a ride on the reversed trade winds back west and Dobbo began to empty. On 2 July, van Waasbergen's prau with Wallace, Ali and the collections aboard sailed for Macassar. They took a slightly different route on the return, going south of Banda. As the winds pushed the prau ever westward at about five knots, Wallace reflected on the weather patterns of the region and concluded that if the Dutch would record weather measurements at all their outposts "various anomalies of climate to be reduced to some dependence on general laws". After nine and a half days of sailing, they reached Macassar on the evening of 11 July 1857.

Macassar, 12 July–19 November 1857

Wallace landed at Macassar on the morning of 12 July and his friend Willem Mesman helped him return to his bamboo house Mamajam. Wallace spent three or four weeks laboriously preparing a shipment of over 9,000 Aru specimens to Singapore.[535] There his consignment was transferred to the 1,000-tonne *Maori* which sailed on 4 September for London which she reached the following January.[536] By March of 1859, Wallace had earned a handsome £360 from his Aru collections.

From about 2 August therefore until he left for his next collecting trip, apparently in the first ten days of September, we have no record of Wallace's activities.[537] During this time, he attended to the seven months' post and the latest scientific publications carried halfway around the world by the coursing lines of steamships. The newspapers were full of the Indian Mutiny that erupted in north and central India in mid-May. He was also reunited with his books and wrote articles about his Aru collections. The second important series of notes on species in *Notebook 4* seems to date from this stay in Macassar. Although the notes are again undated, they begin in a different medium (pencil) after the Aru notes conclude.

Groping into darkness

Many writers have claimed that something called the "species problem" was the goal of Wallace's study and speculations and thus the Ternate essay was "perhaps an inevitable result of an intense search". Historian Barbara Beddall wrote in 1968, "If Darwin had been working on the problem for twenty years, Wallace had been working on it for at least ten, the major difference being that Darwin had long had a theory against which he was collecting facts, while Wallace was still actively searching for one." Or, as biographer Ross Slotten wrote, "Never far from [Wallace's] mind, even in such trying moments, was the main impetus for traveling to the ends of the earth, the problem that had obsessed him for a decade."[538]

But the notes do not support these legendary images of Wallace on a quest for the Holy Grail of biology. Wallace's notes reveal ridicule of divine design arguments, devotion to the rule of natural laws, rationality and uniformitarianism. He accepted that species change from the start; it was not a discovery he sought to make. His notes are largely concerned with the history of life rather than adaptation or ecology. The history of life had seen the major types of living things branching off from one another. Their lineages formed a tree of life. He was sure this must explain the true relationships between the different groups, but there was so much prejudice against it. His notes take issue with arguments against evolution, but they are not questions seeking answers or a hunt for a solution. Someday he planned to write a book.

The 2nd Macassar species notes

His notes begin with whales branching from mammals. These might have been prompted by reading Knight's excellent *Cyclopedia* article on whales.[539] "What are whales?" Wallace asked. Rather than a halfway point between mammals and fishes to complete an abstract philosophical scheme of nature as entomologist William Sharp Macleay had it, whales are surely derived from land mammals because they have almost all the characteristics of land mammals and almost none of fish. Wallace noted that vanished forms were the transitions between what are now isolated groups. He clearly meant evolutionary transitions. This moved beyond the Sarawak paper, because it aimed to understand how groups are related by derivation.

A few pages later are brief but very important notes on reading the German geologist Christian Leopold von Buch's book on the flora of the Canary Islands. "On continents the individuals of one kind of plant disperse themselves very far, and by the difference of stations of nourishment & of soil produces varieties, which at such a distance not being crossed by other varieties & thus brought back to the primitive type, become at length permanent & distinct species."[540] Here, Wallace noted that new varieties could branch off from their parent species and eventually, if isolated, become a new species and not be brought back to type. This sounds almost like the title of Wallace's later essay on natural selection. This was a fundamental step closer to his theory in the Ternate essay. But adaptation was still nowhere in his considerations.

A few passages, written with a different pencil, were added to blank spaces in the old Si Munjon notes on Lyell. Their content seems more similar to the note-taking of this second Macassar stay. "But we have no proof how the varieties of dogs were produced. All varieties we know of are produced at *birth* the offspring differing from the parent. This offspring propagates its kind."[541] Thus, for Wallace, "the varieties of the Primrose adduced by Lyell is complete proof of the transmutation of species". Lyell quoted the following case.

> I raised from the natural seed of one umbel of a highly manured red cowslip a primrose, a cowslip, oxlips of the usual and other colours, a black polyanthus, a hose-in-hose cowslip, and a natural primrose bearing its flower on a polyanthus stalk. From the seed of that very hose-in-hose cowslip, I have since raised a hose-in-hose primrose. I therefore consider all these to be only local varieties, depending upon soil and situation.[542]

Hence, what Wallace called a "complete proof" of evolution was the production of different varieties at one step, in this case by manipulating the soil and fertiliser. This is reminiscent of *Vestiges'* version of new organic forms born quite different from their parents. In the following sentence, Wallace concluded, "Who can declare that it shall not produce a *variety* which process continued at intervals will account for all the facts."

Varieties emerge from a species, and subsequently give rise to further varieties. This model of organic change is very similar to that later expressed in his Ternate essay, though here still bearing a greater resemblance to *Vestiges*. This is redolent of what would be called saltational speciation. Darwin too went through a saltationist phase early in his theorising about species. This is connected with Wallace's belief, expressed in the Sarawak paper and elsewhere, that animal structures may exist prior to their need or a later function. Lyell and von Buch, in contrast, had local conditions causing varieties.[543] Wallace's view was also the opposite of Lamarck, who believed that habit preceded structure.

The Aru paper

His renewed focus on species theory is reflected in an article on the Aru Islands written at this time.[544] "I believe that nearly one-half of the hitherto-described species of passerine birds from New Guinea will be found in my Aru collections, a proportion which we could only expect if all the species of the latter country inhabit also the former." Why were species in Aru like those in New Guinea? Wallace concluded (as Earl had years before) that the two islands were once connected by a land bridge that had since subsided beneath the sea. Like that other notorious independent discovery, natural selection, Wallace may not have been first, but at least he got there on his own.

He used the paper to pose some ambitious questions. "Why are not the same species found in the same climates all over the world?" According to Lyell, "as the ancient species became extinct, new ones were created in each country or district, adapted to the physical conditions of that district". Wallace then quoted from his Si Munjon notes: "'Then,' [Lyell] says, 'the animals and plants of Northern Africa would disappear, and the region would gradually become fitted for the reception of a population of species *perfectly dissimilar in their forms, habits, and organization.*'"[545]

But the case of Aru shows they often are not totally "dissimilar". Therefore, Wallace declared some "other law has regulated the distribution of existing species than the physical conditions of the countries in which they are found". He referred to his Sarawak law as a more likely explanation. History rather than adaptation to the environment was the clue. But once again Wallace concealed what he really thought about where species come from with vague language: "New species have been gradually introduced into [New Guinea and Australia], but in each closely allied to the pre-existing species, many of which were at first common to the two countries. This process would evidently produce the present condition of the two faunas."

The permanent varieties paper

Perhaps the most important of Wallace's theoretical statements on varieties and species up to this point was his brief "Note on the theory of permanent and geographical varieties" which appeared in the *Zoologist* in January 1858.[546] It is one of his most dense and impenetrable writings so its meaning has been hard for commentators to interpret. The paper was written in reaction to an unknown publication or publications in his pile of post that awaited him in Macassar.

At the outset, he noted that "this subject" is now being discussed especially amongst entomologists. Who was discussing this? Could it be the theoretical and evolutionary–philosophical articles of the old radical anatomist Robert Knox appearing in the *Zoologist*?[547] There are few similarities. And Knox was no entomologist.

Another work appeared in London in 1856 that seems to be a likely candidate. The entomologist Thomas Vernon Wollaston published a small volume entitled *On the variation of species with special reference to the insecta; followed by an inquiry into the nature of genera.*[548] It was dedicated to Charles Darwin "whose researches, in various parts of the world, have added so much to our knowledge of Zoological geography". Wollaston was educated at Jesus College, Cambridge. Unfortunately, ill health dogged him. He made several trips to Madeira for his health but ended up making a major study of the insects of the Atlantic island. He found that island forms differed from continental versions, forming permanent or geographical varieties.

Although variation away from parent species was a fact for Wollaston, and more substantial than previously realised, it occurred only within strict limits, which meant species do not evolve. Members of a species are all related to each other by descent from a common parent from "specific centres of creation". Wallace used this phrase in his Aru paper written at the same time.[549] For Wollaston, a variety differed a small amount from its parent species, but this difference was permanent. It was inherited by all subsequent offspring. Even a small distance away from a parent species should be regarded as specific as long as no intermediates are known. This sounds like what Wallace objected to in his paper.

It was just the sort of speculation to stir and at the same time annoy Wallace. Perhaps Stevens sent it knowing Wallace's interest in theories of varieties and species. Wallace's response shows his continuing fascination with varieties. At the start of the paper, he claimed to be impartial, but his argument was clearly not so. He pointed out inconsistencies with "independent creations" of species and hinted for the first time at genealogical descent as the more plausible cause for their origination.

The term "permanent varieties" is now archaic. The anthropologist James Cowles Prichard defined a "permanent variety" as a differing variety that had split off from the parent species.[550] Wallace argued that if permanent varieties are groups differing only slightly from a parent species, not specially created but actually descended from a parent species, on what grounds can we insist that species, differing from each other also only slightly, are specially created? Why should the identical effect of slightly different types of organisms be attributed to radically different causes — descent and creation? It was an ingenious criticism of special creation, but he still refrained from declaring his own belief.[551]

In Sarawak, Wallace published "On the succession of species"; in this note he suggested that new species originated from the direct descent of varieties from earlier species. Wallace's views on the history of life and the relationship between varieties and species had moved on a long way since the Sarawak paper; indeed he was on the cusp of the views that would culminate in the Ternate essay a few months later. But natural selection was still nowhere in sight.

Maros, c. 11 September–8 November 1857

In early September, Wallace obtained a "pass" from the Dutch Resident and set off north by boat for the River Maros to stay with Mesman's elder brother Jacob David Matthijs.[552] The Mesmans' father had been given the estate of Maros by the British in 1816 and was Dutch Resident from 1820.[553] The unfortunate Ali seemed always ill at Macassar. He was left in hospital in the care of Wallace's "friend the Doctor", J. R. Bauer who lent Wallace a copy of Lesson's book.[554] Wallace must have heard more stories of Ida Pfeiffer who had stayed in Macassar with a Dr. Schmitz and his wife in 1853.

Wallace set out with two new servants "utterly ignorant of everything". At daybreak, the boat entered the Maros River and by 3 pm they reached the village. Wallace showed his pass to the Assistant Resident and was given ten porters

Maros. "Mr. Wallace's route" map MA1.

to transport his baggage inland. This gives some idea of how much baggage he must have carried. His *Journal* recorded:

> The country was a uniform plain of burnt up paddy fields but at a few miles distance a range of precipitous hills appeared backed by the lofty central range of the peninsula. Towards these our path lay & after six or eight miles the hills began to close around us & the ground became pierced with blocks & pillars of limestone rock,- while abrupt conical hills rose here & there like islands. Passing over an elevated patch form- ing the shoulder of one of the peaks, a picturesque scene lay before us; a little valley almost entirely surrounded by the hills rising abruptly in huge precipes & forming a succession of knolls & peaks & domes of the most varied & fantastic shapes. In the centre was the thatch & bamboo house of Mr. M. while scattered around were a dozen cottages of his work people. I was received in a saloon, walls of bamboo roof of grass thatch & floor of thin bamboos flattened & plaited....I therefore had a little cottage built in the forest at the base of the mountain about a mile off, & having recovered from a smart touch of fever brought on by too much exertion in the sun I took up my abode there.

He later recalled, "I have rarely enjoyed myself more than during my residence here. As I sat taking my coffee at six in the morning, rare birds would often be seen on some tree close by, when I would hastily sally out in my slippers, and perhaps secure a prize I had been seeking after for weeks."[555]

> Then what delightful hours I passed wandering up and down the dry river-courses, full of water-holes and rocks and fallen trees, and

Maros in 1929. The limestone hills are visible in the background. Tropenmuseum.

overshadowed by magnificent vegetation! I soon got to know every hole and rock and stump, and came up to each with cautious step and bated breath to see what treasures it would produce. At one place I would find a little crowd of the rare butterfly Tachyris zarinda, which would rise up at my approach, and display their vivid orange and cinnabar-red wings, while among them would flutter a few of the fine blue-banded Papilios. Where leafy branches hung over the gully, I might expect to find a grand Ornithoptera at rest and an easy prey.[556]

On 19 September, he borrowed a horse from Jacob Mesman to visit the Bantimurung Waterfalls where the river cascades down a narrow precipice between two nearly vertical limestone ridges close on either side. Here and there, occasional patches of white limestone were visible through the encrusting tropical verdure. It was "a spot much visited & considered very beautiful". It is still an area of remarkable butterflies and Wallace netted as many beautiful specimens as he could. In the wet season, the clear water of the river is stained a mesmerising azure blue. It remains a popular tourist attraction and rows of shacks outside the park entrance sell colourful local butterflies mounted in cases. Wallace stayed until the 22nd.[557]

Bantimurung Waterfall by Woodbury. Tropenmuseum.

In the stillness of the evening in his little cottage near Maros, Wallace could hear a barely perceptible rustling, crunching and popping all around as insects slowly devoured the beams. A fine sawdust rained down on his mosquito net and accumulated in little piles at the base of the wall posts. With this quiet background chorus, Wallace answered many of his letters that had arrived while in Aru. One of these was Darwin's. Only a fragment of Wallace's letter of 27 September 1857 survives, but it offers tantalising clues of species discussion.

Wallace began by thanking Darwin for his letter from May 1857 in which Darwin agreed on "the order of succession of species" in the Sarawak paper. Wallace confessed, "I had begun to be a little disappointed that my paper had neither excited discussion nor even elicited opposition. The mere statement & illustration of the theory in that paper is of course but preliminary to an attempt at a detailed proof of it, the plan of which I have arranged, & in part written, but which of course requires much research in English libraries & collections, a labour which I look" — the remainder of the letter is cut away.[558] This fragment survives because Darwin kept the note about black jaguars on the back. But it is clear that Wallace revealed he would write a book on species theory when he came home. What it might be about was left quite vague. Darwin had seen only the law of succession. No doubt Wallace might be an ally, but the idea of a competitor could never have entered Darwin's mind.

In mid-October, the rains began and insects and animal life began to appear in and around Wallace's cottage. The area was soon crawling with huge millipedes as thick as a finger. Wallace even found one in his bed one morning. Another creature almost made an abrupt end to his collecting career, and his life. "One day when beating in a dead tree for insects & having got my net half full of leaves I was just going to examine them when a rustling & commotion made me stop & up came the head of a long brown snake."[559] No doubt he pulled his hand out of the net in an instant. Despite the better collecting, his hut was too damp to keep his collections dry so he started packing up to return to Macassar.

Wallace's tiger beetles

After his return from Maros, Wallace collected around the edge of Macassar and even on the town's lamp posts. One day during a sunny spell on the edge of town, Wallace found some beetles that would prove the most influential he ever collected. He found two tiger beetles on the shiny brown mud of a saltwater creek. The insects were coloured so exactly like the mud that Wallace could see only their shadow cast by the afternoon sunshine. It reminded him of an insect "singularly agreeing in colour with the white sand of Sarawak", another from Bali that resembled "the dark volcanic sand of its habitat" and the two species from Lombok always found on "the same coloured dark sand".[560] This perfect matching to the environment puzzled Wallace. He recorded the new beetles: "385. on mud of salt creek-? sexes-plain & bk. spots".[561] Their exact match with the mud would make these tiger beetles his equivalent of Darwin's legendary finches. They could be called Wallace's tiger beetles. They would spark the greatest breakthrough of his life. But not quite yet.

Wallace was now convinced that varieties descended suddenly from a species, and themselves could become new species. But how did varieties or new species become adapted to an environment? Wallace never addressed this in his species notes. In fact, he strongly opposed speculation on the utility of structures. This smacked of shallow natural theology where the fit of every structure to a need was supposedly evidence of divine design. But if the tiger beetles were descended from some previous species, how did they come to match the mud so perfectly? He continued to puzzle over it.

Wallace posted his letters and prepared for his next journey. His collections were shipped out on the brig *Corcyra* on 13 December and arrived in Singapore on 16 January 1858.[562] They earned him more than £300 — a very satisfactory result.

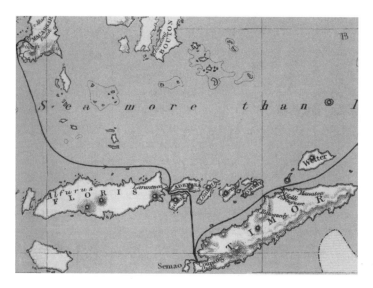

Macassar to Timor. "Mr. Wallace's route" map MA1.

Timor, 24/25–c. 26 November 1857

On 19 November, Wallace and Ali departed Macassar on the new Dutch-built 181-tonne screw steamer *Padang* on her monthly mail run through the Moluccas. Similar to the P&O ships, the Dutch government awarded mail contracts to private companies to facilitate regular mail services. From 1850, the mail service to the eastern part of Java and the outlying posts of the Dutch East Indies was contracted to the company owned by the former naval officer W. F. K. Cores de Vries, which maintained four screw steamers on this route, as well as another one from Batavia to Padang, West Sumatra. In the 1850s, the steamers of Cores de Vries operated a monthly anti-clockwise route through the Moluccas leaving from Surabaya (Java) and calling at Macassar, Coupang (Kupang), Delli (Dili), Banda, Amboyna, Ternate and Menado (Manado). On return to Surabaya, it connected with services to Batavia.

The Batavia–Singapore mail since 1840 was carried by the 516-tonne teak steamer *Koningin der Nederlanden* owned by the Nederlandsch–Indische Stoomboot Maatschappij (Dutch East Indies Steamboat Company). In 1854, the Dutch government entered into a contract with Cores de Vries to operate an additional service from Batavia to Singapore hoping, in vain, that competition would reduce prices.[563] Ida Pfeiffer also used the Cores de Vries steamers to travel through the Moluccas and back to Celebes. There were even advertisements in the Singapore newspapers promoting the Cores de Vries steamers through the Moluccas for tourists.[564] Wallace described the voyage on the *Padang* in his *Journal*:

> The vessel was a roomy & comfortable one, but could only be got to go six miles an hour in the finest weather. However we went along pleasantly & there being only three passengers besides myself we had plenty of room. The arrangements are somewhat different from our steamers. There are no cabin servants as each passenger almost invariably brings his own, the ships' stewards attending only to the saloon & the eating department. At 6 there is tea & coffee, at 7 to 8 a light breakfast, tea eggs, sardines etc etc at 10 Madeira Gin & bitters are brought on deck as a preliminary dejeuner a la fourchette at 11. At 3 pm tea & coffee at 5 Madeira etc and ½ past 6 dinner concluded by tea & coffee at 8. Between whiles beer soda water etc. are called for *ad lib.* so there is no lack of gastronomical excitement to while away the tedium of the days on board.

Once again Wallace was travelling in style. After a 460-mile journey, the *Padang's* first stop, presumably on 25 November, was the island of Timor.[565] The name means "east" in Malay. The island was divided west and east between the Dutch and the Portuguese. After the famous mutiny on the *Bounty* in 1789, Captain William Bligh and his men had sailed for Timor as the closest island occupied by Westerners. The *Padang* dropped anchor at Coupang on the southwest coast. Wallace made a few remarks in his *Journal* on Coupang but it is not clear that he went ashore.[566] The vegetation was only "scrubby". The *Padang's* next stop was the Portuguese port of Delli on the northern side of the island. Here, Wallace was able to put his Brazilian Portuguese to use by speaking to the commandant, Francisco Xavier Lobato de Faria, about local earthquakes and volcanic phenomena.

Timor to Banda. "Mr. Wallace's
route" map MA1.

Gunung Api, Banda c. 1860–1872.
Tropenmuseum.

Banda, c. 28–29 November 1857

The *Padang* steamed north for two days and 540 miles to the island of Banda
arriving around 28 November. The Banda group consists of ten small volcanic
islands south of Ceram. For centuries, Banda was the world's only source for
nutmeg. This valuable spice attracted the first Westerners, the Portuguese, at
the beginning of the 16th century. They were followed by the Dutch who were
long established by the time Wallace arrived. The Dutch maintained a monopoly
on the island's nutmeg, although by this time it was also grown elsewhere. The
Padang steamed into the sheltered harbour of Banda and dropped anchor. The
lush green vegetation of the three little islands made a stark contrast to the dull
yellow colours of Timor. The scene was dominated by the almost perfect volcanic cone of Gunung Api that rises 2,100 feet or 640 metres, forming its own
island. The green cone gave off wisps of steam while the dwarfed steamer below
gave out its own blackish grey smoke from her funnel. In the shadow of the volcano was the lower island where the town nestled vulnerably. Wallace observed
"the little town, with its neat red-tiled white houses and the thatched cottages
of the natives, bounded on one side by the old Portuguese fort".[567] He thought
the natives were particularly of a mixed race. He climbed the highest hill to a
"telegraph station" where flags were posted.[568] He mistakenly concluded that the
island was not volcanic, but instead an ancient continental fragment only later
broken up by volcanic activity.

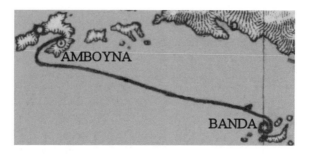

Banda to Amboyna. "Mr. Wallace's route" map MA1.

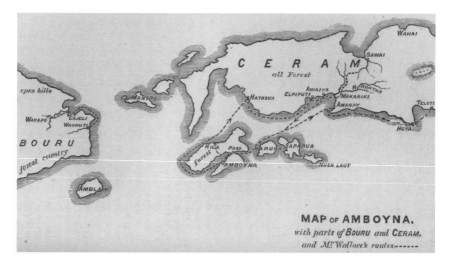

Extract of map of Amboyna. MA1:459.

Amboyna, 30 November 1857–4 January 1858

Wallace continued on the *Padang* as she steamed north towards the island of Amboyna, the Dutch capital of the Moluccas. The journey took twenty hours and they arrived, according to his *Journal*, on 30 November 1857. The town was located midway up the north side of a large peninsula that jutted out parallel to the rest of the island to the north, forming a long narrow bay, about six miles wide at its mouth. There were about 2,000 inhabitants. Amboyna is the original home of cloves whose cultivation was the powerhouse of the economy. Ida Pfeiffer described the trees during her 1853 visit. "The clove tree lives 100 years, and begins to bear at twelve or fifteen; the harvest time is from November to January;

the produce unequal—a tree yielding variously from one to twenty pounds. The cloves are dried in the shade."[569]

Wallace brought a letter of introduction to a forty-four-year-old German doctor, Otto Gottlieb Johan Mohnike, the chief medical officer of the Dutch Moluccas. Mohnike was formerly stationed in Japan where he had attempted to introduce smallpox vaccination.[570] Happily he was a fellow entomologist and an accomplished naturalist. He later published an important book on the natural history of the Dutch East Indies.[571] Wallace found to his chagrin that Mohnike could not speak English as he had heard, but only read it, so the two men were forced to speak awkwardly in French, a language in which neither was very proficient. Mohnike offered Wallace a room in his house and soon the luggage was transported from the *Padang*, her oil no doubt leaking into the sea and starting that process of pollution that has continued to damage tropical coral reefs to this day.

Wallace found Mohnike "a very learned and hospitable man; he has… ascended volcanoes, and made collections: my pleasure may be imagined in looking over his superb collection of Japanese Coleoptera…he has also an extensive collection of Coleoptera made during many years' residence in Sumatra, Java, Borneo, and the Moluccas—a collection that makes me despair."[572] Wallace was also very impressed by the coloured woodcuts and sketches by a Japanese lady. They struck Wallace as the finest illustrations he had ever seen.

The following morning, they went to the hospital where Wallace was introduced to another physician, the thirty-one-year-old Carl Ludwig Doleschall. Wallace described him in a letter to Bates.

> Dr. Doleschall is a Hungarian who studied a year in the Vienna Museum (the Diptera & Arachnida) which he knows well. He also collects the Lepidopt. & Col.ª of Amboyna, and liberally gave me a fine suite for my private colln. He is a delightful young man, but poor fellow's dying of consumption. He can hardly I fear live a year, yet is enthusiastic in entomology.…We conversed always in French, of which I have had to make so much use that I am getting tolerably fluent though fearfully ungrammatical. But we were about equal in that respect & so blundered along gloriously.[573]

Wallace was tragically correct about Doleschall's ill health. He died in 1859.[574]

The same evening, Mohnike took Wallace to the residence of the Dutch Governor of the Moluccas, Carel Frederik Goldman, who had written the polite letter to Aru to assist Wallace. The fifty-seven-year-old Goldman, to Wallace's great relief, spoke excellent English. He offered Wallace letters of introduction and every assistance for a visit to New Guinea. After a few more days with the entomological doctors, Wallace procured a boat and local men to take him up the narrow stretch of water separating the two arms of Amboyna.

They left the town on 3 December. As the boat sailed up the harbour, Wallace marvelled at the crystal clear waters beneath the boat which revealed the dazzling colours of tropical coral reefs. "It was a sight to gaze at for hours, and no description can do justice to its surpassing beauty and interest. For once, the reality exceeded the most glowing accounts I had ever read of the wonders of a coral sea."[575] They ascended a river on the northern side of the bay and then trekked along a good path through the forest to a small plantation in the middle of the northern peninsula. Here, in the middle of a large clearing, was a small thatched hut. He set to work collecting insects the next day.[576] "Only the entomologist can appreciate the delight with which I hunted about for hours in the hot sunshine, among the branches and twigs and bark of the fallen trees, every few minutes securing insects which were at that time almost all rare or new to European collections."[577]

One evening, Wallace was sitting on the veranda of the hut reading. In the distance, one could hear the "loud wailing cry" of the large mound-building scrub fowl. He then heard a noise closer to hand, in the thatch of the roof. The rustling soon ceased. He thought no more about it. The next day he was again reading, this time reclining on a bench. He looked up and saw "a large yellow & black roundish mass between the ridge pole [and] the thatch. I thought at first it was a tortoise shell that had been put there out of the way but looking more closely saw it was a large serpent compactly coiled up in a knot." Wallace called out to his two boys who were skinning birds outside. "'Here's a big snake in the roof;' but as soon as I had shown it to them they rushed out of the house and begged me to come out directly."[578] Wallace found a local man to help. The man poked at the snake with a stick to make it move and then looped a noose over its body and pulled it down. There was a dramatic scuffle as the giant snake writhed amongst the furniture to escape. The man grabbed the snake by its tail and instantly ran out of the hut. The snake was pulled backwards so quickly it was unable to coil around anything or bite its attacker.

Outside, the man lost his grip and the snake slithered under a fallen tree. Again the snake was poked out of hiding and the man took it again by the tail and swinging it through the air dashed it against a tree which left it stunned.

It was then killed with a hatchet. The skin of this snake is now on display at the Linnean Society of London. It was a fully grown Moluccan python (*Morelia clastolepis*). The event was later illustrated for *The Malay Archipelago* by Thomas Baines. Wallace's story is entirely plausible, unlike that of his competitor, the American naturalist Albert Bickmore, who told a tall tale of a fifteen-minute life-and-death battle with a large python. Bickmore managed to defeat his serpent armed only, as he put it, with an axe as depicted in his book of 1869 (see p. 196).

"Ejecting an intruder." MA1:466. Note Wallace's net on the veranda. The European is Wallace.

In about fifteen days of fine weather, Wallace and his team made a very good collection of about 550 species of insects. He noticed that the insects resembled those of Aru. "I also obtained one or two specimens of the fine racquet-tailed kingfisher of Amboyna, Tanysiptera nais [see colour insert], one of the most singular and beautiful of that beautiful family."[579] From his hut, he wrote a letter to Stevens on 20 December, describing Amboyna as "eminently tropical; the number of large and handsome species in all orders of insects is perhaps greater than in any other place I have visited, and the forms far more closely resemble those of Aru than of Borneo or Macassar".

On 24 December, Wallace returned to stay with Dr. Mohnike in Amboyna in order to catch the next mail steamer to Ternate. Wallace dried his collections and studied the entomological collections of the two doctors. He also enjoyed some colonial society. In the evening of New Year's Day 1858, there was a party at the Governor's residence. Ida Pfeiffer had visited it exactly five years before and described it as "about a mile from the town…an insignificant little abode, built of bamboo".[580] Wallace observed that "tea & coffee were presented as is almost universal during a visit as well as cigars for on *no occasion* is smoking prohibited in the Dutch colonies, & the ladies stand the fire most heroically". Wallace was just as interested in a "black parroquet from N. Guinea a pet of the Governors daughter".[581]

Wallace thought the natives of Amboyna "seem a mixture of at least three nations, Portuguese Malay & Papuan or Ceramese, with an occasional cross of Chineese & Dutch".[582] He seems not to have noted that they were mostly Christians. On 4 January, he began a letter to Bates on the Sarawak law paper which is particularly interesting.

> To persons who have not thought much on the subject I fear my Paper "On the Succession of Species" will not appear so clear as it does to you. That paper is, of course, merely the announcement of the theory, not its development. I have prepared the plan & written portions of an extensive work embracing the subject in all its bearings & endeavouring to prove what in the paper I have only indicated.…I have been much gratified by a letter from *Darwin*,[583] in which he says that he agrees with "*almost every word*" of my paper. He is now preparing for publication his great work on *Species & Varieties*, for which he has been collecting information 20 years. He may save me the trouble of writing the 2nd part of my hypothesis, by proving that there is no difference in nature between the origin of species & varieties, or he may give me trouble by arriving at another conclusion, but at all events his facts will be given for me to work upon.[584]

This remark about the "2nd part" of Wallace's theory is quite unique. Darwin would have agreed with the wording, as far as it went, that there is "no difference in nature between the origin of species & varieties". Wallace's permanent varieties paper makes it clear that he now meant genealogical descent. Wallace gives no sense that anything else is needed for his theory to be written. So much for seeking for the "mechanism" or the solution to a "species problem". His book only awaited his return to English libraries and national collections. He had no way of knowing that a new breakthrough awaited him in Ternate one month later.

His collections dried and packed, Wallace sent a consignment of about 1,000 specimens which left Amboyna on a ship bound for Batavia on 6 January 1858. By March 1859, he had earned £80 from their sale. Wallace himself embarked on the monthly Cores de Vries mail steamer, this time the 183-tonne *Ambon*, on 4 January heading for Ternate (although Wallace gave 5 January in his *Journal*).

The name "Ternate" now positively scintillates with the importance of Wallace's sudden scientific revelation there. Even a century ago, biographer James Marchant wrote that Ternate "is now indelibly associated with that particular visit which ended after a trying journey in an attack of intermittent fever and general prostration, during which he first conceived the idea which has made Ternate famous in the history of natural science".[585] And for this we move to a new chapter.

Note on the Theory of Permanent and Geographical Varieties.
By Alfred R. Walllace, Esq.

As this subject is now attracting much attention among naturalists, and particularly among entomologists, I venture to offer the following observations, which, without advocating either side of the question, are intended to point out a difficulty, or rather a dilemma, its advocates do not appear to have perceived.

The adoption of permanent and geographical varieties has this disadvantage, that it leaves the question "What is a *species?*" more indeterminate than ever; for if permanent characters do not constitute one when those characters are minute, then a species differs from a variety in degree only, not in nature, and no two persons will agree as to the amount of difference necessary to constitute the one, or the amount of resemblance which must exist to form the other. The line that separates them will become so fine that it will be exceedingly difficult to prove its existence. If, however, the two things are of essentially distinct natures, we must seek a qualitative not a quantitative character to define them. This may be done by considering the permanence, not the amount, of the variation from its nearest allies, to constitute the specific character, and in like manner the instability, not the smaller quantity, of variation to mark the variety. In this way you define the two things by a difference in their nature; by the other, you assert that they are of exactly the same nature, and differ only in degree.

Wallace's evolutionarily suggestive "Note on the theory of permanent and geographical varieties". *Zoologist*, 1858.

KILLING THE PYTHON

Tall tales. American naturalist Albert Bickmore claimed he fought a Malayan python for fifteen minutes with an axe, before finally killing it. Bickmore, 1869, p. 541.

View of Ternate in the 1850s. S.A. Buddingh, *Neêrlands-Oost-Indië*, 1860, vol. 2, p. 77.

Chapter 8

STRUGGLE AND SPICE

A fter steaming north 547 miles past the string of lush green spice islands, the *Ambon* arrived on the morning of 8 January 1858 on the eastern side of the small volcanic island of Ternate. Ida Pfeiffer had come to Ternate on the *Ambon* five years before. She found the scene "very picturesque" as "the bay is encircled by mountains more than 5000 feet high; amongst them the Tidore and the Ternate; the latter a volcano that is frequently smoking".[586] It was the second time Wallace arrived on a new island on his birthday. He was thirty-five.

The fabled spice islands are strung along the west coast of Gilolo (Halmaheira): Ternate, Tidore, Mortier, Makian and Batchian. But the cloves that had once made Ternate the source of so much wealth, power and dispute were gone. The Dutch had removed them from several islands in order to maintain their monopoly. The people now grew and subsisted on sago.

Ternate, forty-eight miles north of the equator, was tiny, only about eleven and a half square miles. It was made a free port along with Banda and Amboyna only in 1853. The town, with 8,500 inhabitants, was huddled in the shadow of Gamalama, the great conical volcano, which almost covered the whole island. Most of the inhabitants were Malay, but there were about 400 Chinese, 100 Westerners, settlers from Celebes and about 300 slaves.[587] Across a narrow strait to the east lay the sprawling multi-limbed island of Gilolo, like a miniature Celebes.

Ternate was not a Dutch possession but a sultanate. According to Pfeiffer, the Dutch paid the Sultan, Taj ul-Mulki Amir ud-din Iskandar Kaulaini Shah, a pension of 10,800 rupees in order to control the regional spice trade. The Sultan's palace was built on a hillside at the end of the 18th century. It was a European-style

Amboyna to Ternate. "Mr. Wallace's route" map MA1.

stone building with a wooden veranda supported by thick columns with an attap roof; the interior was full of European furniture. It still stands. Before the era of the Dutch, the clove-powered sultanate was one of the wealthiest and most powerful in the archipelago.

There was a Dutch Resident and police magistrate and Fort Orange garrisoned by about 115 soldiers. This together with a small part of the town called Malajoe "and the adjacent Chinese and Makassar camps" were administered by the Dutch.[588] Wallace found the town littered with ruined stone buildings which he took to be remains of Ternate's former greatness. They were however as yet unrebuilt houses, destroyed in the last great eruptions of the volcano in 1840 and 1855.

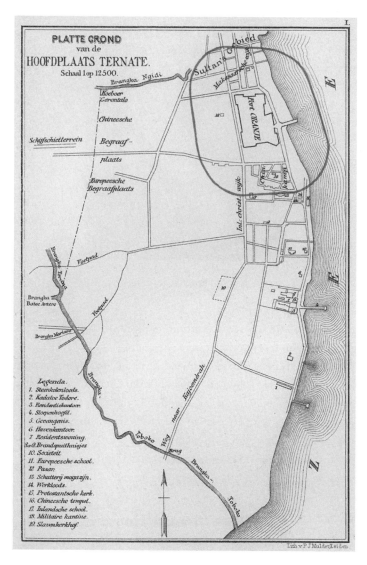

Map of the European part of the town of Ternate from de Clercq, 1890.

Wallace brought letters of introduction to the wealthiest Dutch merchant of Ternate, Maarten Dirk van Duivenbode (1805–1878).[589] He was born on Ternate, educated in England and as Wallace remarked, was "a well educated man fond of literature & science, a phenomenon in these regions". He lived in

Maarten Dirk van Duivenbode. From Heij, 2010.

a luxurious estate about a forty-five-minute walk north of the town.[590] Visitors were impressed by his large pond stocked with saltwater fish.[591] His second wife, the daughter of a Chinese Nakoda named Gim Nio, had been baptised Antoinette Elizabeth Johanna. They had three children. According to Wallace, van Duivenbode was "generally known as the king of Ternate, from his large property & great influence with the native Rajahs & their subjects".[592] He owned 100 slaves and half the town. More importantly, he owned a fleet of three schooners and a barque, making him one of the leading shipowners in the islands. He also operated the island's shipping agency.

Wallace's Ternate House

Through van Duivenbode's assistance, Wallace rented a house from a Chinese man.[593] It was "rather ruinous but well adapted to my purpose being close to the town yet with a free outlet to the country & the mountains. A few repairs were soon made, some bamboo furniture & other necessaries obtained, and after a visit to the Resident [Casparus Bosscher, an amateur botanist] & Police Magistrate, I found myself an inhabitant of the earthquake tortured island of Ternate & able to commence operations & prepare the plan of my campaign for the ensuing year."[594]

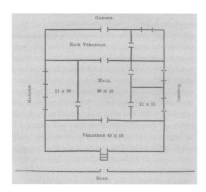

Floor plan of Wallace's house. MA2:3.

A house on Ternate, 1876. A. Raffray, Voyage en Nouvelle-Guinée. *Le Tour du monde* 37 (1878): 225–288.

Ternate houses were somewhat different from the usual Malay style with their floor on ground level, so Wallace described his house in detail and even published a floor plan in *The Malay Archipelago*.

> A description of my house (the plan of which is here shown) will enable the reader to understand a very common mode of building in these islands. There is of course only one floor. The walls are of stone up to three feet high; on this are strong squared posts supporting the roof, everywhere except in the verandah filled in with the leaf-stems of the sago palm, fitted neatly in wooden framing. The floor is of stucco, and the ceilings are like the walls. The house is forty feet square, consists of four rooms, a hall, and two verandahs, and is surrounded by a wilderness of fruit-trees. A deep well supplied me with pure cold water—a great luxury in this climate. Five minutes' walk down the road brought me to the market and the beach, while in the opposite direction there were no more European houses between me and the mountain. In this house I spent many happy days....I had ample space and convenience for unpacking, sorting, and arranging my treasures.[595]

Wallace's house in Ternate is rightly considered a famous site in the history of science, because it was here that he hit on natural selection and composed his famous essay. Many visitors have attempted to locate it over the years but with no apparent success. Most likely the house no longer exists or has been changed beyond recognition. The real difficulty in the quest for Wallace's house is the vagueness of his description of its location. The house was on the outer and uphill side of the town, above a fort, about a five-minute walk to the market and the

Malay part of the town lay to the northeast. "Just below my house is the fort, built by the Portuguese, below which is an open space to the beach, and beyond this the native town extends for about a mile to the north-east." Calling the fort Portuguese has caused some confusion because there are remains of Portuguese forts on Ternate. But the fort Wallace described was Fort Orange, built by the Dutch in 1607 (see map on p. 199). His reference to the distance to the market is less ambiguous and eliminates the Portuguese forts from consideration. His house was probably located along the road to the west of the circle on the map. But where along it can never be known.

The Mystery of February 1858

Just as earlier generations of Westerners had been drawn to the Spice Islands by their valuable natural products of clove, mace and nutmeg, Wallace was drawn by the promise of valuable insects and the beautiful Birds of Paradise. Ternate was the farthest point on the Dutch mail steamer route, surrounded by new islands and in reach of New Guinea, his main target for Birds of Paradise and other rare specimens like those that were so profitable and interesting at Aru.

There was some time before one of van Duivenbode's schooners was due to sail for New Guinea in mid-March. Wallace decided to visit the neighbouring island of Gilolo. In a letter to Bates written on Ternate on 25 January, Wallace mentioned, "In about a week I go for a month's collecting [to Gilolo]."[596] A notebook entry however is dated five days before this letter "*Gilolo - Jan. 20th. 1858*".[597] These fragments are contradictory. Where was he, on Ternate or Gilolo?

It would not matter very much if it were not for the fact that the most important event in Wallace's life — and one of the most pivotal in the history of science — happened about this time, and no one has been able to determine where. It has become one of the most intractable mysteries in the story of Wallace.

Traditionally, it was believed that Wallace discovered natural selection in his house on Ternate. The printed version of his essay is signed "Ternate, February, 1858" and all of his recollections in later years name the same time and place.[598]

In the late 1960s and early 1970s, historian H. Lewis McKinney was a pioneer in using manuscripts rather than just later publications to study Wallace. McKinney assumed that Wallace left Ternate at the end of January (as the Bates letter proposed) and that, as the *Journal* states, "1st. of March I returned to Ternate." So McKinney concluded that Wallace was on Gilolo for *all* of February 1858. Therefore, the famous Ternate essay was actually written on Gilolo! Almost all writers now agree that the Ternate essay was written on Gilolo.[599] In fact, most of the current theories about how Wallace discovered natural selection actually depend on him having his eureka moment on Gilolo because the different human races there supposedly triggered his breakthrough.

If so, why was the essay signed Ternate? Historian James Moore surmised it was "the place of completion and the month of conception of his theory". Entomologist George Beccaloni suggested Wallace "got the month wrong". The most common answer is, as biographer Peter Raby put it, "Wallace simply gave his main residence, and postal base [of Ternate]."[600] This sounds perfectly reasonable.

But there is a serious flaw with this theory. Wallace always signed documents from his actual location, not his postal base. One letter to his mother is signed from "The Jungle, near Malacca" and another from "the mountains of Java".[601]

How do we know that Wallace really did leave about a week after 25 January? McKinney even suggested that Wallace wrote the wrong date and actually meant 15 January. To make things even more difficult, no Wallace manuscripts are dated February 1858. February is mysteriously absent from all of his records. His collecting registry gives only [see colour insert]:

1858	Ternate &c. Insects.
January	
In two	
months.	200 [Coleoptera].[602]

The section is strangely blank compared to the preceding and following locations. It may even have been written later since the September entry next to it is in indistinguishable pencil. The only February 1858 date by Wallace of any kind that survives is the one on the Ternate essay itself! If all of this is not confusing enough, Wallace dated *The Malay Archipelago* chapter on Gilolo as March! No wonder this mystery has been so protracted.

Solving the Mystery

But a few clues have been overlooked. If Wallace was ready to leave "in about a week" of writing Bates on 25 January, that is c. 1–2 February, surely he would wait for the next mail steamer due around 6–9 February. He badly needed a box of collecting supplies that had not arrived on 9 January. Why leave for a collecting trip a few days before the essential supplies might arrive? Also, in a 2 March letter, Wallace said he "received" Bates' letter "a month ago". That was the c. 8 February steamer. Therefore, Wallace probably travelled to Gilolo *after* the c. 8 February mail steamer.

Wallace recounted his stay on Gilolo: "I got some very nice insects here, though, owing to illness most of the time, my collection was a small one."[603] He gave no indication of the duration of his stay except that it was "comparatively short". How long was it then? All modern writers seem to assume it was for a whole month, as Wallace planned in the 25 January letter.

However, two sources show that Wallace only stayed on Gilolo for two weeks, not a month. A previously overlooked note by Wallace gives the stay on Gilolo as "2 weeks?"[604] Similarly, in an 1869 article, Wallace wrote that this and his second trip to Gilolo amounted to one month.[605] Hence, we can be quite confident that Wallace was on Gilolo for about two weeks, not the entire month of February.

If we take these two weeks and subtract them from 1 March, we get a departure from Ternate around 16 February. Therefore, the note Wallace wrote about the area of Gilolo, seen at the start of his stay, was misdated 20 January when he meant to write 20 February.[606] We have no other evidence about when he left Ternate except after c. 8 February when the mail arrived. His consistent practice of signing documents from his actual location will not allow the usual dismissal of the place written on the essay. Therefore, Wallace did write the essay on Ternate in February 1858.

The Eureka Moment

The development of Darwin's theories on the origin of species are some of the best documented in the history of science with thousands of pages of notebooks, reading notes and other manuscripts which have been and continue to be meticulously studied and interpreted by scholars.

The case for Wallace could not be more different. We have little to go on. There are about 400 words related to species in *Notebook 1*, about 7,300 in *Notebook 4* and the passages in shorter publications which have been discussed. This puts historians of Victorian science, normally luxuriating amidst a wealth of evidence, unexpectedly into the situation of Shakespeare scholars or classicists who have so little or sometimes nothing else written by an author of interest but the published text itself.

And yet, Wallace's eureka moment has been described as "one of the most romantic tales of modern science".[607] How did Wallace come up with an essay which struck Darwin to remark, "I never saw a more striking coincidence"? It is now possible to reconstruct what happened knowing where he was, on Ternate, and fully utilising his private notes.

Wallace received his mail on c. 8 February, but he did not leave for Gilolo until about 19 February. Wallace's great flash and the writing of his historic essay probably occurred in this window of time. He left for Gilolo later than his 25 January plan not only because he fell ill, but also because he spent a few days working on the Ternate essay, a task he had not anticipated when writing to Bates.

Wallace was suffering from a bout of intermittent tropical fever (probably malaria) in his house on Ternate. Once again housebound and unable to collect specimens, his books were at hand. Perhaps the latest scientific periodicals that arrived in that post contained something to spur him just as Forbes' polarity thesis had prompted the Sarawak law.[608] The scientific literature from Britain was every bit as influential to Wallace's thinking as the insects and birds flitting around him in the tropical sunshine. Laying on his bed, he may have opened *Notebook 4* and read over his species notes from Si Munjon, Singapore and Macassar. Then, an idea "suddenly flashed upon me".[609]

Malthus

The question that has long been debated is, what prompted this flash? Since McKinney's 1972 book, most writers seem to agree that Wallace was inspired to remember Malthus and the struggle for existence by local human races.[610] As McKinney put it, "Realizing that [the Malthusian process] was even more applicable in the animal kingdom, he made the logical transfer from human races to animal varieties. The search was ended."[611]

Despite how widely accepted these views may be, there is little evidence to support them. First of all, we can now see that Wallace wrote the Ternate essay *before* he went to Gilolo, therefore the races there could not have influenced the essay. There is also no trace of humans in any of his species notes or the essay itself. Wallace's speculations about human races in the Eastern Archipelago concerned classification and biogeography. Are the people of this island Malay or

Papuan or mixed? There is nothing in his interest in human races to connect with Wallace's ideas that the struggle for existence could turn races or varieties into new species. There were no striking instances of racial struggle around him.

Wallace did not have a copy of Malthus nor is there any mention of him in the essay. In his later recollections, Wallace said he thought of Malthus at the moment of insight, but his recollections cannot be fully trusted because they are so distant and not entirely consistent. His recollections from 1887, 1903, 1905 and 1908 all refer to Malthus and transferring the idea of positive checks to population growth from "savage races" to species in nature. But these were all written after Wallace had read Darwin's *Life and letters* (1887) with its emphasis on Malthus as the spur for Darwin. Wallace only mentioned Malthus by name (in a surviving document) after reading Darwin's 1858 paper with its mention of Malthus. In Wallace's first recollection of 1869, he merely stated, "I was led to it by Malthus' views on population applied to animals", that is, with no mention of human races. What Wallace probably meant, at least in 1869, was the theory of superfecundity applied to animals.

The phrase "struggle for existence" was extremely common in the popular and scientific literature Wallace received from Europe. Any of these sources, and of course Lyell, could have reminded Wallace of a struggle for existence. Lyell would seem the most likely source since it was Wallace's most heavily used work for species theorising. But Lyell does not mention the "geometrical ratio" detail that Wallace used. So it remains unclear which source prompted Wallace to recall a struggle for existence.

The spark

Wallace's legacy of reading Combe and *Vestiges* is clear from his devotion, through-out his early writings, to a nature controlled by natural laws and uniformity as espoused by Lyell. Wallace's mechanics' institute legacy is apparent in his ridicule of divine design arguments and zeal for rationality and proposing systemic reforms to the classification of birds, the system of synonyms and his conviction that the origin of species was not miraculous, but due to some form of material-istic evolution.

From his reading and theorising in the Eastern Archipelago, virtually all of the elements of the Ternate essay were already in *Notebook 4* — the implications of fossil succession — that species change without limit and lineages could branch off. As he wrote in his notebook in Macassar six months before, a variety could give birth to new varieties and this "process continued at intervals will account for all the facts".[612] But there was no cause for adaptation. In fact, Wallace had a strong aversion to adaptationist thinking. A theory for adaptation first appears — apparently out of nowhere — in the Ternate essay. Where did it come from?

The final spark to ignite Wallace's powder keg of speculation may have been the tiger beetles from Macassar. It was an intriguing part of his current context — not social or textual, but natural. As Wallace wrote to Frederick Bates less than a month after writing the essay:

> Others [tiger beetles] are sea beach insects as the C. tenuipes & the Baly species - the former singularly agreeing in colour with the white sand of Sarawak, the latter with the dark volcanic sand of its habitat. Others prefer river banks. The two Lombock sp.[ecies] were found always a little way inland on the same coloured dark sand…so also 63 & 126 Macassar, frequent river banks on sand of a lighter col.r than that of Baly & Lomb. but darker than that of Sarawak, *as are the insects.* Another [new species] in the last Mac. coll.[ection] was found in the soft shiny mud of salt creeks, with which its colour so exactly agrees that it was perfectly invisible except for its shadow. *Such facts as these puzzled me for a long time, but I have lately worked out a theory which accounts for them naturally.*[613]

Puzzling over the match between these beetles and the mud or sand they lived on, as Wallace later wrote of conceiving of his theory, "marked out a different line of work from that which I had up to this time anticipated".[614] He mentioned this colour matching twice in the Ternate essay as we will see below.

Wallace already believed that new varieties are constantly appearing. As he noted from Lindley's *Botany*, "New forms, miscalled species, are always starting up in every Botanic Garden."[615] And responding to Lyell: "varieties constantly occur in the same place & under the same circumstances as the original species".[616] Varieties "of many tints" constantly appearing would be coloured differently from their parents and backgrounds. They would be detected more often by predators and destroyed. But those that happened to match their background well enough would survive. Many appearing, and few surviving could lead Wallace to recall Malthusian exploding population numbers kept in check — the struggle for existence.[617]

The astonishing fit between the colour of the tiger beetles and their environments — particularly the one at Macassar — left a lasting impression on Wallace. A decade later he wrote:

> I noticed generally that, whatever the colour of the sand or the soil, the common Tiger beetles of the locality were of the same hue. A most remarkable instance of this was a species which I found only on the glistening, slimy mud of salt marshes, the colour and shine of which it matched so exactly that at a few yards' distance I could only detect it by the shadow it cast when the sun shone![618]

And again, in the context of his discussion of "protective resemblances" in 1867, what since the First World War has been called camouflage: "one which was never seen except on the wet mud of salt marshes was of a glossy olive so exactly the colour of the mud as only to be distinguished, when the sun shone, by its shadow!"[619]

The Ternate Essay

Soon after his feverish eureka moment, Wallace wrote out his famous essay "On the tendency of varieties to depart indefinitely from the original type". A fresh interpretation, using only contemporary materials and the new insights from his notebooks, reveals that we have never completely understood Wallace's original theory. Despite its fame, it has remained obscure for over 150 years. Part of the problem is that Wallace left out some crucial pieces. We can supply the missing pieces of the puzzle from his notebooks and thus reconstruct Wallace's original theory.

As the title of the essay suggests, varieties were again the focus — as well as species. Like the Sarawak paper, this title was also taken silently from Lyell.[620] The lineal descent of varieties from parent species was widely accepted, but this was not taken as evidence of evolution since purported limits to change prevented a variety from becoming any more different from the parent species. Lyell argued that if domesticated varieties were released to the wild, they returned to the form of the parent species or died out.[621] Hence, wild varieties must also have a tendency to return to the original type.

This tendency to return suggested that species must have fixed limits. It was as if, when released, a domesticated variety would snap back to the form of the original species like a stretched rubber band. It was also believed that "permanent or true varieties" might exist in nature, but these probably had fixed limits, so the stability of species remained true.

The Ternate essay begins by opposing this argument. Wallace reiterated the problem of distinguishing "which is the *variety* and which the original *species*", the focus of his permanent varieties paper. But the Ternate essay would show that wild varieties are *not* the same as domesticated ones. In fact, "there is a general principle in nature" which will cause varieties to become ever more different from the original species, not snap back to it. And this process would go on and on. Thus, species could change or evolve indefinitely.

The struggle for existence

Wallace next discussed a "struggle for existence" in nature. Species have a tendency or ability to increase at a geometrical rate, but they do not because the struggle for existence keeps population sizes in check. It does so because the weakest or "the very young, the aged, and the diseased" tend to die, whereas "the most perfect in health and vigour" tend to survive. Just as this process works for individuals within a species, a similar process acts between different species.

To modern readers, this sounds like evolution by natural selection, and that Wallace is referring to the selection of individuals just as Darwin did. But that was not Wallace's meaning in this early section of the essay. He was merely introducing the point about the struggle for existence (primarily food supply and predation) limiting population sizes. No species change resulted from this part of the discussion. It was a balance mechanism. This sort of "struggle for existence" was, after all, nothing new. Wallace had seen it in the writings of Blyth, Lyell, Darwin and others.[622] Whereas earlier writers used this principle to argue that species could not change, Wallace had something very different in mind.

In the next section, he offered an analogy. Having established how the struggle for existence works amongst individuals in a species and between separate species, he would "proceed to the consideration of *varieties*, to which the preceding remarks have a direct and very important application". Just as the weakest individuals in a species tend to die and the strongest tend to survive because of their characteristics, it was the same with the daughter varieties of a species.

The origin of varieties

As a few commentators have pointed out, how varieties were "produced" in the first place is never stated in the essay.[623] It is the key to unlocking Wallace's full theory of evolution as it stood in February 1858. It was, Wallace mentioned, an "undisputed fact that varieties do frequently occur" in the same environment. But they *did not* arise from a slow gradual process of natural selection. As we saw above, for Wallace, varieties arose quite independently of need or the environment. And, as we will see shortly, some of his varieties were "inferior" to the parent species, not something that could be produced by Darwin's natural selection of individuals. In *Notebook 4*, we find the answer, "All varieties we know of are produced at *birth* the offspring differing from the parent. This offspring propagates its kind."[624] If a variety was produced "at birth", this means not only that its members were descended from the parent species, but that they appeared fully formed as a distinctly different form worthy of the name "variety".

What were varieties?

But what exactly was a "variety" for Wallace at this time? This is a question that has long been debated. Did Wallace mean variant individual or variant populations/races?[625] This is important for teasing apart how similar Wallace and Darwin's views originally were. The fact that scholars have disagreed so much about what Wallace meant by "variety", "varieties" and "variations" is due partly to vagueness on Wallace's part and because he withheld some of his private views from the essay.

In his most recent paper on permanent varieties, Wallace mentioned that varieties were so-called "where a smaller amount of difference exists" than between two species.[626] How much difference constituted a variety for Wallace? All of the examples I have found are greater than the slight individual variations Darwin considered in his scheme of individual competition. Varieties named by Wallace in his other writings up to this time were either a completely different colour from the parent species, of different size or otherwise different enough to be classed as a variety, but not so different as to be classed as a distinct species.

In his Amazon narrative (1853), Wallace mentioned a type of umbrella-bird that was white instead of the usual black: "a mere white variety, such as occurs at times with our blackbirds and starlings at home". Similarly, a butterfly species on the Amazon occurred in varieties of blue and orange. If any female orangutans with cheek pads (so far known only in males) were found, these must be an "accidental variety". On Amboyna, there were insects that were either "very closely allied" species or varieties to those on other islands because they were of "larger size and more brilliant colours".[627]

Wallace later wrote, presumably in 1860, in his copy of *Origin of species*, "The old objection that *albino* animals do not increase & form distinct species is now well answered since they cannot do so unless *albinoism* is profitable to them which there seems no reason to think it can possibly be."[628] Albinos, which appear from time to time from normally coloured parents, are a variety that is not increased numerically by selection. Hence for Wallace, a variety could be either a group or a single individual that differed markedly from the parent species and whose differences were passed on to its offspring.

Since a variety began as an individual sport of nature, even when Wallace clearly referred to varieties as races, he could also mean an individual, and vice versa. This does not mean that Wallace believed that individuals were selected and not groups. His discussion is primarily about varieties as races — i.e., groups. But it means that in some cases his language cannot be further resolved. A variety could be either a single individual or many individuals. "Varieties" is used in the essay synonymously several times and in the same sentence as "races". In his earlier publications, the term "varieties" almost always refers to races.[629]

So in the Ternate essay, a "superior variety" was well suited to the environment and therefore numerous and perhaps numerically increasing, but other varieties were "inferior" and remained less numerous than their parent species or perhaps were numerically declining. It was in this important sense that Wallace developed the analogy of traditional struggle for existence between individuals within a species. Because over the very long term, the strongest varieties would outsurvive, not only the weaker ones, but even perhaps the parent species.

The causes of varieties appearing in the first place were due to Wallace's higher law which he had hinted at in the Sarawak law (1855) and Orang-utan papers (1856). Perhaps he was not entirely satisfied with his current understanding of the origin of varieties. For whatever reason, he left it out. After reading *Origin of species*, Wallace never mentioned these early ideas again.[630] This is why part of the meaning of the Ternate essay has remained hidden for so long.

The higher law and the origin of varieties at one blow, I believe, finally settle the old debate about the differences between the theories of Wallace and Darwin in 1858. Darwin had varieties themselves gradually formed by a long slow process of struggle and selection between individuals with slight differences. Wallace had the sudden appearance of full-blown varieties. But the causes for the appearance of these different forms were, in effect, random in the sense that they were not a response to need or the environment.

The role of the environment

We can illustrate Wallace's original theory with the Macassar tiger beetle. The parent species was, for example, brown. Imagine the mud was also brown. From time to time, daughter varieties (that is, individual beetles) of varying tints were born, some grey, black, olive or white and so forth. The brown parent species and a numerous variety of olive beetles co-exist with a less numerous black variety. If the mud/environment slowly changed from brown to olive, the black variety would no longer be so well concealed. Predators would prey on them more. So the black variety would go extinct. If the process continued, the brown parent species would also go extinct. But the olive variety, which happened to match the new olive-coloured mud very closely, prospered in this altered environment. It could never revert to brown as Lyell assumed because any brown ones now appearing would be inferior. The olive variety would be classed as a new species.

This new olive species might itself give rise to "new varieties, exhibiting several diverging modifications of form". The same tendency might see some of these varieties in turn superseding their parent, and so on. Wallace concluded grandly, "Here, then, we have *progression and continued divergence* deduced from the general laws which regulate the existence of animals in a state of nature, and from the undisputed fact that varieties do frequently occur."

A divergence digression

Something called "divergence" in the writings of Wallace and Darwin has been hotly debated.[631] McKinney, Brackman, Brooks, Davies and others have claimed that Wallace discussed divergence in the Sarawak law paper and/or in the Ternate essay, and that Darwin somehow derived his own "principle of divergence" from them. These claims were demolished in an important essay by historian David Kohn in 1981.[632] Kohn showed that the supposedly singular "divergence" in Darwin and Wallace are actually two very different things. Kohn distinguished them as "taxonomic divergence" and a "principle of divergence". Taxonomic divergence is the observation that "taxa can be arranged in a branched hence diverging scheme".[633]

Darwin made this observation in his famous 1837 tree diagram which depicts daughter species diverging off a common ancestral trunk. He published an implicit reference to this in the second edition of *Journal of researches* (1845) and an explicit reference to a divergent tree in his monograph on fossil barnacles (1851).[634] Taxonomic divergence was mentioned in one passage of Wallace's Sarawak paper, but no explanatory principle was given.

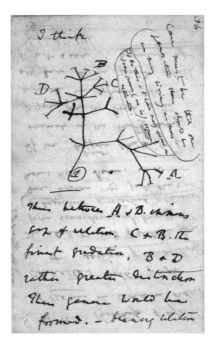

Sketch from Darwin's *Notebook B*. 1837. An ancestral species was at the base of the tree (1). Many lines ended in extinction. The letters represented existing species. Some were very similar, such as B and C. Less similar would be D. A would be very different because its ancestors had split off and shared less of the common heritage of B, C and D. Cambridge University Library.

A "principle of divergence", according to Kohn, explains "how divergence occurs". Darwin developed this by the mid-1850s.[635] Historians continue to debate the exact date. It was written into Darwin's table of contents and draft chapters of his big book by March 1857, and described in detail in his September 1857 enclosure to Asa Gray, Fisher Professor of Natural History at Harvard University.

An 8 June 1858 letter from Darwin to Hooker has been used to suggest that Darwin came to his principle of divergence after purportedly receiving the Ternate essay, "I will try to leave out all allusion to genera coming in & out in this part, till when I discuss the 'principle of Divergence', which with 'Natural Selection' is the key-stone of my Book & I have very great confidence it is sound."[636] The confusion comes from the misinterpretation that Darwin was announcing the discovery of a new idea (around the time Wallace was in the field); he was actually referring to the structure of his book draft.[637] He could not mean a new discovery since his "principle of divergence" was already described in the note to Asa Gray the year before. Historian Janet Browne has traced the development of Darwin's principle of divergence in great detail, leaving no possibility for a borrowed inspiration from Wallace or anyone else.[638]

There were fundamental differences between Wallace's brief references to divergence and Darwin's "principle of divergence". As Kohn noted, "[Wallace] offered an explanation that is ecologically static, where a new species forms only by the extinction of its parent. There is none of the creation of new evolutionary opportunities by the subdivision of the environment that characterized Darwin's principle of divergence."[639]

Hence, there was no Wallace principle of divergence. Lineages branched in the sense that one parent species can give rise to multiple varieties, not because there was an ecological bias to make this happen. Wallace argued at length against the idea that varieties were brought about in response to the environment. Darwin's principle of divergence was driven by ecological specialisation, leading to adaptive radiation. There simply was no similarity to borrow or steal.

Back to the essay

Wallace stressed that domestic varieties are artificially kept alive by humans. So in the wild they must revert to their ancestral form or go extinct. Wallace then mentioned Lamarck's theory, though in the usual English-language characterisation of his theory as driven by the will of individual organisms. Wallace made the logical point that even though Lamarck's views were frequently refuted, evolution *per se* was not.[640]

Wallace switched to a smaller focus on adaptations in a passage inspired in part by the tiger beetles. "Even the peculiar colours of many animals, especially insects, so closely resembling the soil or the leaves or the trunks on which they habitually reside, are explained on the same principle; for though in the course of ages varieties of many tints may have occurred, *yet those races having colours best adapted to concealment from their enemies would inevitably survive the longest.*"[641]

Many writers have said that Wallace compared natural selection to "the centrifugal governor of the steam engine". Some have even said that this shows how much clearer his vision or terminology was compared to Darwin's. But this passage is not about the differential survival of individuals or varieties, but an additional outcome of selection, namely "that balance [of features in animals] so often observed in nature".[642] If some parts become weak, Wallace argued, others will become strong in their place: "a deficiency in one set of organs always being compensated by an increased development of some others". It was this subsidiary organic balancing tendency that was likened to the governor of a steam engine. This balancing tendency also explained what anatomist Richard Owen called a "more specialized structure" or the increasing specialisation of some structures over geological time.[643]

Wallace then drew his essay to a close. His tendency would cause some varieties to supersede their parent species and become a new species. Parent species constantly give rise to multiple daughter varieties. And by extension, this process of gradual genealogical descent could explain the succession of forms in the history of life "and all the extraordinary modifications of form, instinct, and habits which they exhibit". And so species give rise to varieties before going extinct, their varieties do the same and so on. Thus, Lyell was wrong, varieties can and do depart indefinitely from the original type. Life evolves.

The essay was a remarkable tour de force, in places the reasoning sparkles with genius. And it clearly contains an expression of what Darwin called "natural selection". But it was not the same as Darwin's theory.

What was it about?

But many writers, including Wallace himself, tend to treat this whole matter ahistorically, as if the origin of species was a problem towards which Wallace was intentionally seeking a solution. For example, as he put it forty-seven years later in his autobiography: "the great problem of the origin of species had been continually pondered over".[644] But there is no evidence for this. During his voyage, Wallace actually referred to the Ternate essay as "on the subject of 'Varieties'".[645]

It is easy to see the advent of a scientific innovation in retrospect as striving for a goal, much harder to banish this from one's mind while trying to trace what actually happened. But it is only with this banishing that one can understand the complex process of innovations and seemingly irrelevant twists and turns which later form solutions to problems not formally seen or defined beforehand. Many discoveries are made before their use or application is known or imagined. The idea that new discoveries are just solutions to "problems" is a grossly oversimplified way of understanding how scientific "discovery" actually happens.

Varieties were ever before Wallace's eyes in his collections. Row upon row of insects pinned out on wooden trays drying on his veranda — their similarities obvious but they also varied on different islands, from Singapore, Borneo, Celebes, Aru and now Ternate.

But another curious aspect of the essay is how apparently detached it is from local context and examples. The essay never mentioned plants, just animals. It bears more signs of his reading about wildlife overseas. His examples of animal types are not drawn from the Eastern Archipelago, but instead include lions, antelopes (from Knight, 1854), "the wild asses of the Tartarian deserts" (from Lyell, 3:59), horses and passenger pigeons from the Americas. These international examples presumably present Wallace as more than a local expert — someone who was *au fait* with the international literature and of the worldwide facts of animal life.

What was it for?

What did Wallace write it for? What did he intend to do with it? These questions are shrouded in conflicting claims and conspiracy theories. Many writers have been rather incautious. Some say that Wallace wrote the essay with the intention of sending it to Darwin. But how do we know that? Not only is there no contemporary evidence, but his recollections from 1869 to 1903 do not say so, merely that *it was* sent to Darwin. Only in 1905 did Wallace add the detail that he wrote the essay "in order to send it to Darwin". The essay may have been originally written for another purpose. Wallace did not, after all, post the essay to Darwin, or for publication, or to anyone else on the next mail steamer as will be shown below.

One previously overlooked source throws light on Wallace's intentions *before* he sent the essay to Darwin. Wallace recollected to his ornithologist friend Alfred Newton in 1887 that when writing the essay, "I *had* the idea of working it out, so far as I was able, when I returned home".[646] We know from letters to Darwin and Bates from September 1857 and January 1858 that Wallace planned to write a book on species when he returned home.[647] Wallace scholar Charles Smith has suggested that Wallace was apparently not finished with his evolutionary speculations with the Ternate essay.[648] It was not a complete theory. It was the next step, after Sarawak and succession, after permanent varieties and descent. Now, selection extinguished parent forms and enabled varieties to adaptively replace them. These would form the core of his book on species one day. Perhaps a refinement of the higher law explaining how varieties were born would have formed the following part.

Hence, if Wallace had *not* sent his essay to Darwin, it, or a later version of the theory it contained, would not have been published by Wallace until *after* his return home in April 1862. If Darwin had not published his theory by then and Wallace had not sent him the Ternate essay, we might roughly estimate that a volume by Wallace could have appeared between 1864 and 1869. This is, of course, impossible to estimate since his *Malay Archipelago* only appeared in 1869 and much of it was already drafted in his *Journal*. The Ternate essay, together with the notes in his notebooks, would have required at least as much time and probably even more to prepare for publication. At any rate, after finishing this brilliant and historic essay, Wallace put down his pen, closed it up in his house and prepared to visit Gilolo for more collecting.

Gilolo, c. 19 February–1 March 1858

A brother of Wallace's Chinese landlord provided a boat and a slave crew to visit Gilolo. Wallace was joined by two of van Duivenbode's sons eager to hunt deer. They started early in the morning and rowed and sailed three hours across to the village of Sedingole (Sidangoli), where a house belonging to the Sultan of Tidore was available to them.[649] Wallace thought it was "a dirty ruinous shed". While the others hunted and fished, Wallace explored inland but was disappointed to find the country was "a plain covered with coarse high grass thinly dotted here & there with trees".[650] This was very poor collecting ground for birds and insects. He commented on the grassy plains in a mis-dated notebook entry *"Gilolo-Jan. 20th. 1858."*[651] The entry was Wallace's first use of his new balance by the struggle for existence theory:

> Plains in the tropics. Why are some covered with lofty forests, - others with grasses only? This for a long time puzzled me, but I think I have found the explanation…Ground once taken possession of by grasses cannot be reconquered by forest even if surrounded by it.
>
> A clearing for a few years only, will if left become forest, from roots & seed left in the earth, but if once covered with grass all woody growth is kept down.[652]

The phrase "for a long time puzzled me, but I think I have found the explanation" is reminiscent of the phrase in the 2 March letter to Frederick Bates where tiger beetle colouration "puzzled me for a long time, but I have lately worked out a theory". Both puzzles were solved by the struggle for existence.

According to his *Journal*, his friends returned to Ternate after two days and Wallace travelled by boat down the coast and then up a little river to the village of Dodinga. The village, surrounded by hills, consisted of about twenty houses and the ruinous remains of a Dutch fort which Wallace again mistakenly called Portuguese.[653] The fort was occupied by a "Dutch corporal & four Javanese soldiers". Most of the inhabitants were "Ternate men" (Malays). The forest soon revealed some new insects. But Wallace was more interested in learning about a race of people called Alfuros. He was informed that these "true indigenes of Gilolo" lived on the opposite side of the island and in the interior.[654]

As he eventually recorded, "The natives of this large & almost unknown islands were examined by me with much interest, as they would help to determine whether, independent of mixed races, there is any transition from the Malay to the papuan type: I was soon satisfied by the first half dozen I saw that they were of genuine papuan race."[655] Once again, Wallace made a snap judgement about racial groups, as with the Papuans at Ké. This was despite his own criticisms of travellers. "I am convinced no man can be a good ethnologist who does not travel, and not travel merely, but reside, as I do, months and years with each race, becoming well acquainted with their average physiognomy and their character, so as to be able to detect cross-breeds, which totally mislead the hasty traveler, who thinks they are transitions!"[656] Ternate was visited a few years later by the American naturalist Bickmore, who told the tall tale about fighting a

"Scene in the Moluccas." Wallace, *Australasia*, p. 396.

python with an axe, and concluded that the "Alfura" he saw were "strictly of the Malay type, and have not the dark skin and frizzly hair of the Alfura of Ceram and Buru".[657]

Wallace and "savages"

A classic and long-standing distinction made between Wallace and Darwin is their attitudes to non-European races. Darwin is said to have been negative, dismissive and rather arrogant, based on his reactions to the people of Tierra del Fuego during the voyage of the *Beagle*, whereas Wallace is represented as much more sympathetic.[658] Their class backgrounds are usually said to explain this. The high-born Darwin was horrified by his first encounter with degraded savages whereas the low-born Wallace, as Desmond and Moore put it, "viewed his Dyaks, not as Darwin had his bestial Fuegians, but in an egalitarian socialist light".[659]

But there are fundamental problems with this good cop, bad cop story. First of all, it is not acceptable to equate the people of Tierra del Fuego seen by Darwin with the Brazilian rainforest Indians first encountered by Wallace (or the peoples of the Eastern Archipelago either). This is treating all non-Westerners as if they were the same. They were not.[660] Had Wallace first encountered the Fuegians, we would have a meaningful comparison.

Second, a candid survey of Wallace's writings reveals a much wider range of attitudes to other races.[661] Speaking of what he described as Papuan natives of Matabello:

> What a contrast between these people & such savages as the hill dyaks
> of Borneo or the Indian of the Uaupes in S. America… There exist in
> fact almost as great differences in savage as in civilised life,- & we may
> safely affirm the better specimens of the former are very far superior
> to the lower examples of the latter.

And there was no doubt where he rated those of Matabello. "The people in fact are wretched ugly dirty savages clothed in unchanged rags & living in the lowest state of misery." Similarly, when Wallace encountered the Papuans of the Aru Islands, his reactions sound quite similar to Darwin's at Tierra del Fuego: "They are on the whole a miserable set of savages. They live much as all people in the lowest state of human existence & it seems to me now a more miserable life than ever I have thought it before." The peoples of Minahassa were formerly "naked savages, holding obscene festivals". And the Papuans of Dorey struck Wallace as in "the lowest state of civilisation".[662] Even in 1906, Wallace wrote, "On a calm consideration of the whole problem it must be admitted that the former point of view—that of inherently superior and inferior races—of master and servant, ruler and ruled, is the most consistent with actual facts and perhaps not the less fitted to ensure the well-being, contentment, and ultimate civilisation of the inferior race."[663]

My point is not to represent Wallace as a racist — his views were quite typical for the age — but to refute the constantly repeated and facile stereotype that he was sympathetic to non-Westerners while Darwin was not. In fact, both men wrote at times highly sympathetically and at other times disparagingly of other races.

Ternate, 1–25 March 1858

On 1 March, Wallace returned to Ternate and his house. It was time to prepare for the main event, his great six-month expedition to New Guinea where the French naturalist Lesson had found so many lucrative Birds of Paradise. Van Duivenbode's trading schooner for New Guinea was due on the 10th. Wallace badly needed more collecting supplies from Stevens which were long overdue. Surely they would arrive on the next steamer in time for his expedition to New Guinea.

The *Ambon* duly arrived at Ternate again on 9 March. But alas, Wallace "was disappointed in not receiving a box from England (due two months) & containing fresh arsenic & many other necessaries for my voyage. I had therefore to get what substitutes I could, & as the schooner arrived the next day I was very busy packing up the collections I had already formed to remain at Ternate, & making the best preparations I could in boxes bottles & ammunition for this long looked forward to & interesting journey."[664] The combination of packing up his collections from Ternate and Gilolo to survive uneaten for six months and equip himself for this major six-month expedition without the proper supplies must have made 9 March one of the busiest days of Wallace's voyage.

Although his supplies were not on board, the *Ambon* brought his mail. The arrival of the mail boat was like nowadays getting back online after a holiday and suddenly downloading a torrent of new emails. There were magazines, newspapers and 140 specimens of small beetles from his friend Willem Mesman at Macassar. There was also a letter from Darwin, written 22 December 1857.[665] This letter would be as pivotal for Wallace as his next letter so famously was for Darwin. But we have now reached one of the most contentious, contorted and intractable mysteries in the history of science. When did Wallace send his essay and when did Darwin receive it?

Until recently, no one has been able to clarify some uncertainties that have fuelled the Wallace–Darwin conspiracy industry. This too goes back to McKinney who discovered a letter from Wallace to Henry Bates' brother Frederick that was sent from Ternate on 9 March 1858. The letter still bears postmarks showing that it arrived in London and Leicester on 3 June. But Darwin claimed to receive Wallace's Ternate essay on 18 June. How could that be if both were sent at the same time? There was only one mail route.[666] All of these mysteries can now be cleared up and the conspiracies dispelled.

According to the conventional view (extrapolating from the date on the essay and a literal reading of his recollections), Wallace had already written his Ternate essay *and* letter to Darwin in order to send them by the 9 March steamer. But this cannot be correct. One of the few details known about the letter is that it was a reply to Darwin's 22 December letter.

Darwin wrote, "But you must not suppose that your [Sarawak] paper has not been attended to: two very good men, Sir C. Lyell & Mr E. Blyth at Calcutta specially called my attention to it. Though agreeing with you on your conclusion in that paper, I believe I go much further than you; but it is too long a subject to enter on my speculative notions."[667] Wallace learnt here for the first time that the great Lyell had been interested in the Sarawak law paper. Both it and the recent Ternate essay were almost conversations with Lyell. Wallace recalled in his following letter, "I asked [Darwin], if he thought [the Ternate essay] sufficiently important, to show it to Sir Charles Lyell, who had thought so highly of my former paper."[668]

Wallace could not have written a reply to Darwin before receiving the letter on 9 March.[669] Conspiracy theorist Roy Davies knew well that Wallace would have to reply to Darwin by the same steamer in order for the Ternate essay to be sent on 9 March. So Davies imagined that Wallace "must have opened and read it on the quayside".[670] Letters were, however, delivered in a mail bag to the post office, not distributed to the public from the ship. The steamer was not at a jetty but anchored out in the bay. Even worse for Davies' imaginary scenario, the steamer crews were forbidden by contract to accept mails bound for Europe.[671]

It is not clear when the mail bag to be sent *from* Ternate was closed, but it would have been some fixed time before the departure of the steamer in order for post office staff to complete sorting and franking. Some mail was bound for remaining stops on the way to Batavia, but the mail was not sorted on board these small mail steamers. Wallace's letters were probably deposited in a bag in the Ternate post office bound for Batavia or Europe.

Third, of all surviving letters of Wallace from the Moluccas, none provide evidence that Wallace could or ever did reply to a letter via the same steamer. One surviving letter to Wallace at Ternate from the mining engineer and naturalist James Motley is addressed "Alfred R. Wallace Esq | Messers Duivenboden | Ternate".[672] So at least some letters to Wallace on Ternate were directed to the van Duivenbode office, where Wallace retrieved his mail. At any rate, it was probably not possible for Wallace to reply to his letters received on 9 March before the following steamer in April.

Brooks used mail schedules to estimate the transit times of Wallace letters between London and Ternate. These estimates were highly inaccurate. For example, Brooks estimated, assuming the letter to Darwin was sent on 9 March, that it could have arrived in London by 14–20 May 1858.[673] Yet, the extant letter to F. Bates which actually was posted on 9 March arrived in London and Leicester on 3 June! Assuming that the letter to Darwin was sent on the same day, McKinney, Brackman, Davies and apparently Quammen therefore conclude that Darwin received his letter on 3 June.[674] This would make Darwin a liar, and maybe worse.

2 March 1858 letter from Wallace to F. Bates. Natural History Museum, London.

So much for assumptions and estimates. Wallace's correspondence that actually survives today reveals transit times of 75–119 days between Ternate and London.[675] The famous 2 March 1858 letter is the only known Wallace letter from Ternate with postmarks. It was not in an envelope as usual, but folded and sealed with wax in the traditional manner. Maybe Wallace ran out of envelopes.

"The next post"

Wallace never explicitly said he sent the essay in February or March 1858, just vaguely that he sent it by "the next post". He must have been mistaken in his recollections, written between eleven and fifty years later, that he sent it by the next one after composing the essay — that was the 9 March steamer as the essay was referred to in the 2 March letter.

Almost all writers accept that Wallace was incorrect in all five of his recollections of composing the essay on Ternate, saying instead it was Gilolo. Yet, we have even less reason to accept Wallace's "next post" recollection. There is no contemporary evidence at all for this detail — it first appeared in Wallace's 1869 recollection. So much confusion rests on the unquestioned acceptance of "the next post" as both an accurate recollection and indicating 9 March. Why has there not been a similar readiness to consider that this too could be a mistaken retrospection not to be taken so strictly just as composition on Ternate itself? Furthermore, as we have seen so abundantly, Wallace was terribly inaccurate about dates, even when only weeks away.

No one has considered if the arrival at Darwin's home on 18 June actually connects to one of the monthly mail steamers from Ternate, and if so, which one? The only tangible *contemporary* evidence of any kind about the transit of this letter is its purported receipt by Darwin on 18 June 1858.

To doubt the receipt of Wallace's letter on 18 June (something purportedly written on the day) on the basis that *it is assumed* to have been sent in March 1858 is not only a weak argument, but ignores a great deal of converging contextual evidence.[676] Darwin's letter to Lyell enclosing Wallace's essay was dated only "18" as usual for friends not far away (one day for correspondents in London) and referred to Wallace's letter as received "to day". The letter has since been endorsed with "June" by Lyell or Francis Darwin.

Darwin had been engaged in writing his study of pigeons since 14 June, as part of his big book on species. However, uniquely in his "Journal", where he recorded the progress of his work and publications, this was "interrupted".[677] The next line in this notebook is "July 20th to Aug 12th at Sandown, begun abstract of Species book", which became the *Origin of species*.

Therefore, professional historians tend to agree that Darwin received Wallace's letter and essay on 18 June 1858. Yet previous writers assumed, indeed many stated as if it were a historical fact, that Wallace's fateful letter to Darwin was sent on the 9 March steamer like the surviving letter to F. Bates.[678] But there is no evidence at all that Wallace *sent* the essay on 9 March. As biographer Michael Shermer rightly pointed out, sending it on 9 March "is not a historical fact but an inference".[679] The inference is derived from two sources — only one of them contemporary: the date on the essay "February, 1858" and Wallace's later recollections that he sent it "by the next post". Taken alone, without any other evidence, these indicate the 9 March steamer.

Wallace's recollections were written at the distance of between eleven and fifty years later.[680] Brooks wrote that, "The crucial elements of all are, naturally, the same."[681] This is incorrect. Suffering from fever is a part of the story from the beginning but other details differ. Many writers cite that Wallace sent his essay by "the next post, which would leave in a day or two" — a detail that only appeared in the 1905 recollection.[682] In the preceding recollections, Wallace maintained simply "by the next post", which could mean transit as much as a month later.

Another inconsistency in his recollections is the amount of time spent on the essay. In 1869, Wallace wrote and copied it all in a single evening. In 1887, it became "[I] finished the first draft the next day". In 1895, this was increased to: "the same evening I sketched the draft of my paper, and in the two succeeding evenings wrote it out in full". And finally in 1908, he recounted that all of this, including posting it to Darwin, occurred "all within one week".

Yet, these retrospective accounts of what was by then a well-rehearsed story of events long ago should not be taken as literal recordings of what occurred in 1858. Wallace himself had only the date on the published essay to show when he wrote it or when it was sent. He did not treat the essay with such gravity in 1858.

How the Ternate essay reached Darwin

The editors of the Darwin correspondence pointed out: "It is of some significance to note that the schedules in Brooks 1984 show that another mail from the East Indies arrived in London on 17 June, a delivery date that is consistent with the arrival of Wallace's communication at Down on 18 June."[683] No one investigated this arrival date further. Postal historians and philatelists routinely reconstruct the itineraries of historical letters. The same can now be done for Wallace's fateful letter to Darwin.

- Wallace deposited his letter to Darwin at the Ternate post office before leaving for New Guinea on 25 March 1858.
- The mail steamer *Makasser* left Ternate c. 5 April and arrived in Surabaya on 20 April.[684]
- The *Banda* left Surabaya on 20 April and arrived in Batavia on 23 April.[685]
- The *Banda* left Batavia on 26 April and arrived in Singapore on 30 April.[686]
- The *Pekin* left Singapore on 1 May and arrived in Galle on 10 May.[687]
- The *Nemesis* left Galle on 14 May and arrived at Suez on 3 June.[688]
- Overland transfer for the mails between Suez and Alexandria took two days.[689]
- The *Colombo* left Alexandria on 5 June and arrived at Southampton on 16 June. Her letters arrived in London on the 17th.[690]
- Darwin received the letter on 18 June 1858.

Thus, the mystery of how Bates' letter could arrive on 3 June and Darwin's on 18 June is solved.[691] They were not, and could not have been, posted on the same day as previously assumed. Bates' letter actually took longer to arrive, eighty-seven days, as compared to c. seventy-five days for Darwin's letter because the latter caught a fortuitous steamer connection.[692]

"In order to send it to Darwin"

It has often been asked, why did Wallace send the Ternate essay to Darwin? But before we try to answer this question, we need to take note of a new realisation — Wallace *did not* send the essay for publication, or Darwin, or anyone else on the 9 March steamer, even though the essay was already written. Neither did he mention it explicitly to Bates or anyone else. Why? In an earlier letter, Wallace mentioned his disappointment that no notice was taken of the Sarawak law paper. Darwin's letter which arrived on 9 March revealed that not only Darwin, one of the most eminent men of science Wallace had ever corresponded with, but the far more eminent Sir Charles Lyell, were highly impressed with the Sarawak paper. And Darwin had revealed that he was an evolutionist. He was a uniquely sympathetic possible recipient.

Hence, the letter from Darwin suggested a new possibility. If the Sarawak paper impressed Darwin and Lyell, the Ternate essay was likely to impress them even more. Rather than further reworking the essay over the next few years, Wallace could use his greatest intellectual achievement to date to gain the recognition and "acquaintance of these eminent men on my return home". Perhaps a scientific position might be his, and thus his ultimate dream at the time, to work on his collections in a rural English home?[693]

We tend to look too much to Darwin because of what happened next. Wallace's main target for the essay was not Darwin, but his scientific hero Lyell.[694] Wallace read and noted Lyell more than any other figure. It was Lyell that he challenged in his notes, in the Sarawak paper and again in the Ternate essay. Wallace used Lyell's own principles to show that "the struggle for existence" would indeed cause a "tendency to departure to an indefinite extent from the original type of the species". These words are from Lyell, not Wallace. Maybe Lyell could be convinced.

And as we now know from the letter from Sir James Brooke, Wallace was quite aware of the prejudice against evolutionary views. Perhaps sending the essay to the only eminent naturalist he knew who was an evolutionist makes sense. And we can now further appreciate that Wallace was sticking to what he had done all along, refraining from public avowal of belief in evolution. He had concealed his belief in evolution in the Sarawak, Aru, arrangement of birds, permanent varieties and other papers. The Ternate essay was the first to openly declare evolution. And it alone was not sent directly for publication. Surely this is not a coincidence. Wallace also concealed his religious scepticism.

Wallace had the prospect of returning to England to unemployment or continuing as a land surveyor as he wrote to his brother-in-law the following year, "I have not yet made enough to live upon, and I am likely to make it quicker here than I could in England. In England there is only one way in which I could live, by returning to my old profession of land-surveying. Now, though I always liked surveying, I like collecting better."[695]

Sending the essay to another man of science would equally establish the date of Wallace's ideas. Priority was not only settled by publication as it is today, but by composition and sharing. By sending it to other men of science, Wallace established the date of his ideas just as respectably as publishing. Shermer noted this point in his highly commendable treatment of the conspiracy theories.[696] Darwin adopted the same technique when referring to Hooker's reading of the 1844 essay and the 1857 enclosure to Gray.[967]

And Wallace *did* secure the patronage of Lyell, Darwin and other eminent scientific men. In addition to his Ternate essay being read at the prestigious Linnean Society and published in their proceedings (something not open to Wallace who was not a Fellow), Wallace was later supported by these men in job applications, offered paid editorial work by Lyell and Darwin, and eventually Darwin, Hooker, Huxley and others arranged for a state pension of £200 a year for Wallace in 1881. Wallace wrote in 1903, "My connection with Darwin and his great work has helped to secure for my own writings on the same questions a full recognition by the press and the public; while my share in the origination and establishment of the theory of Natural Selection has usually been exaggerated."[698]

At least one contemporary's opinion supports the interpretation that, without backup, the Ternate essay might have been just as ignored as the Sarawak paper. As Samuel Haughton, president of the Geological Society of Dublin, said of the Darwin and Wallace publication of 1858, "This speculation of Messrs. Darwin and Wallace would not be worthy of notice, were it not for the weight of authority of the names under whose auspices it has been brought forward [Lyell and Hooker]. If it means what it says, it is a truism; if it means anything more, it is contrary to fact."[699]

At any rate, commentators who claim that Wallace could have or would have (without the Linnean affair) published before Darwin should keep this in mind. Wallace could have sent it for publication on the 9 March steamer, but he did not do so. He could have sent it for publication via the April steamer, again he did not. He intended to work it up after he returned home. Wallace had, just like Darwin in 1842 and 1844, written a draft essay.

By this time, Darwin had written 70,000 words in his transmutation notebooks between 1837 and 1839,[700] a pencil sketch of 13,500 words in 1842, a 54,000-word essay in 1844,[701] compiled between c. 1840 and 1858 about a dozen subject specific portfolios bulging with thousands of notes, clippings and correspondence bearing on the theory (including one on divergence),[702] and as Kohn has written, "completed ten and a half chapters of his book, that is, over 250,000 words of well-articulated argument supported by a masterly array of facts. Darwin had virtually completed the plan that Wallace was just contemplating."[703]

Darwin would probably have finished his book *Natural selection* (if uninterrupted by Wallace) by 1860.[704] Wallace, on the other hand, had produced about 7,700 words of notes on species in his notebooks, the Sarawak paper on succession, the permanent varieties note and the 4,200-word Ternate essay. Wallace's projected species book would have been forestalled by Darwin about two years before Wallace even returned to Britain. Therefore, without the Linnean affair, Wallace would probably never have been credited as co-discoverer of evolution by natural selection. In light of this, there is nothing to regret about the Linnean reading from Wallace's perspective.

Dorey, New Guinea, 11 April–29 July 1858

Wallace wrote his historic letter to Darwin sometime between c. 11–25 March 1858, enclosed the fateful essay and left them at the post office in Ternate. For some reason, van Duivenbode's "very fast sailing" schooner, the *Esther Helena*, did not leave for New Guinea until 25 March.[705] Although Ternate was the main port from which trading ships sailed to New Guinea, there were only about two ships per year.[706] Hence, Wallace had to take this one.

For this major expedition, he took four servants with him: "my head man Ali, and a Ternate lad named Jumaat (Friday), to shoot; Lahagi, a steady middle-

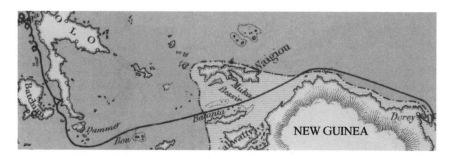

Ternate to Dorey. "Mr. Wallace's route" map MA1.

aged man, to cut timber and assist me in insect-collecting; and Loisa, a Javanese cook".[707] Collecting even started on the schooner itself when Wallace found five or six small beetles in the ship's biscuits.[708]

The *Esther Helena* sailed south, touching the next day at the small island of Makian and then into the straits between Batchian and Gilolo. On the 28th, they anchored and took on water at Ganeh on the southern end of Gilolo. Wallace collected insects and sent a boy to shoot birds. The following day, they sailed on though often becalmed. Eventually, on 11 April, they dropped anchor off the small island of Mansinam near the coast of Dorey (Manokwari) Harbour, in Cenderawasih Bay, on the north coast of the Vogelkop Peninsula of New Guinea.

Since 1855, Mansinam was the base of two German missionaries, Carl Ottow and Johann Geissler. Their efforts to Christianise the natives and establish European farming won the financial backing of the Dutch government.[709] They were the first Christian missionaries in the region and are locally remembered to this day by the Christian community in the region. Uncharacteristically snooty, Wallace noted that they were "working men…being more useful among savages than persons of a higher class".[710]

No Wallace letters survive from the New Guinea expedition, and perhaps none were written there since there was no way to post or receive letters. At the village of Dorey near the end of a narrow promontory on the bay of the mainland, Wallace was intrigued not only by a different race of people with dark skin and tight curly hair but different cultural forms. "The houses all stand completely in the water, & are reached by long rude peers from the high water mark."[711] Their woodwork was adorned with what Wallace thought grotesque figures and shapes. With his boys, and a collection of somewhat baffled locals, Wallace built a rough house near the village overlooking the white sandy beach and the sea.

Dorey village. Wallace, *Australasia*, p. 440.

His collecting records start on 19 April 1858.[712] Despite having come up with his new theory for the origin of species only two months before, his collecting and note-taking were entirely unchanged. There was no new research programme or method as yet.

The rough and muddy jungle tracks soon revenged themselves on the pillaging foreigner when Wallace wounded his ankle amongst some fallen trees on 2 May. It became infected. He was housebound for many days. He passed the time by re-reading Laurence Sterne's *Tristam Shandy*. For company, and fresh eggs, he had a young chicken sitting in a basket in the house. Outside there was a continual drizzle. He might have used some of this time off to purchase some local artefacts like carved human figures and a scoop made of coconut shell.[713]

An old Dutch paddle steamer, the *Etna*, arrived at Dorey with faint wisps of grey smoke trailing from her funnel. She had been sent by a Dutch government commission. Wallace recorded in his *Journal* that she arrived on 5 May, but in *The Malay Archipelago*, he gave 15 May. According to the expedition's official narrative, the *Etna* arrived on 3 May 1858.[714]

"The captain [Georg Roijer], doctor, engineer and some other of the officers paid me visits; the servants came to the brook to wash clothes, and the son of the Prince [Amir] of Tidore, with one or two companions, to bathe; otherwise I saw little of them, and was not disturbed by visitors so much as I had expected to be." As much as Wallace enjoyed the company of some Westerners, he was annoyed that so many unique specimens from the area went to them, Prince Amir and the Resident of Banda, H. D. A. van der Goes. For example, "on board the steamer they had a pair of the curious tree kangeroos alive".[715] Wallace and his men never even saw one in the forest.

The *Etna*'s tender, the barque *Atie Atul Barie*, reached Dorey on 14 May.[716] On board, Wallace was delighted to find "a brother naturalist" Hermann von Rosenberg (1817–1888) serving as "draughtsman to the surveying staff. He had brought two men with him to shoot and skin birds, and had been able to purchase a few rare skins from the natives." Von Rosenberg's detailed account of his travels between 1839 and 1871 was published in German as *Der malayische Archipel* (1878) [The Malay Archipelago].

Wallace the spy?

Unknown to Wallace, the Dutch were somewhat suspicious of an Englishman based in remote New Guinea just as they were seeking to cement their influence in the region. The writer of the Dutch Commission's narrative suspected Wallace might be up to more than just natural history collecting. He might be a spy reconnoitring for an English trading station.

> It has not escaped our notice that we found at Doreh an Englishman by name of Russell Wallace, who told us that he for several years had visited several countries in the eastern hemisphere, to keep himself occupied with catching and collecting birds, butterflies and insects.... And without judging whether the desire to investigate natural history was the only reason for the visit to Doreh by Mr. Russell Wallace, we believe it right not to keep silent about his presence there.[717]

Wallace's ankle finally began to heal but then a secondary infection bothered his foot. The Dutch doctor, Johan Hendrik Croockewit, advised poultices and other medications which kept Wallace in his house for several more days.[718] Poor Wallace was distraught that so much of his short and valuable time in this exotic locale was being lost sitting in his hut. He had come to Dorey because of Lesson's gorgeously illustrated 1835 book with hand-coloured plates of Birds of Paradise acquired here.[719] As usual, Wallace spent some time writing, this time a short essay on "Synonyms & the quotation of Authorities by Naturalists".[720] On 27 May, Wallace sent his boys by boat to Amberbaki, a village about 100 miles to the west. Their mission was to shoot, find or buy Birds of Paradise at all costs.

Horned flies. MA2:314.

One interesting discovery Wallace made were some bizarre little flies, which appeared to have antlers on their heads and "about half an inch long, slender-bodied, and with very long legs, which they draw together so as to elevate their bodies high above the surface they are standing upon....The horns spring from beneath the eye...In the largest and most singular species...these horns are nearly as long as the body, having two branches, with two small snags near their bifurcation, so as to resemble the horns of a stag."[721]

On 5 June, the Dutch coal barque *Ydroessie* returned. She had been at Dorey from 21 February to 16 April (Wallace recorded the 17th in *Notebook 4*). Soon thereafter, Wallace's boys returned from Amberbaki with more bad news. The collecting was no better there and no other species of Bird of Paradise were to be found.

The thunderbolt strikes

At 4 pm on 17 June, both the *Etna* and *Atie Atul Barie* left for Humboldt Bay to the east. Wallace had a fever. Dr. Croockewit left medicine for him. The following day, Friday 18 June 1858, Wallace's fateful letter and Ternate essay arrived at Darwin's Down House. The newspapers of the day were full of news of the bloody Indian Mutiny which was finally coming to an end. For Wallace, it was "wet all day — nothing — still unwell".[722] On the other side of the world, Darwin opened the historic letter. As Wallace later put it, "without any apparent warning, my letter, with the enclosed essay, came upon him, like a thunderbolt from a cloudless sky!"[723]

Much has been written about Darwin's reaction or feelings on seeing Wallace's essay. It is important to remember how speculative these descriptions necessarily are because we have no long diary entries or discursive letters describing the moment. All we have is the letter Darwin penned to Lyell that day.

> Down Bromley Kent 18
>
> My dear Lyell
>
> Some year or so ago, you recommended me to read a paper by Wallace in the Annals, which had interested you & as I was writing to him, I knew this would please him much, so I told him. He has to day sent me the enclosed & asked me to forward it to you. It seems to me well worth reading. Your words have come true with a vengeance that I shd be forestalled.

But, as he thought Lyell may not remember the conversation, Darwin reminded him of the circumstances in April 1856:

> You said this when I explained to you here very briefly my views of "Natural Selection" depending on the Struggle for existence.— I never saw a more striking coincidence, if Wallace had my M.S. sketch written out in 1842 he could not have made a better short abstract! Even his terms now stand as Heads of my Chapters.
>
> Please return me the M.S. which he does not say he wishes me to publish; but I shall of course at once write & offer to send to any Journal. So all my originality, whatever it may amount to, will be smashed. Though my Book, if it will ever have any value, will not be deteriorated; as all the labour consists in the application of the theory. I hope you will approve of Wallace's sketch, that I may tell him what you say.
>
> My dear Lyell
>
> Yours most truly
>
> C. Darwin[724]

The Dutch steamer *Etna* at anchor at New Guinea. Rosenberg, 1878, p. 441.

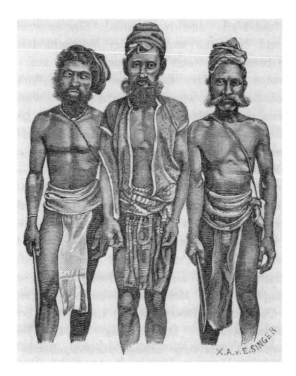

Alfuros from Elpaputi Bay, Ceram. Rosenberg, 1878, p. 288.

Chapter 9

THE LONGEST DAY

Ever since 18 June 1858 when Darwin first read Wallace's Ternate essay and forwarded it to Lyell as a "striking coincidence", there has been a tendency to refer to "the same theory" conceived by Darwin and Wallace. Darwin even wrote in his autobiography that Wallace's essay "contained exactly the same theory as mine".[725] But in recent decades, several scholars have stressed some of the differences. For example, historian Peter Bowler pointed out in 1984 that "the two men were certainly arguing along different lines: Wallace did not deal with selection of individual differences, postulated only an episodic selection of varieties, and had no concept of divergence through ecological specialization".[726]

But there is a more fundamental problem — and that is talking about "the theory" as a single thing in the first place. Just what exactly is "it"? Of course as shorthand language, it is a useful and efficient way of speaking in most contexts. But if we want to be more exact, the shorthand becomes a hindrance to our thinking.

If we compare what Darwin and Wallace wrote, we see many details that are not the same. How can we then conclude that the written words are just varying descriptions of what was *really* the same idea in their heads? It is a poor solution to imagine that their differing words are somehow actually derived from the same idea. How can we know that the idea is the same when the only evidence we have are their differing words?[727]

There is no single idea or theory that both Darwin and Wallace had in their heads. And that is part of the reason why what they wrote is not the same. Instead, we should always refer to a plurality — their ideas. It is not that difficult to do and it allows for more of the underlying complexity and saves us from all sorts of blunders.

We are investigating a whole population of ideas, not just one idea. When looked at this way — which is more accurate and correct — of course Wallace did not have all of Darwin's twenty-year work at once; much he already had, some new things he first added at Ternate, but not all was in his essay by a long shot. So there were some similarities between their ideas. This is both more accurate and allows for the necessary gaps in our understanding rather than glossing them over with a singular thing we imagine we have before us called "the theory".

Lyell was perhaps the first, and alas maybe one of the last, to refer to the ideas of Darwin and Wallace in this pluralistic sense. In his *Antiquity of man* in 1863, Lyell wrote that Wallace "thought out, independently for himself, one of the most novel and important of Mr. Darwin's theories".[728] Darwin's theories, after all, included many components not in Wallace's. These included the analogy of man's selective shaping of domesticated plants and animals, laws of variation, transitional varieties, inherited instincts, family selection, hybridism, embryology, taxonomic classification, no inherent progress, sexual selection, vestigial organs as remnants, natural dispersals rather than former land bridges and answers to objections to evolutionary theories. Wallace himself made similar points to Lyell, listing details, in his February 1860 notes on reading *Origin of species* (see p. 280, Chapter 11).

Charles Darwin, 1855.　　Charles Lyell, 1860s.　　Joseph Dalton Hooker, 1864.

The Big Three.

What Happened Next

The events that unfolded after Darwin received Wallace's essay were far less dramatic or sinister than so often portrayed. Darwin's letters from the days after receiving Wallace's essay that survive reveal little out of the ordinary. I discovered one to Murchison about the British Museum and there is another about his daughter Etty who had a sore throat since the 18th of June. The Darwins were using their new dining room and the grey horse had cut its knee. Two days after Wallace's letter arrived, Darwin mentioned to his cousin Fox, "I am getting on very well about Bees' cells & my theory, I feel pretty sure, will hold good."[729]

At some time during the ensuing week, Lyell drafted a letter to Wallace and sent it to Darwin to read. But Lyell then reconsidered and wrote again to suggest that his letter to Wallace might have been precipitous. Darwin replied the following Friday, "I confess it never did occur to me" that Wallace "could have made any use of your letter".[730] What did Lyell say? We don't know. He may have commented on the similarities between the Ternate essay and Darwin's views or given further details. In the next lines, Darwin outlined his manuscripts that showed by how many years his unpublished views preceded Wallace's unpublished views.

Darwin was subtly reminding Lyell that he had long preceded Wallace. Darwin conceded that he would now consider publishing a sketch of his theory to be published after Wallace's essay, but Darwin had scruples that this might be improper because of privately knowing of Wallace's views — and assuming that Wallace would soon publish them. Darwin reiterated that Wallace's views were independent. "I do not in least believe that he originated his views from anything which I wrote to him."

As the days passed, the subject consumed Darwin's thoughts more and more. He wrote to Lyell again on the following day, enclosing Wallace's letter itself (instead of the essay already sent) for the first time.

Down. 26th

My dear Lyell

Forgive me for adding P.S. to make the case as strong as possible against myself.

Wallace might say "you did not intend publishing an abstract of your views till you received my communication, is it fair to take advantage of my having freely, though unasked, communicated to you my ideas, & thus prevent me forestalling you?" The advantage which

I should take being that I am induced to publish from privately knowing that Wallace is in the field. It seems hard on me that I should be thus compelled to lose my priority of many years standing, but I cannot feel at all sure that this alters the justice of the case. First impressions are generally right & I at first thought it wd be dishonourable in me now to publish.—

Yours most truly

C. Darwin

I have always thought you would have made a first-rate Lord Chancellor; & I now appeal to you as a Lord Chancellor[731]

Darwin *was* concerned about priority. Who wouldn't be? But the game was far from an all or nothing one. At some point, Darwin wrote half a letter to Wallace offering to publish the Ternate essay and give Wallace priority for publishing natural selection first.[732] Before this letter was finished, however, a new possibility emerged.

Apparently, Darwin's friend Hooker proposed a solution to the conundrum. Concerned that his friend's generosity would mean losing his twenty years' priority in natural selection, but also trying to be fair to Wallace, Hooker proposed to present Wallace's essay along with extracts from Darwin at the next meeting of the Linnean Society in about a week's time.

On the 28th, Darwin's baby son Charles Waring tragically died of scarlet fever. The next day Darwin replied to a now lost letter from Hooker. "I have received your letters. I cannot think now on subject, but soon will. But I can see that you have acted with more kindness & so has Lyell even than I could have expected from you both most kind as you are. I can easily get my letter to Asa Gray copied, but it is too short."[733]

Later that night, Darwin received another note from Hooker and answered it. Darwin had already sent Wallace's letter to Lyell and by now Darwin had the Ternate essay back. He now forwarded it to Hooker, along with the fair copy of his 1844 essay with Hooker's comments, to remind Hooker that he had read it.[734] Hardly something that needed doing if we believe that everyone regarded this as the theory of theories. Darwin did not supply the sketch to form part of the Linnean Society presentation. He also sent his copy of the Asa Gray enclosure. All were sent via a servant to Hooker at Kew.

Meanwhile in New Guinea, Wallace and his team were still beset with illnesses. Jumaat died on Saturday 26 June from "Diarrhaea".[735] "He was about eighteen years of age, a native, I believe, of Bouton, and a quiet lad, not very active, but doing his work pretty steadily, and as well as he was able. As my men were all Mahometans, I let them bury him in their own fashion, giving them some new cotton cloth for a shroud."[736] Wallace felt that respecting his Malay servants and their beliefs was important. As he later remarked, "Malays make admirable servants if due regard be paid to their feelings and prejudices; they readily become attached."[737] Respecting others also felt natural for Wallace. After all, the London scientific toffs treated him as more junior than he felt he really deserved, and this probably rankled.

The Main Event

Thursday 1 July 1858 was a mild summer day in London. The usual array of activities and entertainments was on offer in the bustling, smoky, soot-stained Victorian capital. The Grand National Rose Show met in St. James's Hall with fifty principal rose growers. The Bradford Festival Choral Society performed at the Crystal Palace on Penge Common. Charles Dickens read some of his short stories at St. Martin's Hall at 8 pm.[738] At exactly the same time a few blocks away at Burlington House, Piccadilly, was a meeting of the Linnean Society, one of the dozen or so scientific societies in London.

The Linnean was one of the more prestigious learned societies and regarded as the representative of the natural history sciences in the United Kingdom. It was founded for "the cultivation of the Science of Natural History in all its branches" in 1788 after the purchase of the library and collections of the famous Swedish naturalist Carl Linnaeus (1707–1788), who pioneered the system of hierarchical biological classification still used today.[739] Prospective members or Fellows of the Society had to be nominated by three existing Fellows. The admission fee was a hefty £6 and the annual subscription £3. The Linnean met every other Tuesday from the beginning of November to mid-June. Some of the papers presented at meetings were selected for publication in the Society's journal. The last meeting of the 1858 season was 17 June. However, the former president of the society, the Scottish botanist Robert Brown, had died aged eighty-four the week before. Therefore, the current president of the Society, sixty-six-year-old reptile expert Thomas Bell, suggested the meeting should adjourn "in consideration of [Brown's] long connexion with and eminent services to the Society and to Natural Science".[740] A new vice-president and member of council needed to be elected to replace Brown.

So the Society met again on 1 July for a "special meeting" to elect a replacement and conduct other business deferred from the previous meeting.[741]

About twenty-seven fellows and two guests were present. Bell was the chairman for the evening. The minutes of the previous meeting were first read, followed by a list of a dozen donations to the library and museum. The botanist George Bentham was elected in Brown's place. Lyell spoke for a few minutes about the contributions of Brown during sixty years of association with the Society.

The meeting moved on to the scientific papers to be presented. The minutes recorded:

> Read 1st, a letter from Sir Charles Lyell, F.L.S., and Dr. Hooker, F.L.S., addressed to the Secretary, as introductory to the following Papers, on the laws which affect the production of varieties, races, and species, viz.:
>
> An abstract from a MS. work on species, by Charles Darwin, Esq., F.R.S. L.S., sketched in 1839 and copied in 1844.
>
> An abstract of a letter addressed by Mr. Darwin to Prof. Asa Gray of Boston, U.S., in Oct. 1857.
>
> An essay on the tendency of varieties to depart indefinitely from the original type; by A. R. Wallace, Esq.[742]

The contributions were probably read by the Society's secretary, botanist John Joseph Bennett. The introductory letter by Lyell and Hooker contained 650 words, Darwin's contributions 2,800 and Wallace's 4,200. Lyell and Hooker, themselves only partially convinced by Darwin at that point, added a few verbal remarks about the importance of the subject.[743] It was very appropriate that the anti-evolutionist Lyell should have thus served as the godfather to the theories of Darwin and Wallace. Not just because his status might have helped gain an attentive hearing for these radical views, but because he was the single most influential figure for both Darwin and Wallace.

Five more papers followed: on a marine worm, the larvae of lampreys, a new genus of tropical American cucumbers, a manuscript by a deceased naturalist on the South American shrub cinchona (once used to treat malaria) and two letters on the vegetation of West Africa.[744] There was no discussion at the end of what was a very long meeting, neither was there any fuss at the tea which followed.

At that very moment, it was 5 am on the shores of New Guinea and Wallace was just waking in his hut with another fever. Incessant ants were crawling over him and mosquitoes sang their annoying bloodthirsty song in his ears. He could not have imagined what was then unfolding in London. Later in the day, still unwell, Wallace limped out for a short walk and collected forty-two insect species. Darwin and his family attended to the melancholy funeral of their infant son in the Down churchyard.

Little notice was taken of the reading of the Darwin–Wallace papers.[745] This was probably made worse by the fact that the meeting was after the usual season. For this reason, there was no report of the meeting in the usual places such as the *Athenaeum* or *The Times*.

Over the ensuing weeks, mobs did not run through the streets. No brimstone descended from the pulpits. Darwin later recalled, "Our joint productions excited very little attention, and the only published notice of them which I can remember was by Professor Haughton of Dublin, whose verdict was that all that was new in them was false, and what was true was old."[746] Wallace recalled that "it received little attention till Darwin's great and epoch-making book appeared at the end of the following year".[747]

The response that was, as Janet Browne put it, "destined to become known as one of the most unfortunate misjudgements in the history of science" was that of Thomas Bell, in his presidential report to the Linnean Society of May 1859. He famously said, "The year which has passed has not, indeed, been marked by any of those striking discoveries which at once revolutionize, so to speak, the department of science on which they bear."[748]

Darwin, again rather embarrassed about the selfish motive, wrote to Gray a few days later about the date of the enclosure.

> It is very unlikely, but if by any chance you have my little sketch of my notions of "natural Selection" & would see whether it or my letter bears any date, I sh^d. be very much obliged. Why I ask this, is as follows. M^r. Wallace who is now exploring New Guinea, has sent me an abstract of the same theory, most curiously coincident even in expressions. And he could never have heard a word of my views. He directed me to forward it to Lyell.— Lyell who is acquainted with my notions consulted with Hooker, (who read a dozen years ago a <u>long</u> sketch of mine written in 1844) urged me with much kindness not to let myself to be quite forestalled & to allow them to publish with Wallace's paper an abstract of mine; & as the only very brief thing which I had written out was a copy of my letter to you, I sent it and, I believe, it has just been read, (though never written, & not fit for such purpose) before the Linnean Soc^y.[749]

It is not clear when Darwin first heard about the Linnean Society meeting. He thanked Hooker for his care but was still embarrassed at having cared about priority. It is clear Darwin assumed that Wallace's essay would be published in the *Linnean proceedings* with a note from Hooker saying that Darwin had similar views before. On 13 July, Darwin told Hooker, "I am *much more* than satisfied at what took place at Linn. Soc.ʸ— I had thought that your letter & mine to Asa Gray were to be only an appendix to Wallace's paper."[750] On the same day, Darwin sent a letter (now lost) to Wallace giving details of his book in progress, *Natural selection*, and forwarded a letter from Hooker explaining the Linnean Society reading to Wallace.

Darwin was expecting to prepare a separate article on his views for a subsequent issue of the *Linnean proceedings* and asked Hooker how many pages would be allowed. If it was not accepted, Darwin planned to publish it as a pamphlet. On 15 July 1858, Hooker sent the proofs of the *Linnean proceedings*. The publication was printed just as presented at the meeting. Lyell and Hooker presented the three documents "in the order of their dates" an extract from Darwin's 1844 sketch, the 1857 note to Gray and Wallace's 1858 essay.

These documents were published together in the Society's proceedings in August 1858. Both the reading and the publication, despite their retrospective importance, were too brief to engender any scientific revolution.[751] The opening address by the Scottish naturalist Andrew Murray to the Botanical Society of Edinburgh on 10 November 1859 underlines this point: "The hints thrown out in these Linnean Society Papers are very brief, and merely indicate their author's views on one or two points of a great question, leaving the greater part untouched on."[752] Even Lyell and Hooker themselves were not fully convinced of Darwin's views and hence neither believed that they were unveiling a revolution in science in favour of their friend at the expense of Wallace. But we should not go to the extreme of Darwin's grandson Sir Charles Darwin who said in 1958, "These papers were completely and absolutely stillborn!"[753]

Richard Owen, usually represented as the anti-Darwinian villain, at first approved of the paper.[754] The zoologist Alfred Newton, at least according to his later account, sat up late to read the papers. "Never shall I forget the impression it made upon me. Herein was contained a perfectly simple solution of all the difficulties which had been troubling me…after reading these papers more than once, I went to bed satisfied that a solution had been found. All personal feeling apart, it came to me like the direct revelation of a higher power; and I awoke next morning with the consciousness that there was an end of all the mystery in the simple phrase 'Natural Selection.'"[755]

Was It Fair to Wallace?

Some modern commentators voice strong opinions about fairness and credit in the arrangement. Many of these opinions and accusations are not only anachronistic judgements of the actions of Victorian men of science by current (or rather the writer's own pro-Wallace) standards, but also quite uninformed about the standards of acceptable practice in the mid-19th century. A more rigorous historical approach can clarify most, if not all, of these.

Priority

Priority is one of the main concerns of recent writers. As usual with questions of credit, opinions vary widely, but the word "priority" is seldom absent from books and articles about Wallace. Conspiracy theorists such as Brackman and Davies are in no doubt that despite having conceived of natural selection twenty years after Darwin, Wallace was somehow "robbed in 1858 of his priority". Even Darwin biographer Janet Browne has written, "Yet Darwin had no priority over Wallace, except in manuscript form....In actuality, the joint reading of the two papers in 1858 deprived Wallace of his priority."[756] I don't see how, since both men had produced unpublished manuscripts, "Darwin had no priority" but Wallace was "deprived...of his". Presumably, Browne assumed that Wallace's paper was sent for solo publication and therefore was otherwise destined to take priority.

Order

Many commentators are concerned with the order in which the papers were printed. The reasons offered differ but they all come down to the same thing. Many people think Wallace should have been published alone or at least printed before rather than after Darwin.

So, for example, Charles Smith claimed that "as the initiating work Wallace's paper rightfully should have been read first, but instead was presented third".

David Quammen opined that "Darwin's contribution [was] given priority on the grounds of prior composition—although it was only a set of excerpts cobbled together". Even Bill Bryson has claimed "protocol required Darwin to step aside and allow Wallace full credit for the theory".[757]

In order to make these claims, some evidence from the time must be provided; otherwise, there is no way to distinguish the writer's opinion from contemporary standards of fairness. I have been unable to find any contemporary evidence to support either the claim that Wallace's essay ought to have been placed first or should have been published alone. Multi-authored contributions appeared in the *Linnean proceedings* from time to time with no clear system about order.[758] Even today, the order of names on scientific papers can be a grey area.

Credit

It seems that most modern writers on Wallace assume that being first in print with natural selection would result in Darwin's fame. On the one hand this is a natural enough assumption. But given that it has been known since 1860 that earlier authors *did* publish versions of natural selection, particularly the Scottish writer Patrick Matthew in 1831, this should not matter.[759] But as Darwin's and Wallace's descriptions of natural selection were published jointly, many argue they should enjoy identical status and fame today. Since Wallace is nowhere near as famous as Darwin, surely something has happened to diminish Wallace's rightful share?

But as Darwin noted, even on the very day he first saw Wallace's essay, "My Book…will not be deteriorated; as all the labour consists in the application of the theory." It is not that Darwin was the first to think of natural selection; he conceded that Matthew and others preceded him, and if Wallace had published the Ternate essay first, Darwin would have added Wallace to the top of the list. Hence, even if the Ternate essay had been published by itself in 1858, and Darwin published the *Origin of species* in 1859, the reputations of Darwin and Wallace would not be any different today.

It was Darwin's *Origin of species* which, within the scope of ten to fifteen years, changed the views of the international scientific community to accept that evolution was a fact. Natural selection was less widely accepted at first. This seems to dampen even further the urgency of the impression that Wallace has been diminished. Although he formulated a version of natural selection, he had thought of a small percentage of the web of arguments and evidence that made the *Origin of species* convincing. I cannot see that it would have made any difference either way whose paper was printed first, unless the dates of their composition were concealed — but this would have been very odd.

Wallace, in his reply to Hooker, indicated how he thought priority in science was normally recognised: "to impute *all* the merit to the first discoverer of a new fact or a new theory, & little or none to any other party".[760] In this statement, Wallace attributed priority not to being the first to publish, but being the first to discover. Wallace was grateful that his later independent discovery had been published together with Darwin, rather than it being pointed out that Darwin long preceded him.

When Darwin later admonished Wallace for referring to natural selection as solely Darwin's, Wallace replied, "I shall always maintain it to be actually yours & your's only. You had worked it out in details I had never thought of, years before I had a ray of light on the subject...All the merit I claim is the having been the means of inducing you to write & publish at once."[761] Even decades later, Wallace said in an interview:

> Many people have asked me if I was not a little disappointed to find that Darwin had anticipated me, because he had his materials gathered and ready? For that very reason I was not disappointed. I felt that while I could only have written a slight volume, he was able to launch the discovery backed, first, by a mass of carefully compiled evidence, and secondly, by his already very high reputation as a naturalist. Everything was for the best, as perhaps it usually is.[762]

Was it proper to publish?

The historian Barbara Beddall may have been the first, in 1968, to opine that Wallace did not give his "consent" to the reading or publication and that the arrangement was not quite fair to Wallace.[763] David Quammen went so far as to write, "The conventional version claims that 'Wallace heartily agreed to' the joint presentation...but that's bullshit. Nobody waited for his permission." Several other writers have even claimed that Lyell and Hooker were being less than truthful when they stated in the introductory letter: "both authors having now unreservedly placed their papers in our hands".[764]

We have seen that Darwin was uncertain about the proper course. He wrote to Lyell, "Please return me the MS which he does not say he wishes me to publish; but I shall of course at once write & offer to send to any Journal." Did Darwin write this because it was not acceptable to publish the Ternate essay without Wallace's say-so or because Darwin felt he was not a neutral party and this made him uncomfortable? It seems like the former to those whose fund of contemporary evidence is only these Darwin–Wallace materials.

One should not go so far as to say that there were explicit rules, and this case was not typical. Nevertheless, we can only understand and assess the way Lyell and Hooker acted with reference to the standards of the time. What was normal practice in the 1850s?

Every meeting of the London scientific societies had private correspondence read publicly. Every issue of the *Linnean proceedings* and similar scientific journals published private correspondence, "without permission". Wallace's own letters were routinely published, not all by his instruction.

When Darwin was travelling with the *Beagle,* he too had scientific writings published by a correspondent at home without his knowledge in 1835.[765] When Darwin found out, he reacted similarly to Wallace, pleased his writing was considered worthy of publication and discussion by his seniors and somewhat embarrassed he had not been able to correct the proofs (which did contain typographical errors). But it was perfectly normal practice. Similarly, two letters from the astronomer and doyen of British science Sir John Herschel to Lyell and Murchison, one of which contained the now famous phrase "the mystery of mysteries", were published without "consent" in Charles Babbage's *Ninth Bridgewater treatise* (1838). No one objected.

When Darwin wrote to Asa Gray about the details of his theories in 1857, Darwin asked Gray *not* to let the contents be known; "the reason is, if anyone, like the Author of the Vestiges, were to hear of them, he might easily work them in". How could the author of *Vestiges* hear about what Darwin wrote in a private letter to Gray? Because Darwin knew that Gray might refer to or even publish Darwin's words. Without explicitly declaring them private, it was acceptable to do so. Wallace used the same convention, for example when he asked Sclater, "Please do not print this gossiping letter, except an extract or two."[766] This is how to understand the line "both authors having now unreservedly placed their papers in our hands".

What Did Wallace Say?

Another misconception is that Wallace expressed dissatisfaction or regret that the Ternate essay was published without his knowledge. George Beccaloni claimed, "In several of his publications Wallace *complained* that he was not given the opportunity to correct the proofs of his 1858 essay."[767] But Wallace made a qualification, not a complaint. In fact, none of Wallace's statements indicate any dissatisfaction or disappointment. They contain only disarming qualifications that the work before the public had not been checked by him in proof. We could not expect a clearer or more unguarded indication of how Wallace received the news of the arrangement than the letter to his mother after learning the news. He told her that "Dr. Hooker and Sir C. Lyell… thought so highly of it that they immediately read it before the Linnean Society". They thought so highly of it they had it immediately read! And that's that.

No matter how many times Wallace said how happy he was with the Linnean arrangement (and we have many instances), and how much he thought he benefited more than he deserved, this does not deter some Wallace fans from feeling aggrieved. Indeed, given how overwhelmingly advantageous the joint publication was for Wallace, it is hard to see how he could have regarded it as anything but positive and fortunate — which is how he described it in all of his later recollections. Wallace remarked in 1903, "My connection with Darwin and his great work has helped to secure for my own writings on the same questions a full recognition by the press and the public; while my share in the origination and establishment of the theory of Natural Selection has usually been exaggerated."[768] "It was really a singular piece of good luck that gave to me any share whatever in the discovery."[769] He felt he had received "ample recognition by Darwin himself of my independent discovery of 'natural selection'".[770] And in his autobiography, Wallace stated that he "obtained full credit for its independent discovery".[771]

What Did Darwin Say?

A long overlooked fragment in the Darwin Archive sheds some interesting light on Darwin's views on a parallel but separate case of convergence with another naturalist mentioning both plagiarism and priority. His daughter recorded:

> Feb 1871 Just before publication of [Descent of] Man, my Father told
> me "I have just heard that a German book has come out apparently

the very same as mine, "Sittlichkeit & Darwinismus"; whereupon I said "Well, at any rate nobody can say you've plagiarized." "Yes, that is the only bother, that is very disagreeable Otherwise I never have cared abt the paltry feeling of priority & it doesn't signify a bit its coming out first It is sure to be not exactly the same." It is a good thing it is coming out when two men hit upon the same idea it is more likely to be true."[772]

Cleaning Up the File

Another myth is that historical documents relating to these events were selectively destroyed. This comes from Beddall's claim that there appeared to her to be intentionally missing correspondence relating to the Linnean affair and so "somebody cleaned up the file".[773] This discredited assertion has unfortunately been repeated by recent writers.[774] But Kohn long ago showed:

> The historical record is inevitably imperfect....there is the lamentable paucity of extant Lyell letters in Darwin's papers in Cambridge. The bulk of these should be among the over 3900 mostly post-1862 letters to Darwin arranged in alphabetical order, scores of which show signs of damp. Worst hit was the letter L, for which only seven miscellaneous letters, 82 Lubbock letters, and three Lyell letters survive. One may infer that a packet of Lyell letters disintegrated in good uniformitarian fashion along with the bulk of the L's.[775]

Much has also been made of the loss of Wallace's letter and essay. Is this evidence of a conspiracy or cover up? Shermer noted, "Eight months later Darwin received another paper from Wallace...that he also forwarded to the Linnean Society for presentation and publication. The originals of this letter and paper are also missing, but no one has concocted a conspiracy about that fact."[776] The same is true for almost all of Wallace's articles published from the Eastern Archipelago as well as the letters he received. Once again, silly conspiracies have been concocted without first checking the evidence — is this unusual or not?

The Ternate essay, after its epic journey of 9,240 miles from Ternate to London by steamship and steam train, and then back and forth between Darwin, Lyell, Hooker and the Linnean Society, ended up at the printers Taylor and Francis at Red Lion Court, Fleet Street. Here, it was set in type for publication in the *Journal of the Linnean Society*. The Ternate essay was unceremoniously discarded by the printers like other manuscripts that went into the journal.

Chapter 10

DARWIN'S DELAY

After twenty years, Darwin's views were now public. This raises one of the biggest questions in the entire story of the discovery of evolution. If Darwin conceived of the theory twenty years before 1858, why had he not published it? If he had, then Wallace could not have conceived of his explanation independently and surprise Darwin with the Ternate essay. There would have been no Linnean Society affair or change of Darwin's book plan giving us the *Origin of species*. Wallace would not have achieved co-discoverer status and much later we would not have so many conspiracy theories about it all. In fact, there are as many myths and misunderstandings surrounding the question of Darwin's so-called delay as the Ternate essay affair.[777]

For decades, the story of Darwin and evolution has revolved around this twenty-year delay.[778] It has been called the greatest puzzle about Darwin. What were his reasons or motives for postponing for so long? He is said to have delayed because he was afraid of the reactions of his scientific colleagues, offending his religious wife, upsetting the captain of the *Beagle*, Robert FitzRoy or even upsetting the social order. Desmond and Moore's popular biography, appropriately subtitled "the life of a tormented evolutionist", revolves around Darwin's delay.[779]

In fact, the evidence is overwhelmingly against the view that Darwin postponed publishing his theory because he was afraid. The very idea of "Darwin's delay" dates only to the mid-20th century. Darwin hardly veered from his original plans for working out and publishing his species theory when he had completed his other projects.

Darwin's Secret?

Many writers claim that Darwin's theory was kept secret before publication. Probably the most insistent are Desmond and Moore. For them, Darwin's theory was secret because they believe Darwin and his contemporaries must have seen "evolution as a social crime". "Darwin could expect a furore among his geological friends if they discovered his secret. No more 'hail fellow, well met'. He could be labelled as a traitor. His respectability would be compromised. Not only would his science be impugned. He himself would be accused of reckless abandon."[780] They describe Darwin's transmutation and expression notebooks as "secret", "clandestine" and "covert".[781] Darwin preferred, in their view, "living a lie".[782]

Was Darwin's theory really a secret? Did he lie? I know of no evidence that Darwin or anyone who knew him referred to it *or treated* it as secret.[783] The editors of Darwin's correspondence observed, "Darwin is usually depicted as having been very careful to keep secret his heretical views on species, but the correspondence does not bear out this view, if what is meant is that Darwin was afraid to divulge his conviction that species had evolved."[784] In addition to there being no evidence that Darwin's theory was a secret, there is a lot of evidence that shows that he did not keep it secret.

Darwin told *many* people about his interest in evolution during the years before publication.[785] Darwin's very first known recorded doubts about the stability of species were written in the now famous passage in his ornithological notes in 1836: "If there is the slightest foundation for these remarks the zoology of Archipelagoes — will be well worth examining; for such facts would undermine the stability of Species." What is so seldom realised is that these notes were not private. They were prepared to give to another naturalist along with the bird collection when the *Beagle* returned.[786] So from the very advent of our detailed paper trail of the development of Darwin's theory, the evidence flatly contradicts the secrecy theory.

Darwin's letters and notes demonstrate that many people knew of his belief in evolution. These included his wife, his father, his brother, his children, his cousins Hensleigh and Elizabeth Wedgwood, Julia Wedgwood, as well as

F. M. Wedgwood	J. S. Henslow
E. Dieffenbach	L. Horner
E. Cresy	L. Jenyns

J. Lubbock	R. Owen
H. Falconer	G. R. Waterhouse
E. Forbes	J. D. Hooker
E. Blyth	W. D. Fox
W. Lonsdale	C. Lyell
H. E. Strickland	C. J. F. Bunbury
G. H. K. Thwaites	A. R. Wallace
S. Covington	A. Gray
L. Edmondston	Mr. Fletcher
S. P. Woodward	E. Norman
J. E. Gray	T. H. Huxley
G. Grey	H. C. Watson
T. C. Eyton	J. D. Dana
P. H. Gosse	E. L. Layard
G. Bentham	G. M. Craik
J. Quatrefages de Bréau	M. Butler
T. V. Wollaston	G. Tollet
C. A. Murray	

and probably W. Herbert, H. Cuming and W. Yarrell.[787]

Given the nature of interpreting historical evidence, a few of these identifications could be disputed. Nevertheless, more than fifty people certainly knew first-hand that Darwin believed in evolution before publication. There were, no doubt, others who were told in conversation or in the perhaps half of his letters that no longer survive. As the editors of his correspondence point out, "It is clear from the correspondence that his close friends were not outraged by Darwin's heterodox opinions."[788]

Finally, we have at least three explicit statements by Darwin that he discussed evolution with "very many" people. In the first years of his theorising, Darwin wrote in his *Notebook C* (1838), "State broadly scarcely any novelty in my theory, only slight differences, «the opinion of many people in conversation.» the whole object of the Work is its proof."[789] In the sixth and final edition of the *Origin of species* (1872), Darwin responded to critics who suggested that he exaggerated his originality. He answered by pointing out that before publishing, "I formerly spoke to very many naturalists on the subject of evolution, and never once met with any sympathetic agreement."[790] In his autobiography he recalled, "I occasionally sounded not a few naturalists, and never happened to come across a single one who seemed to doubt about the permanence of species. Even Lyell and Hooker, though they would listen with interest to me, never seemed to agree. I tried once or twice to explain to able men what I meant by Natural Selection, but signally failed."[791] Notice the distinction Darwin made between discussing evolution "occasionally", but only "once or twice" natural selection.

Contrary to the secrecy view, Darwin was quite open about his belief in evolution with his family, friends and colleagues. It does not indicate that he revealed natural selection so openly, however. This, after all, was his unique solution. Lyell was similarly private about the details of his *Principles of geology* before publication.[792] A more appropriate term for Darwin's belief that species evolved is "private".[793]

The Darwin Don Quixotes

In light of this overwhelming evidence, how anyone can still maintain that Darwin kept his belief in evolution secret is inconceivable to me. Nevertheless, Desmond, Moore and Kohn make the Quixotic effort. In their book *Darwin's sacred cause* (2009), Desmond and Moore repeat their views that Darwin postponed and kept his theory strictly "secret" because he was afraid. Their mode of argumentation remains the same, the social and political atmosphere *must* have made it so. The radical press *must* have influenced him because of his tone on human arrogance. This is weak "guilt by association" logic. Although Desmond and Moore never mention my work, in a few passages they address one of my arguments.

> As [Darwin] explained in his letters, he was working on 'the variation & origin of species', or 'the origin of varieties &c species'; and given the contemporary brouhaha over the zoning of life and aboriginal place of creation of each species, a catch-all term like 'origin of species' could easily be interpreted to mean the *place* of each origin (especially as he was writing to such distant outposts). He wasn't giving anything away - he certainly wasn't declaring himself an evolutionist. To study the 'origin' of species in 1855 was permissible in implying the place and time of local creation.[794]

These gymnastic contortions to explain away evidence simply do not work. Darwin's language cannot have been vague concealment. He used the identical language in his own drafts of his theory. In 1841, Darwin told his cousin W. D. Fox, "I continue to collect all kinds of facts, about 'Varieties & Species' for my some-day work to be so entitled."[795] Even as the *Origin of species* was going to press in 1859, Darwin still proposed the title "An Abstract of an Essay on the Origin of Species and Varieties through Natural Selection".[796] Darwin used the same wording "my work on varieties and species" with confidants like J. D. Hooker who knew all about Darwin's theory.[797] It sounds like special pleading to insist that using this language to other correspondents was a way of concealing his true meaning.

Darwin might not have fully explained his views on evolution in many letters. But this is not the same as concealing that he believed in evolution. This was normal language to mean evolution at the time. Wallace used the phrase "origin of species" in an 1847 letter to Bates — and no one doubts that he meant evolution by it.[798] Similarly, when Wallace wrote to Bates that Darwin was preparing a book "on species and varieties", both Wallace and Bates understood this to mean evolution.[799] Wallace also referred to evolution as "the subject of varieties and species" in the Ternate essay itself.

If Darwin was not really revealing what he seems to be saying in so many letters and in so many requests for information and assistance, what did all those people think when the "secret" was revealed with the *Origin of species* in 1859? Darwin stated on the opening page that he had been working on this for twenty years. This would be an admission of dishonesty if he had concealed his views from so many people. Where are the reactions by all those deceived by Darwin for so many years by this extraordinary tactic? As far as we know, there are none. This objection to Darwin's delay seems unanswerable.

The Down village schoolmaster, Mr. Fletcher, was paid £2 to make a fair copy of Darwin's 1844 essay and the copyist Ebenezer Norman made a duplicate of the abstract sent to Gray in 1857.[800] Paying to have his evolutionary theory commercially copied makes no sense if the contents were secret and considered capable of undermining Darwin's respectability and reputation. On the other hand, a secret or at least very private document for Darwin was his autobiography. This was not sent out, but was copied instead at home by his son Francis.

Let us compare this to a document that really was secret. Robert Chambers, in order to keep his authorship of *Vestiges* secret, had the whole manuscript copied out by his wife so that his handwriting could not be recognised. This was then sent from Scotland to a friend in Manchester who then sent it to the publisher in London.[801] If the authorship of *Vestiges* was kept secret because the social and political climate was impossible, as Desmond and Moore insist, then why did its authorship remain a secret long after the *Origin of species* appeared and that climate was no more?

Another way to consider the question of whether or not something was a secret for Darwin is to compare it with unequivocal secrets. How did Darwin treat these? He clearly said so! Writing to Lyell about a conversation with Owen, Darwin wrote, "Please repeat nothing."[802] In a letter to Agassiz reporting early findings on barnacles, Darwin asked, "I should be glad if you would not mention my present results."[803] This was not a secret, but nevertheless a request not to repeat Darwin's not-yet published views. And of course, according to the convention of the day, letters were marked "private" if they were meant to be secret or were not to be published. Darwin sometimes used this practice. But none of his many letters discussing evolution in the gap years is marked private. Darwin's work was in fact so far from a secret that he expressed no surprise or regret when writing to Fox that *Vestiges* "has been by some attributed to me".[804]

Nevertheless, Moore and Kohn continue to claim that Darwin was very afraid and kept his belief in evolution secret. One tactic they employ to give the impression of fear and secrecy is to stress the small number of people Darwin told about natural selection while implying that this is the number of people he told about his belief in evolution. Moore emphasised in 2009 that Darwin told only "three" people, and Kohn claimed in a 2009 documentary that Darwin "hides his theory, he really delays, keeps it virtually secret for many years. I mean he only tells one person [holding up index finger] during a 15 or 16 year period about natural selection."[805]

Stressing how taboo evolution was in Victorian Britain may make for dramatic documentaries and sell more books, but it is not an accurate reflection of historical reality. At any rate, Moore and Kohn are mistaken even about the number who knew about natural selection. At least nine people knew: Emma Darwin, Erasmus Darwin, Hooker, Lyell, Gray, Hensleigh and Elizabeth Wedgwood, Mr. Fletcher and Norman.

Like Confessing a Murder?

Probably the most persuasive evidence that Darwin felt fear is the famous line "it is like confessing a murder". It is now one of the most oft-quoted passages by Darwin and is even the title of several books about him. Many scholars since the 1950s have felt this passage means that Darwin was very afraid. Perhaps scholars in the 20th century were so imbibed with Freudian expectations that they became convinced there must be some buried tension and anguish. Freud's biographer Ernest Jones wrote about "the psychology of discoverers" in 1959:

> [Discovering] the relation of Natural Selection to Evolution…meant displacing God from His position as a detailed Creator specially concerned with mankind…Darwin, the one who stood in such awe of his own father, said it was 'like committing murder [sic]'—as, indeed, it was unconsciously; in fact, parricide. He paid the penalty in a crippling and lifelong neurosis[806]

The "confessing a murder" line comes from an 1844 letter to Hooker.

> I am almost convinced (quite contrary to opinion I started with) that species are not (it is like confessing a murder) immutable. Heaven forfend me from Lamarck nonsense of a 'tendency to progression' 'adaptations from the slow willing of animals' &c,—but the conclusions I am led to are not widely different from his—though the means of change are wholly so—I think I have found out (here's presumption!) the simple way by which species become exquisitely adapted to various ends.[807]

In their biography, Desmond and Moore wrote, "When Darwin did come out of his closet and bare his soul to a friend, he used a telling expression. He said it was 'like confessing a murder.' Nothing captures better the idea of evolution as a social crime in early Victorian Britain." More recently, Bill Bryson repeated this image: "Darwin never ceased being tormented by his ideas. He referred to himself as 'the Devil's Chaplain' and said that revealing the theory felt 'like confessing a murder.' Apart from all else, he knew it deeply pained his beloved and pious wife." [808]

Many readers today will find it hard to imagine that there can be any other interpretation for this passage. Yet it is undeniable that it can be read differently. This quotation has been in print since 1887. I have been unable to find any writer before the 1950s who interpreted this passage as evidence of fear. And this of course includes everyone who actually knew Darwin.

Darwin barely knew Hooker at the time (they had been corresponding for only two months) and was, as so often, humorously melodramatic in telling his correspondent, probably somewhat embarrassed, that he held an unorthodox view.[809] He never asked Hooker to keep the matter confidential. Hence, ironically, "confessing a murder" is quoted out of context, or at least the wrong context, to make it sound like an expression of fear. It is in fact typical humorous language for Darwin. Considered next to the language Darwin used all the time, the correct interpretation becomes obvious.

When work on his books felt overwhelming, he would write "the descent half kills me" or "I am ready to commit suicide".[810] When Hooker was preparing to travel overseas, Darwin wrote, "I will have you tried by a court martial of Botanists & have you shot." Darwin once playfully remarked, "May all your theories succeed, except that on oceanic islands, on which subject I will do battle to the death."[811] Even the word "murder" was typical Darwin hyperbole:

> "If [the plant] dies, I shall feel like a murderer."
>
> "You ought to have seen your mother she looked as if she had committed a murder & told a fib about Sara going back to America with the most innocent face."
>
> "I fear that I shall kill the splendid specimen of Sarracenia, which Hooker sent: it is downright murder, but I cannot help it"

And the best for last: "When I saw your bundle of observations, I felt as if I had committed theft, arson or murder." This remark, over embarrassment that a correspondent had taken too much trouble for Darwin, is much stronger than the famous "confessing a murder" quote about evolution.[812]

What Did People Say?

If Darwin spent twenty years postponing his famous evolutionary theory, one would expect this to have been mentioned by his friends or other contemporaries. Instead, not only is this postponement *never* mentioned, but it was said at the time that Darwin was widely known to have been working on the subject for many years. In 1859, the *Saturday review* referred to "the work upon which [Darwin] was known to have been long engaged".[813] The *Westminster review* noted, "It has long been known to the friends of Mr. Darwin, that he had arrived at a mode of accounting for the diversity of the specific forms of organic life."[814] Lyell, in his *Antiquity of man* (1863), wrote that Darwin had been "patiently" working on the theory for twenty years and "some of the principal results were communicated to me on several occasions".[815]

It cannot be emphasised enough that Darwin himself *never* stated that he postponed or withheld his theory for any reasons, nor did he ever refer to keeping it a secret. Instead on several occasions, he said that he was working on the theory during the years after 1837.[816] Consider this letter to Asa Gray from 1857:

> It is not a little egotistical, but I sh^d. like to tell you, (& I do not *think* I have) how I view my work. Nineteen years (!) ago it occurred to me that whilst otherwise employed on Nat. Hist, I might perhaps do good if I noted any sort of facts bearing on the question of the origin of species; & this I have since been doing.[817]

This, to me, is an explicit contradiction of the entire delay thesis in Darwin's own words.

Not Finished

Darwin did not regard his species theory as finished or complete in 1844. There was a great deal of thinking to be done, many problems to be solved and books to be read in addition to the five or so years of actual composition Darwin envisaged.

One difficulty which Darwin called in 1848 "the greatest *special* difficulty I have met with"[818] and described rather grandly in the *Origin of species* as "one special difficulty, which at first appeared to me insuperable, and actually fatal to my whole theory. I allude to the neuters or sterile females in insect-communities: for these neuters often differ widely in instinct and in structure from both the males and fertile females, and yet, from being sterile, they cannot propagate their kind."[819] Only in November 1854 did Darwin feel he had solved this to his lasting satisfaction with what he called "family" or what is now called kin selection.

He experimented extensively in the 1850s with natural means of dispersal and the individual variations and crossings of pigeons and other domesticated and wild varieties. He also experimented with hivebees, attempting to explain their geometrical constructions with natural selection.

But most of all he worked on the classification and morphology of barnacles. The barnacle studies led Darwin to think less of species originating in geographical isolation, and more towards omnipresent variations ever sifted by natural selection. Nothing but a study like that of the barnacles could have given Darwin so well founded an understanding of natural variation. The experience Darwin gained in taxonomy, morphology and ontogenetic research can hardly be overestimated.[820] Nevertheless, the barnacle work should not be justified in terms of its use for a later project, either in terms of skills gained or reputation earned.[821] To do so would be to ahistorically restrict Darwin to a man interested only in his theory of evolution when in fact, as his wide range of publications demonstrates, he had interests in a huge spectrum of natural phenomena.

One last point. There has been so much stress on the twenty years that elapsed between starting to work on evolution and publishing that many people are convinced that this simply must be the result of postponement. But once again, we need only compare this case with others. A long gestation for books was typical for Darwin. The *Origin of species* was perhaps the *least*, not the longest. His inheritance theory of pangenesis was not published for twenty-seven years. His orchid work c. thirty years. The notes on his baby son were not published until thirty-seven years later; his work on cross-fertilisation after thirty-seven years; and his book on earthworms was published forty-two years later!

Chapter 11

CROSSING BACK

In mid-July 1858, the Dutch steamer *Etna* returned to Dorey where Wallace was still stranded, sick and disappointed. Flies covered his drying bird skins, laying thousands of eggs that soon became maggots. "In no other locality have I ever been troubled with such a plague as this," he lamented. The *Esther Helena* finally returned on the 22nd. She departed taking Wallace and his two surviving boys on 29 July.[822] Wallace left Dorey "without much regret for in few places I have visited have I encountered more disagreeables & annoyances. Continual rain, continual sickness, little or nothing to eat with a plague of ants & flies surpassing everything I have yet met with, required all a naturalists ardour to encounter, & when they were uncompensated by success in his collections, became all the more insupportable."[823]

They returned to Ternate on 15 August 1858. Wallace thought the trip to Dorey was not much of a success: "never have I made a voyage so disagreeable, expensive and unsatisfactory". He was disappointed not to have bagged more of the lucrative Birds of Paradise he had hoped for. He had not researched the area well beforehand and most of the thirteen known species eluded him. But he was delighted to be back in his Ternate house with products such as milk and varied food again available. A torrent of letters and packages had accumulated in his absence. He also had a lot of work to do packing up his collections for shipment to London. Van Duivenbode's third son Constanijn married and Wallace was invited to the ball in the evening. The mail steamer did not arrive at the expected time at the beginning of September. Wallace waited a week for it in vain and then decided to try further collecting on Gilolo.

Gilolo, 14 September–1 October 1858

This time Wallace travelled to the village of Djilolo on Gilolo at the end of a bay to the north of Ternate. Casparus Bosscher, the Resident of Ternate, sent orders ahead to prepare a house for Wallace. He took Ali and Lahagi along. His collecting notes span from 15 September to 1 October.[824] Yet again, the locale was poor due to deforestation so he moved to the village of Sahoe (Sahu) about twelve miles to the northwest. This was also not very productive collecting ground. He returned to Ternate with 400 insects.

Ternate, 2–9 October 1858

Back on Ternate, the mail steamer arrived rather early at the start of October. As usual, Wallace had a pile of new newspapers, magazines and letters to read through. Probably in response to a letter from Stevens, on 5 October Wallace drafted a short note "Direction for Collecting in the Tropics by A.R. Wallace". The front of the note says "for *Mr. Foxcroft*" but the outside has "for Mr H. Squires".

James Foxcroft (dead by 1861) was a commercial entomological supplier and collector. He advertised for subscribers to invest in his collecting expeditions for which they received a share of the resulting collection.[825] Foxcroft set out for Sierre Leone in 1858 and Wallace was a subscriber.[826] Foxcroft exhibited insects from Sierre Leone at the Entomological Society in December 1859.[827] Henry Squire (also dead by 1861) was another insect collector. He collected in England, Scotland and Rio de Janeiro. Presumably, Wallace subscribed to Squire also. These were additional financial investments for Wallace. The pale blue English paper on which the note is written is watermarked "Towgood 1855".[828] It is likely that the Ternate essay was written on the same paper, despite Wallace's oft-quoted c. 1902 recollection of writing it on "thin foreign note paper".[829]

The following day, Wallace wrote at least two more letters. One was to his mother informing her that he was well, recovered from the New Guinea expedition and about to set off for Batchian. But right away he informed her of his exciting news received in the latest batch of letters.

> I have received letters from Mr. Darwin & Dr. Hooker two of the ~~greatest~~ most eminent Naturalists in England which has highly gratified me I sent Mr Darwin an essay on a subject in which he is now writing a great work He shewed it to Dr. Hooker & ~~Mr Darwin~~ Sir

C. Lyell, who thought so highly of it that they immediately read it before the 'Linnean Society' This insures me the acquaintance and assistance of these eminent men on my return home.[830]

The other good news was financial. "Mr. Stevens also tells me of the great success of the Aru collection, of which £1,000 worth has actually been sold. This makes me hope I may soon realise enough to live upon and carry out my long cherished plans of a country life in old England." It was the stereotypical middle-class dream.

On the same day, Wallace answered Hooker's letter from July.

Ternate, Moluccas, Oct. 6. 1858.

My dear Sir

I beg leave to acknowledge the receipt of your letter of July last, sent me by Mr. Darwin, & informing me of the steps you had taken with reference to a paper I had communicated to that gentleman. Allow me in the first place sincerely to thank yourself & Sir Charles Lyell for your kind offices on this occasion, & to assure you of the gratification afforded me both by the course you have pursued, & the favourable opinions of my essay which you have so kindly expressed. I cannot but consider myself a favoured party in this matter, because it has hitherto been too much the practice in cases of this sort to impute *all* the merit to the first discoverer of a new fact or a new theory, & little or none to any other party who may, quite independently, have arrived at the same result a few years or a few hours later.

I also look upon it as a most fortunate circumstance that I had a short time ago commenced a correspondence with Mr. Darwin on the subject of "Varieties", since it has led to the earlier publication of a portion of his researches & has secured to him a claim to priority which an independent publication either by myself or some other party might have injuriously effected;— for it is evident that the time has now arrived when these & similar views will be promulgated & must be fairly discussed.

It would have caused me much pain & regret had Mr. Darwin's excess of generosity led him to make public my paper unaccompanied by his own much earlier & I doubt not much more complete views on the same subject, & I must again thank you for the course you have adopted, which while strictly just to both parties, is so favourable to myself.

Being on the eve of a fresh journey I can now add no more than to thank you for your kind advice as to a speedy return to England;— but I dare say you well know & feel, that to induce a Naturalist to quit his researches at their most interesting point requires some more cogent argument than the prospective loss of health.

I remain

My dear Sir

Yours very sincerely

Alfred R. Wallace[831]

The style of this letter has been noticed before. Shermer seemed to think it typical language for the time.[832] In fact, it was the most obsequious and deferent letter Wallace ever wrote. Together with the letter to his mother, there could be no clearer evidence as to how Wallace felt about the public reading of his essay when he first learnt of it. In a letter to Silk the following November, Wallace confessed, "As I know neither of them [Lyell and Hooker] I am a *little* proud."[833] The great Lyell did not deign to send Wallace a reply.

This letter to Hooker is often quoted, but what has been overlooked is that Hooker's letter (now lost) apparently informed Wallace of the reading of his essay but not its imminent publication. When Wallace wrote to Stevens at the end of October (see p. 265), Wallace was unsure if his essay was to be published or not.

Darwin's letter is also lost, but Wallace copied something extraordinary from it into a blank page of *Notebook 4* between old beetle wings. Darwin shared the table of contents of his species book in progress. In doing so, he left a snapshot of his thinking that has survived nowhere else.

Sketch of Mr. Darwin's "Natural Selection"

Chap. I. On variation of animals & plants under domestication, treated generally.

" II. do. do. treated specifically, external & internal structure of Pigeons & history of changes in them.

" III. On intercrossing, principally founded on original observations on plants

" IV. Varieties under Nature.

" V. Struggle for existence, malthusian doctrine, rate of increase,—checks to increase &c.

" VI. "Natural Selection" manner of its working

" VII. Laws of variation. Use & disuse reversion to ancestral type &c. &c.
" VIII. Difficulties in theory. Gradation of Characters
" IX. Hybridity.
" X. Instinct.
" XI. Palæontology & Geology.
" XII. and XIII. Geog. distribution.
" XIV. Classification, Affinities, Embryology.[834]

Darwin had written as far as Chapter X on Instinct when Wallace's Ternate essay arrived. As he had in March, Wallace deposited these letters at the Ternate post office before leaving on his next expedition.

Tidore, 9–10 October 1858

Wallace hired a boat to take him and his servants south to the larger island of Batchian. His team included "my Bornean lad Ali, who was now very useful to me; Lahagi, a native of Ternate, a very good steady man, and a fair shooter, who had been with me to New Guinea; Lahi, a native of Gilolo, who could speak Malay, as woodcutter and general assistant; and Garo, a boy who was to act as cook" and "Latchi, as pilot. He was a Papuan slave, a tall, strong black fellow, but very civil and careful. The boat I had hired from a Chinaman named Lau Keng Tong, for five guilders a month."[835]

They sailed south and stopped that evening on a beach of Tidore for dinner at sunset before pushing off again. There followed an extraordinary sight.

> The sun set behind the rugged volcanic hills which rose above us & soon after venus was shining in the twilight with the brilliancy of a new moon & casting a very distinct shadow. As we left a little before 7, & got out from the shadow of the mountain I observed a bright light over a portion of the ridge & soon after what seemed a bright fire of remarkable whiteness on the very *summit* of the mountain. I called the attention of my men to it, & they too thought it merely a fire. A few moments afterwards however as we got further off shore the light rose clear up from the ridge of the hill, & some slight clouds clearing away from it discovered a most magnificent comet far more brilliant than any I had hitherto seen.

This was Donati's Comet, which was just at this time nearest the earth and so at its most brilliant. Wallace drew a sketch of the comet in his *Journal*:

Wallace's sketch of Donati's Comet. *Journal 3*:155. Linnean Society of London.

Kaióa, 13–20 October 1858

They sailed on south touching at the islands of March, Motir and Makian before arriving at Kaióa Island (Kayoa) on 13 October. Wallace collected insects as usual while his servants were sent out to shoot birds. He found a beautiful new species of longhorn beetle. As usual, the spectacle of a foreigner so intently working on utterly worthless dead things was a great puzzle to the local people.

On the 16th, Wallace went for a long walk and came upon a recent clearing in the forest. The abundance of fallen trees and branches made for excellent insect collecting. "It was a glorious spot, and one which will always live in my memory as exhibiting the insect-life of the tropics in unexampled luxuriance."[836] In all he bagged about 100 species of beetles.[837] With such a good sample of the insect life of the island, it was time to continue towards Batchian.

Batchian, 21 October 1858–13 April 1859

On the evening of 21 October 1858, Wallace and his team arrived on the north-west coast of the mountainous island of Batchian (Bacan), the largest of the group strung along the western side of Gilolo.[838] Batchian was another sultanate, but essentially belonged to the Dutch who took control of it from the Spanish in 1610. Wallace brought a letter of introduction to the Sultan.

Travelling in his own boat was good experience "which enabled me afterwards to undertake much longer voyages of the same kind".[839] The village where they arrived was "a place where there are some soldiers, a doctor and engineer who speak English".[840] Wallace set up in "a house outside of the

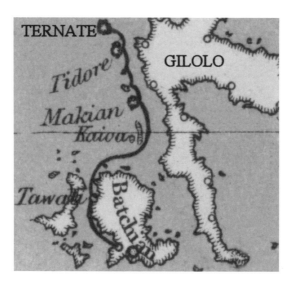

Ternate to Batchian. "Mr. Wallace's route" map MA1.

village on the road to the coal mines".[841] Once the house was in order, the collecting began on the 24th.[842] On their first day of collecting, Ali made a spectacular discovery.

> Just as I got home I overtook Ali returning from shooting with some birds hanging from his belt. He seemed much pleased, and said, "Look here, sir, what a curious bird," holding out what at first completely puzzled me. I saw a bird with a mass of splendid green feathers on its breast, elongated into two glittering tufts; but, what I could not understand was a pair of long white feathers, which stuck straight out from each shoulder. Ali assured me that the bird stuck them out this way itself, when fluttering its wings, and that they had remained so without his touching them. I now saw that I had got a great prize, no less than a completely new form of the Bird of Paradise, differing most remarkably from every other known bird.[843]

Five days later, Wallace wrote a letter to Stevens in which he could not resist sending the news, the excitement resulted in probably the largest number of exclamation points Wallace ever squeezed into a letter.

> Birds are as yet very scarce but I still hope to get a fine collection though I believe I have already the finest & most wonderful bird in the island. I had a good mind to keep it secret but I cannot resist telling you. I have got here a new Bird of Paradise!! of a new genus!!!

quite unlike any thing yet known, very curious & very handsome!!! When I can get a couple of pairs I will send them overland to see what a new Bird of Paradise will really fetch. I expect £25 each! Had I seen the bird in Ternate, I would never have believed it came from here, so far out of the hitherto supposed region of the Paradiseidae. I consider it the greatest discovery I have yet made & it gives me hopes of getting other species in Gilolo & Ceram. There is a species of monkey also here much further east than in any other island so you see this is a most curious locality combining forms of the East & West of the Archipelago yet with species peculiar to itself....The Dutch are working the coals & there is a good road to the mines which gives me easy access to the interior forests.[844]

George Gray, the ornithological curator at the British Museum, named the new bird Wallace's Standardwing because of its outstretched white feathers like banners or standards [see colour insert].[845] It was later displayed at the British Museum in "two nicely-prepared square glass cases, in which a complete series of each of the Paradise-birds obtained by Mr. Wallace is beautifully mounted and arranged".[846] Ali's role as the actual observer and collector was, as usual for the time, almost completely lost.[847]

In the same letter to Stevens, Wallace mentioned the reading of his Ternate essay:

An Essay on varieties which I sent to Mr. Darwin has been ~~presented~~ read to the Linnaean Soc. by Dr Hooker & Mr. C. Lyell on account of an extraordinary coincidence with some views of Mr. Darwin, long written but not yet published, & which were also read at the same meeting. If these are published I dare say Mr. Kippist will let you have a dozen copies for me. If so send me 3, & of the remainder send one to *Bates, Spruce*, & any other of my friends who may be interested in the matter.[848]

As Peter Raby pointed out, Wallace "wrote, much as he might have done with a paper on the Aru Islanders or the orang-utan - and this was request number three in his letter, following a complaint about the length of the No. 14 pins Stevens had shipped out: 'perfectly useless'".[849]

The rains began at the end of November. Wallace was often housebound. He wrote to Silk, "Malays and Papuans, beetles and birds, are what now occupy my thoughts, mixed with financial calculations and hopes for a happy future in old England, where I may live in solitude and seclusion, except from a few choice friends."[850] As Wallace signed the letter, a big spider fell on the paper and he smudged the signature.

Wallace's Standardwing. MA2:41.

Wallace enclosed a 350-word draft article "Note on the smoke nuisance" for the *Athenaeum*. It was never published. A gold medal had been offered by the Society of Arts for the best essay on the prevention of the smoke nuisance, then a hot topic of debate. The speed of steamships allowed Wallace to comment on issues in London almost as if he was there, if a little delayed. This note was a forerunner of the sort of social reform issues that would interest him increasingly in later life. Or maybe it was the temptation for a gold medal.

On 19 December, he recorded in his collecting notes: "Walk to mines. Mr. Huguenin's books", a name never mentioned in his published writings. Otto Fredrik Ulrich Jacobus Huguenin (1827–1871) was a mining engineer who joined the Dutch East Indies government service in 1850 and by 1859 was head of the Mining Service and oversaw the mine work on Batchian. He was also a naturalist and seems the likely source for the next book noted in Wallace's *Notebook 4*, Louis Agassiz's *Lake Superior* (1850).[851] It is the second most heavily noted work in Wallace's notebooks after Lyell's *Principles*, but the notes reveal little if any new theoretical views.

On 20 December, Wallace moved to the village. Again he thought he saw Portuguese everywhere in local people when there was none. During his stay, Wallace acquired one of the most remarkable insects he ever collected — a huge black bee, the biggest in the world — fully an inch and a half long! It was named *Megachile pluto*, and is still called Wallace's Giant Bee. The entomologist Frederick Smith called it "the giant of the genus to which it belongs, and is the grandest addition which Mr. Wallace has made to our knowledge of the family Apidae".[852] Wallace apparently did not realise it was a bee since he described it as "a large black wasp-like insect, with immense jaws like a stag-beetle". But it was one of the "treasures" that "sweetened my residence in this little-known island".[853] Amazingly, the bee was not sighted again for over 100 years. It was believed extinct until it was found in 1981 building nests inside living termite mounds.[854]

Ever since his first walk on Batchian, Wallace had occasionally seen a very large "'bird-winged butterfly,' the pride of the Eastern tropics" but they eluded

Wallace's Giant Bee (now *Chalicodoma pluto*). Smith, 1860, pl. 1.

his net. Finally, on 6 January 1859, he staked out a bush where he once found a female. At last he captured a brightly coloured male butterfly. He recorded in his notebook "Took 1st ♂ of grand *Ornithoptera* !!!" In *The Malay Archipelago*, he described this capture in one of the most breathtaking passages in his book:

> [It is] one of the most gorgeously coloured butterflies in the world. Fine specimens of the male are more than seven inches across the wings, which are velvety black and fiery orange, the latter colour replacing the green of the allied species. The beauty and brilliancy of this insect are indescribable, and none but a naturalist can understand the intense excitement I experienced when I at length captured it. On taking it out of my net and opening the glorious wings, my heart began to beat violently, the blood rushed to my head, and I felt much more like fainting than I have done when in apprehension of immediate death. I had a headache the rest of the day, so great was the excitement produced by what will appear to most people a very inadequate cause.[855]

But he did not neglect the little things. One day, he saw a swarm of small black ants on the leaves of a tall grass with clusters of minute aphids. Wallace carefully cut off a section of leaf. One ant remained with some aphids. Wallace carried the leaf back to his house where he could observe the tiny creatures under a lens. He sat quietly for a long time observing the ant running back and forth to the aphids, stroking them with its antennae. Occasionally, an aphid would excrete a tiny droplet of honeydew which the ant would eagerly collect and drink. Wallace recorded the details of this little performance in his notebook.[856]

On 8 January 1859, Wallace turned thirty-six. On the 30th, Ali returned alone to Ternate to ship a consignment with letters to Stevens via the Ternate steamer. The reason for the hurry was the eight specimens of the new Bird of Paradise which Wallace was eager to share, and to discover how much they were worth. The box reached London on 24 May. Wallace later scribbled in the margin of his notebook "£88" — a very profitable discovery.[857]

Perhaps finding a new Bird of Paradise so far from their previously known range prompted Wallace to think again about geographical distribution. Even before travelling to the island, Wallace was aware that Batchian was "the most easterly spot on the globe on which any monkey is found in a wild state [the crested macaque (*Macaca nigra*)]".[858]

Wallace now wrote a letter to Sclater mentioning, "Your paper on 'The Geographical Distribution of Birds' has particularly interested me…With your division of the earth into six grand zoological provinces I perfectly agree."[859] Wallace later used Sclater's scheme in *Geographical distribution of animals* (1876). Wallace believed that the sort of geological forces of crustal subsidence and uplift expounded by Lyell, Darwin and many others could explain the puzzles of geographical distribution in the Eastern Archipelago. "Here then is the key to the problem:—Sumatra, Java, Borneo, and the Philippines are parts of Asia broken up at no distant period" whereas "Celebes, Timor, the Moluccas, New Guinea, and Australia are remnants of a vast Pacific continent in part marked out by coral islands (see Darwin)". Yet Darwin did not agree in the case of oceanic islands. He saw them not as isolated remnants of sunken continents, but often as fragments of new land never connected together. These were populated by "accidental" natural dispersal via wind, sea, or birds.[860]

Near the end of March, Wallace took a boat to Kasserota Island to the north and visited a place he called Langundi. After ten days, he went south and visited an unnamed area of Batchian. On 6 April, he returned to Batchian village. A government boat arrived bringing rice and troops. The boat was a four-tonne twin-outrigger kora kora with twenty rowers. Wallace with his men and baggage took a passage back to Ternate departing on the 13th. A couple of days later, they anchored at Makian in the afternoon and evening as they made their way north, too leisurely for Wallace. As he settled down in his part of the cramped cabin that night, he blew out his candle and reached for his handkerchief. His hand unexpectedly touched something cold and smooth. It moved! He instantly jerked his hand away.

> "Bring the light, quick," I cried; "here's a snake." And there he was, sure enough, nicely coiled up, with his head just raised to inquire who had disturbed him. It was now necessary to catch or kill him neatly, or he would escape among the piles of miscellaneous luggage, and we should hardly sleep comfortably....On examination, I found he had large poison fangs, and it is a wonder he did not bite me when I first touched him. Thinking it very unlikely that two snakes had got on board at the same time, I turned in and went to sleep; but having all the time a vague dreamy idea that I might put my hand on another one, I lay wonderfully still, not turning over once all night, quite the reverse of my usual habits.[861]

Ternate, 20 April–1 May 1859

The kora kora arrived at Ternate on 20 April 1859. Five days later, Wallace penned a letter to Thomas Sims mentioning the plan that had emerged for the voyage. "I am engaged in a wider and more general study—that of the relations of animals to space and time, or, in other words, their geographical and geological distribution and its causes. I have set myself to work out this problem in the Indo-Australian Archipelago, and I must visit and explore the largest number of islands possible, and collect materials from the greatest number of localities, in order to arrive at any definite results."[862] A letter from Darwin was waiting — in answer to Wallace hearing the news of the Linnean Society reading.

Down Bromley Kent

Jan. 25th

My dear Sir

I was extremely much pleased at receiving three days ago your letter to me & that to Dr. Hooker. Permit me to say how heartily I admire the spirit in which they are written. Though I had absolutely nothing whatever to do in leading Lyell & Hooker to what they thought a fair course of action, yet I naturally could not but feel anxious to hear what your impression would be. I owe indirectly much to you & them; for I almost think that Lyell would have proved right & I shd. never have completed my larger work, for I have found my abstract hard enough with my poor health, but now thank God I am in my last chapter, but one. My abstract will make a small vol. of 400 or 500 pages.— Whenever published, I will of course send you a copy, & then you will see what I mean about the part which I believe Selection has played with domestic productions. It is a very different part, as you suppose, from that played by "Natural Selection".—

I sent off, by same address as this note, a copy of Journal of Linn. Soc. & subsequently I have sent some 1/2 dozen copies of the Paper.— I have many other copies at your disposal; & I sent two to your friend Dr. Davies(?) author of works on men's skulls.…

Everyone whom I have seen has thought your paper very well written & interesting. It puts my extracts, (written in 1839 now just 20 years ago!) which I must say in apology were never for an instant intended for publication, in the shade.

You ask about Lyell's frame of mind. I think he is somewhat staggered, but does not give in, & speaks with horror often to me, of what a

thing it would be & what a job it would be for the next Edition of
Principles, if he were "perverted".—But he is most candid & honest
& I think will end by being perverted.—Dr. Hooker has become
almost as heterodox as you or I.—and I look at Hooker as by far the
most capable judge in Europe.—

Most cordially do I wish you health & entire success in all your pur-
suits & God knows if admirable zeal & energy deserve success, most
amply do you deserve it.

I look at my own career as nearly run out: if I can publish my abstract
& perhaps my greater work on same subject, I shall look at my course
as done.

Believe me, my dear Sir

    Yours very sincerely

    C. Darwin.[863]

    In London, Charles Dickens' *A tale of two cities* was starting serial publica-
tion. Wallace left Ternate on the evening of 1 May in a steamer heading south.
Ali seems to have stayed behind in Ternate — he was newly married. After
three and a half days, they reached Amboyna where Wallace called on his friend
Dr. Mohnike and marvelled at his growing beetle collection. Wallace continued
south with the steamer to Banda on 7 May. The Dutch Resident gave him a pass
to visit the island of great Banda the following morning. The steep hills and
valleys were exhausting walking.[864]

Coupang, Timor, 13–27 May 1859

Wallace continued his journey on the steamer for another three and a half days
south to Coupang, on the western end of Timor. He found it "even more burnt
& parched than when I had visited it in 1857". May is the beginning of the dry
season when there is virtually no rainfall on Timor, yellowing the vegetation.

> The Government doctor, Mr. Arndt, kindly offered me a room in his
> house, which as I intended going some distance from the town as
> soon as I could get boat & men, I accepted. We began talking French
> but his hesitation in that language was so great that we soon broke
> into Malay. Captn. Draysdale a scotchman long settled here & mar-
> ried to a Portuguese lady from Delli, was also at home & gave me
> much information about the country to the East & south of whose
> capabilities he thinks very highly.[865]

Around this time, Wallace wrote an important article "On the zoological geography of the Malay Archipelago". He sent the article to his "eminent" contact Darwin who communicated it to the Linnean Society where it was read at a meeting on 3 November 1859. In it Wallace elaborated on his ideas about the division of the archipelago into Asian and Australian regions.

Semao, c. 17–20 May 1859

Around 17 May, Wallace sailed about twenty miles west in a "large dug-out boat with outriggers" to the island of Semao (Semau) and stayed at the village of Oeassa (Ui-Assa) which had remarkable mineral springs that appeared to bubble with soap. The collecting was poor so he decided to return to Coupang. On the return voyage, the wooden trading vessel sprung a leak and they narrowly escaped drowning at sea. Wallace collected birds for a week at Coupang before leaving on the mail steamer on 27 May.

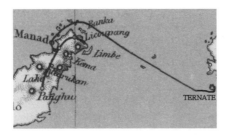

Ternate to Menado. "Mr. Wallace's route" map MA1.

Menado, Celebes, 10 June–23 September 1859

The Cores de Vries steamer took Wallace back north via Banda, Amboyna and Ternate to Menado on the northwestern tip of the long crescent-shaped peninsula of Celebes on 10 June 1859. Again, Ali seems to have stayed in Ternate. Menado was the second principal Dutch town on Celebes, surrounded by a small Dutch territory in the district of Minahassa. The northern peninsula was more agricultural than Macassar, which was primarily a port. A free port since 1849, the main export of Menado was coffee over which the Dutch government maintained a monopoly. Two or three ships of 800 tonnes collected coffee

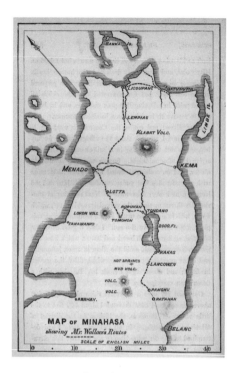

Map of Minahasa. MA1:386.

for shipment back to the Netherlands every year. Wallace described Menado as "by far the prettiest I have yet seen in the east. It had the aspect of a garden containing rows of rustic villas with broad paths between them forming the streets."[866]

He was met by an English businessman and long-time resident named R. Tower who invited Wallace to stay at his house. Tower introduced Wallace to two other local merchants who assisted him during his stay. One was the fifty-year-old Lodewijk D. W. A. van Renesse van Duivenbode, the son (from the first marriage) of Wallace's Ternate friend. Wallace was glad to find that the son had "much taste for natural history". The other gentleman was a Menado-born Dutch merchant named Johannes Wilhelmus Neys who "was educated at Calcutta, and to whom Dutch, English, and Malay were equally mother-tongues".[867]

Twelve days after arriving, Wallace set off into the volcanic mountainous interior of Minahassa. He followed a steadily ascending road up the Tondáno plateau and stayed the night at the village of Lotta where he was joined by the

Dutch Controlleur of the district of Tondáno, Charles Ferdinand Bendsneijder, who offered to act as his guide. As the road climbed through the mountainous region, the villages impressed Wallace with their order and tidiness.

On the 23rd of June 1859, Wallace arrived at the village of Rurukan which he estimated was at 3,500 feet. All around, the volcanic peaks stood sentinel above the coffee plantations. The following day was wet but Wallace collected anyway. He stayed at Rurukan until 1 July.[868] He moved on to the village of Kerubas and the following day to Tondáno at the head of a large lake of the same name. He crossed to the southern side of the lake on the 4th and reached the village of Kakas in a forested area. Without Ali or his usual assistants, Wallace hired local help. At some point, he noted that "both my hunters were sick and left me" and at Langowan, four miles beyond Kakas, he "made arrangements for a man to shoot & for a guide to accompany me to the forest".[869]

Six miles farther south, he stayed for three weeks at the village of Panghu. Here, he set about his usual collecting forays into the surrounding forests. "After some time I found out that by ascending the torrent…Tiger beetles were to be found, & I succeeded greatly beyond my expectations in obtaining several new & most beautiful species."[870]

On 2 August, he left Panghu and travelled a few miles farther south to the small port of Belanc (Belang). By 4 August, he had returned to Menado by sea where he spent a fortnight. He "collected in that neighbourhood and on to the eastern extremity of the peninsula".[871]

On 18 August, he set out again to the north going by Mount Klabat through Lempias where he stayed for ten days. Then on 30 August, he went with "Goldmann the son of the Governor of Amboyna" who organised a hunting trip for "wild pigs, Babirusa, and Sapi-utan [dwarf buffalo]" so Wallace could accompany them to the bay between the islands of Limbé and Banca. Wallace was particularly interested in collecting the island's endemic Maleo or brush-turkey.

In a letter published in *Ibis*, Wallace later noted how the birds buried their very large eggs (about five times as large as a chicken's) in the sand of secluded beaches.

> They come down to the beach, a distance often of ten or fifteen miles, in pairs, and, choosing either a fresh place or an old hole, scratch alternately, throwing up a complete fountain of sand during the operation, which I had the pleasure of observing several times. When a sufficient depth is reached, the female deposits an egg and covers it up with sand, after which the pair return to the forest.…The appearance of the birds when walking on the beach is very handsome. The glossy black

and rosy white of the plumage, the helmeted head and the elevated tail, roofed like that of the common hen, form a tout ensemble quite unique, which their stately and somewhat sedate walk renders still more remarkable.[872]

In this letter, he discussed natural selection in a publication for the first time since the Ternate essay. It was also the first time Wallace used the phrase "natural selection" in print. The feet of the maleo are "slightly webbed at the base, and thus the whole foot and rather long leg are well adapted to scratch away rapidly a loose sand".[873] He mentioned one of his favourite theoretical concerns — whether behaviour precedes and thus determines the structure of animals or vice versa. "For a perfect solution of the problem we must, however, have recourse to Mr. Darwin's principle of 'natural selection,' and need not then despair of arriving at a complete and true 'theory of instinct.'" In his *Journal,* Wallace wrote that it was "a case in which the habits of a bird can be very fairly traced to its organization". This mention of natural selection is also remarkable as the first time Wallace omitted himself and referred to the theory as Darwin's — even before *Origin of species* was published. Wallace would continue what biologist Andrew Berry has called "almost pathological modesty" for the rest of his long life.[874]

Despite its central location in the midst of the Eastern Archipelago, Wallace was unsure whether the species on Celebes were more allied to the Asian or

Babirusa. Wallace, *Australasia*, p. 382.

Australian regions. He later concluded that its many unique species, and a smaller percentage of those shared with surrounding islands, made Celebes a unique region of its own. He attributed this to Celebes having risen out of the sea earlier than other parts of the archipelago. But the African-like appearance of some of its unique creatures, like the "Babirusa or Pig-deer", the "Cyno-pithecus, a genus of Baboons" and the dwarf buffalo, all suggested that a conti-nent might once have existed that connected it to Africa![875] It sounds like an extraordinary proposal, but was quite logical if the species there really were most closely related to those of Africa. It was later found that these were just coinci-dental appearances.

On 23 September 1859, Wallace took a passage on a mail steamer that, unu-sually, was travelling in clockwise direction for Amboyna. The steamer stopped for two days at Ternate where Wallace packed up all the collections that remained in his house there as well as "a nice collection of birds brought by my two boys from E. Gilolo".[876] Wallace arrived back on Amboyna on 29 September. He rented a small house near the beach for a month to pack his collections, write letters and prepare for his next expedition to the large island of Ceram to the north.

Amboyna to Ceram. "Mr. Wallace's route" map MA1.

Ceram, 31 October–28 December 1859

With plans laid and the usual letters of introduction and instructions to provide assistance, Wallace was ready to explore the biological riches of the large island of Ceram to the north. He was offered a passage by a Dutch Nakoda, Mr. van der Beek (misspelled "Beck" by Wallace), who had a tobacco plantation at Hatosua.

They landed on 31 October. Wallace and his servants were soon installed in one of van der Beek's sheds and began collecting on 2 November.

Wallace described van der Beek as "one of the most remarkable men and most entertaining companions I had ever met with" and a remarkable linguist.[877] A year later, some of van der Beek's property was destroyed during an episode of violent social unrest later quelled by the Dutch.[878] Ali remained behind again on Ternate, but another servant was "an Amboyna Christian named Theodorus Matakena".[879] Despite collecting 264 species of beetles in twenty days, Wallace found the area poor so resolved to move on.

On 21 November, Wallace and his team set out in a prau for the village of Elpierputih farther east along the coast. This was a poor collecting area with no visible virgin forest so Wallace continued on to Awaiya on the 25th. Insect bites all over his legs confined him for a time to the house so he wrote letters. He lamented to Stevens, "Ceram is a wretched place for birds. I have been here a month and have got literally not a single pretty or good bird of any kind, except the small Lory I sent before from Amboyna." Wallace also complained about the help. "My three best men have all left me—one sick, another gone home to his sick mother, and the third and best is married in Ternate, and his wife would not let him go: he, however, remains working for me, and is going again to the eastern part of Gilolo."[880]

Ceram is now known to have an unusually large number of endemic birds including parrots, lories, kingfishers and cockatoos. But as he travelled along coastal villages, Wallace barely penetrated into the interior of the island. Nevertheless, he concluded that the island was "very deficient in all forms of animal life".[881] After three weeks, his friend von Rosenberg "who is now the Government Superintendent of all this part of Ceram, returned from the other side (Wahai)" and showed Wallace some butterflies.[882] Following von Rosenberg's advice, Wallace crossed the bay around 17–18 December to Makariki. With two men and six porters, he set out on an inland expedition along von Rosenberg's route through the valley of the River Tanah across a ridge to the "centre of the western half of the island" where they stayed two or three days. Wallace later wrote:

> Never in the whole of my tropical wanderings have I found a luxuri-
> ant forest so utterly barren of almost every form of animal life.
> Though I had three guns out daily, I did not get a single bird worth

having; beetles, too, were totally wanting; and the very few butterflies seen were most difficult to capture. Those who imagine that a tropical forest in the very midst of so rich a region as the Moluccas *must* produce abundance of birds and insects, would have been woefully disillusioned if they could have been with me here.[883]

The wet conditions and hardships of jungle travel confirmed his preference for a fixed residence to make day trips.

Amboyna, 31 December 1859–24 February 1860

Wallace returned to Amboyna and on the 31st arrived at the village of Passo on the narrow isthmus that connects the two great arms of the island. Ida Pfeiffer visited Passo exactly six years previously.[884] The isthmus at that point was so narrow that at high tide, praus could be dragged over and continue their voyage to Ceram without sailing all the way out of Amboyna Harbour and around the island. Wallace was led there by information that there were plentiful common green birdwing butterflies, long-horned beetles, racquet-tailed kingfishers and the yellow-necked lory. He rented a house from the local rajah for a month. As so often, Wallace left a lasting impression. Fifteen years later, the rajah still remembered him.[885]

In London, Darwin's *Origin of species* was published in November 1859. This event is of course now treated as a watershed in history. True, all copies were sold on the first day, but to the bookseller trade, not in the bookshops. Darwin called his theory "descent with modification through natural selection". It followed from the

> Struggle for Existence amongst all organic beings throughout the world, which inevitably follows from their high geometrical powers of increase…As many more individuals of each species are born than can possibly survive; and as, consequently, there is a frequently recurring struggle for existence, it follows that any being, if it vary however slightly in any manner profitable to itself, under the complex and sometimes varying conditions of life, will have a better chance of surviving, and thus be naturally selected. From the strong principle of inheritance, any selected variety will tend to propagate its new and modified form.
>
> Nothing is easier than to admit in words the truth of the universal struggle for life, or more difficult…than constantly to bear this

conclusion in mind....We behold the face of nature bright with gladness,...we do not see, or we forget, that the birds which are idly singing round us mostly live on insects or seeds, and are thus constantly destroying life; or we forget how largely these songsters, or their eggs, or their nestlings, are destroyed by birds and beasts of prey.[886]

The book became hugely controversial. Some thought Darwin was opposing God. Others, like the irascible botanist H. C. Watson, were immediately convinced that Darwin had revolutionised science:

My dear Sir,—Once commenced to read the 'Origin,' I could not rest till I had galloped through the whole....Your leading idea will assuredly become recognized as an established truth in science, i.e. "natural selection".—(It has the characteristics of all great natural truths, clarifying what was obscure, simplifying what was intricate, adding greatly to previous knowledge. You are the greatest Revolutionist in natural history of this century, if not of all centuries.[887]

The debate would rage for more than a decade. But despite some initial ridicule and abuse, the scientific community soon came round to Darwin's main conclusions.

Meanwhile, back on Amboyna, poor Wallace was still covered in insect bites from Ceram that became boils. "I had them on my eye, cheek, armpits, elbows, back, thighs, knees, and ankles, so that I was unable to sit or walk, and had great difficulty in finding a side to lie upon without pain. These continued for some weeks, fresh ones coming out as fast as others got well; but good living and sea baths ultimately cured them." So to Wallace, it was "a not very pleasant memento of my first visit to Ceram".[888]

Thus afflicted, Wallace passed his thirty-seventh birthday. Amongst his post was a complimentary copy of Darwin's *Origin of species*, sent care of Hamilton, Gray and Co. in Singapore.[889] Its perusal must have deeply interested Wallace. He wrote a brief note at the back of an offprint of their Linnean paper:

1860. Feb.

After reading Mr Darwin's admirable work "*On the Origin of Species*", I find that there is absolutely nothing [in the Ternate essay] that is not in almost perfect agreement with that gentlemans facts & opinions.

His work however touches upon & explains in detail many points which I had scarcely thought upon, — as the *laws of variation, correlation of growth, sexual selection, the origin of instincts & of neuter insects, & the true explanation of Embryological affinities*. Many of his

facts & explanations in Geographical distribution are also quite new to me & of the highest interest —
ARWallace .. Amboina.[890]

On 16 February, Wallace wrote a letter to Darwin, now lost. From Darwin's reply, it is clear that Wallace said essentially the same things as he wrote other correspondents at the time. Wallace greatly admired Darwin and the book. Wallace felt he could never have written such a work. He also commented on its strengths and weaknesses, considering "the imperfection of Geolog. Record…the weakest of all".[891]

Wallace wrote to George Silk about the *Origin of species* in September 1860, "It is not one perusal which will enable any man to appreciate it. I have read it through 5 or 6 times, each time with increasing admiration. It is the Principia of Natural history. It will live as long as the Principia of Newton.…Mr. Darwin, has given the world a new science, and his name should, in my opinion, stand above that of every philosopher of ancient or modern times. The force of admiration can no further go ! ! !"[892] In a December 1860 letter to Henry Bates, Wallace expressed himself in similarly panegyric tones:

> I know not how or to whom to express fully my admiration of Darwin's book. To him it would seem flattery, to others self-praise; but I do honestly believe that with however much patience I had worked up and experimented on the subject, I could never have approached the completeness of his book—its vast accumulation of evidence, its overwhelming argument, and its admirable tone and spirit. I really feel thankful that it has not been left to me to give the theory to the public. Mr. Darwin has created a new science and a new philosophy, and I believe that never has such a complete illustration of a new branch of human knowledge been due to the labours and researches of a single man. Never have such vast masses of widely scattered and hitherto utterly disconnected facts been combined into a system, and brought to bear upon the establishment of such a grand and new and simple philosophy![893]

To increase his specimen intake, Wallace hired his old assistant Charles Allen to work for him again as a freelance collector. Allen arrived at Amboyna around the end of January. "He had grown to be a fine young man, over six feet." Allen would be sent "by a Government boat to Wahai on the north coast of Ceram, and thence to the unexplored island of Mysol". Once the preparations for both men, including collecting materials and native servants, were complete, Wallace left Amboyna on 24 February 1860.[894]

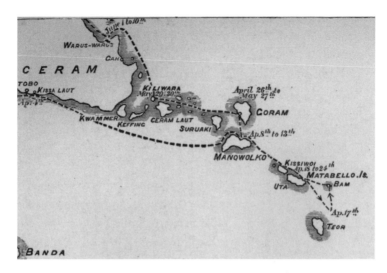

Extract of map of Ceram to Matabello. MA2:95.

Ceram, 26 February–17 June 1860

Wallace arrived at the village of Amahay on 26 February 1860. There followed a series of voyages in borrowed native boats skirting the southern coast of Ceram on his way, he hoped, to revisit Ké. They veered southeast to small islands protruding from the sea as dark silhouettes in the distance, but adverse winds blocked their way and on 17 April they put about. Passing the island of Manowolko on the way back north, Wallace purchased a small prau "for 100 florins (£9)".[895] Settling in at Ondor, the chief village of Goram (Gorong) with its encircling coral reef "visible as a stripe of pale green water", his prau was brought over so that expert boat builders from Ké and other local men could be hired to refit the craft.[896]

Now the skipper of his own vessel, on 27 May Wallace sailed north back to the coast of Ceram. The sea was dangerously rough. The brilliant blue colour turned to a leaden grey. The waves tossed the prau about breaking crockery. Wallace was more than a little alarmed. They anchored off the village of Warus-warus (Waru) on the northeastern corner of Ceram to wait for better weather. At anchor they settled in for the night.

The following morning, Wallace was even more alarmed to find that his entire crew had deserted in the night taking his small boat. Wallace was puzzled why the men would sneak away. "I can impute their running away only to their being totally unaccustomed to the restraint of a European master, and to some undefined dread of my ultimate intentions regarding them."[897] After getting a temporary crew to take him to Wahai, he set out again.

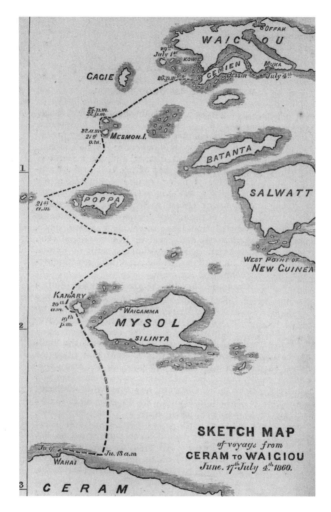

The voyage from Ceram to Waigiou. MA2:332.

Waigiou, 4 July–29 September 1860

At Wahai, Wallace was met by von Rosenberg who lent money to pay off the replacement boat crew. Another crew was eventually procured. There was a letter from Charles Allen on the island of Mysol to the north: "He was out of rice and other necessaries, and was short of insect-pins. He was also ill, and if I did not soon come would return to Wahai."[898]

Wallace and a crew of five sailed on 17 June for the island of Waigiou (Waigeo) in search of the exquisite and lucrative Birds of Paradise. He was ill-prepared with no guide, no experience and no real nautical map. He got in many scrapes, grounded on reefs and even got lost. The winds carried his prau right by Mysol. Poor Allen was left to fend for himself.

At one small island, they anchored to make new rope so "the two best men, a Papuan and a Malay, now swam on shore, each carrying a hatchet, and went into the jungle to seek creepers for rope".[899] But then the anchor dragged loose and the two men were left stranded on the uninhabited island. Unable to turn about to rescue them because of the winds, Wallace continued on to Waigiou.

Meanwhile, back in England, one of the most famous episodes in the wake of *Origin of species* was unfolding. On 30 June, during the Oxford meeting of the British Association for the Advancement of Science, came the now famous encounter between T. H. Huxley and Bishop Samuel Wilberforce or "Soapy Sam" over Darwin's recent work.[900] As countless works have told, supposedly Wilberforce asked Huxley, before a large public audience, if he was descended from an ape on his grandmother or grandfather's side. Huxley, again according to legend, replied that he would rather be descended from an ape than a man who used his talents to bring ridicule into a scientific discussion. Wallace, lost in the vastness of the Eastern Archipelago, would learn of these events only later.

Waigiou is located only about sixty-five kilometres west of New Guinea. They settled at the village of Muka on the southern coast on 4 July.[901] It "consists of a number of poor huts, partly in the water and partly on shore, and scattered irregularly over a space of about half a mile in a shallow bay. Around it are a few cultivated patches, and a good deal of second-growth woody vegetation; while behind, at the distance of about half a mile, rises the virgin forest, through which are a few paths to some houses and plantations a mile or two inland."[902] He thought the local people a mixed race of Malay and Papuan. In three days, he had built a hut for himself and established his specimen production line. From the nearby forest, Wallace could hear the tantalising cries of Birds of Paradise. His first attempt to send a boat for his lost crewmen failed. Wallace did not interrupt his pursuit of Birds of Paradise to rescue them himself. He sent a second rescue party which finally returned his men on 29 July. They were thin but well and grateful for their rescue.

Wallace kept his gun by him though the Birds of Paradise were wary and the trees often too high or the jungle too thick to shoot one. Finally he got one. It was a stunningly beautiful bird with an emerald green face, bright yellow neck and shoulders and tail feathers a sumptuous deep crimson between long black ribbons that curved gracefully outwards from either side [see colour insert]. Despite spending a further month in search of these birds, Wallace only managed to shoot one more male.

Bessir, 7 August–25 September 1860

Wallace decided to travel to the village of Bessir (Besir) on Gam Island, just south of Waigiou where he heard Birds of Paradise were caught by Papuans. He rented a small outrigger and arrived at his new collecting station on 7 August 1860. Wallace procured a tiny hut.

> It was quite a dwarf's house, just eight feet square, raised on posts so that the floor was four and a half feet above the ground, and the highest part of the ridge only five feet above the floor. As I am six feet and an inch in my stockings, I looked at this with some dismay; but finding that the other houses were much further from water, were dreadfully dirty, and were crowded with people, I at once accepted the little one, and determined to make the best of it.[903]

By creeping underneath, he could just sit in his wicker chair and work at a table with his head almost touching the floorboards. He eventually got used to bending down to get into his snug workplace under the hut "to and from which I had to creep in a semi-horizontal position a dozen times a day; and, after a few severe knocks on the head by suddenly rising from my chair, learnt to accommodate myself to circumstances".[904] Here, Wallace lived for six weeks. When not working, he read a recent work in French on the history of prostitution.[905] He wrote to Silk of events back home but noted, "I care not a straw and scarcely give a thought as to what may be uppermost in the political world."[906]

Wallace offered payment to locals to bring him Birds of Paradise which several did, though often in a tattered or filthy state. They caught the birds with snares on long poles. He tried keeping some alive in a bamboo cage he made, but they all died within days.

Wallace's house at Bessir. MA2:359. From a sketch by Wallace who is seated in his wicker chair underneath. Note his gun hung across two poles on the right.

There was almost no surplus of food at Bessir and Wallace became malnourished and suffered from illness and fever. By the end of September, he felt he had to return to Muka for his health and before the monsoon season ended. On the 25th, he returned to Muka and after packing up set off again on the 30th.[907] He arrived back at Bessir the following day from where they finally set out for Ternate on 2 October 1860. Wallace had not achieved what he hoped at Waigiou, but of the seventy-three species of birds collected, twelve were new to science and there were twenty-four fine specimens of the Red Bird of Paradise.

Their voyage back was again beset with difficulties caused as much by the inexperience of Wallace and his crew as by unfavourable winds and currents. They lacked an experienced pilot. The voyage should have taken twelve days, but instead lasted an incredible thirty-eight! Much of the time Wallace was seasick: "10 times aground mostly on coral reefs, four anchors lost or broken…my sails eaten up by rats, my small tender lost astern, many times short of food & water & no oil to burn in the compass lamp at night".[908]

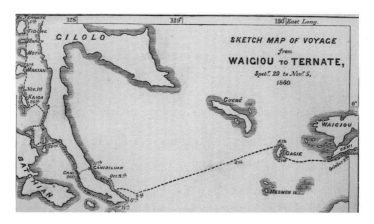

The voyage back to Ternate. MA2:369.

With much difficulty, they beat around the southern point of Gilolo, stopping at Kaioa, Makian, Motir, Mareh and Tidore. Finally on 5 November 1860, they reached Ternate. On the following day, Abraham Lincoln was elected. On 1 December, *Great expectations* started serial publication.

On the 7th, Wallace wrote to Stevens, "My collections are immense, but very poor, when it is considered that they are the result of nine months' collecting by two persons in East and North Ceram, Mysol, and Waigiou. Ceram is a wretched country; and the Papuan Islands, now that the cream is taken off by Aru and Dorey, are really not worth visiting, except for the Birds of Paradise….C. Allen starts in a week or two for N. Guinea."[909] The mail steamers had continued bringing his post during his absence.

> For two months I was stupefied with my year's letters, accounts, papers, magazines, and books, in addition to the manipulation, cleaning, arranging, comparing, and packing for safe transmission to the other side of the world of about 16,000 specimens of insects, birds, and shells. This has been intermingled with the troubles of preparing for new voyages, laying in stores, hiring men, paying or refusing to pay their debts, running after them when they try to run away, going to the town with lists of articles absolutely necessary for the voyage, and finding that none of them could be had for love or money, conceiving impossible substitutes and not being able to get them either,—and all this coming upon me when I am craving repose

from the fatigues and privations of an unusually dangerous and miserable voyage[910]

Wallace lent his copy of *Origin of species* to van Duivenbode who read it three times before returning it, as Wallace told Darwin, "expressing himself so much pleased & interested that he wished to master the whole argument".[911]

Delli, Timor, 7 January–25 April 1861

Having repacked and shipped his consignments to Stevens, Wallace set out on 2 January for Delli, the principal town of Portuguese Timor. He arrived on the steamer on 7 January, the day before his thirty-eighth birthday.[912] No servants were mentioned so he apparently travelled alone. The small town was sparsely shaded with a few scattered palm trees. From the parched hills above the town, strings of pack horses brought coffee beans down to the harbour.

Wallace was received by an old English resident, Alfred Edward Hart, a merchant captain and coffee planter. He invited Wallace to stay with him at his place at Malua about a mile outside the town in the foothills, above the mellifluous influence of the swamps.[913] Wallace also met an easy-going young Cornish mining engineer, Frederick F. Geach, with whom he struck up a friendship. Geach had been hired from England two years before to open up copper mines. The local officials had not conducted a proper mineral survey first.[914] It was a local scandal that Geach found no workable quantities of copper despite the persistent belief that it must be abundant.

Wallace was unwell at first and then the collecting was poor due to the dry vegetation and climate. In the nearby valleys, Wallace collected insects. One day, he observed local men collecting wild honey high in a tree. "I obtained, however, some rare birds and a few very rare and beautiful butterflies by the side of a stream in a little rocky valley shaded by a few fine trees and bushes. Of beetles, however, there were absolutely none worth collecting."[915]

But Wallace enjoyed the pleasant company and conversation of Hart and Geach. Wallace lent his now well-thumbed copy of *Origin of species* to Hart who, as Wallace later wrote to Darwin, "kept it all the time, was constantly reading it & we made it a subject of conversation almost whenever we met, & when I was leaving he did not return it till the steamer arrived going over the recapitulations of the chapters & the conclusion to get the most of it he possibly could".[916] Wallace also wrote a lengthy letter to Thomas Sims about the book.

It is a book in which every page and almost every line has a bearing on the main argument, and it is very difficult to bear in mind such a variety of facts, arguments and indications as are brought forward. It was only on the fifth perusal that I fully appreciated the whole strength of the work, and as I had been long before familiar with the same subjects I cannot but think that persons less familiar with them cannot have any clear idea of the accumulated argument by a single perusal....The evidence for the production of the organic world by the simple laws of inheritance is exactly of the same nature as that for the production of the present surface of the earth—hills and valleys, plains, rocks, strata, volcanoes, and all their fossil remains—by the slow and natural action of natural causes now in operation.

Sims apparently made references to his belief that accepting Christianity was essential for the future of Wallace's soul. Wallace answered Sims' remarks in a lengthy post script (marked not to be seen by Wallace's mother).

I have since wandered among men of many races and many religions. I have studied man, and nature in all its aspects, and I have sought after truth. In my solitude I have pondered much on the incomprehensible subjects of space, eternity, life and death. I think I have fairly heard and fairly weighed the evidence on both sides, and I remain an utter disbeliever in almost all that you consider the most sacred truths.[917]

Early in February, Wallace and his friends made a horseback excursion for a week to a tiny village called Baliba four miles away in the mountains. On 24 February, Geach left Timor for Singapore where he would continue working as a mining engineer. Wallace moved into Geach's former cottage.

Bouru, 4 May–3 July 1861

After almost four months near Delli, Wallace left Timor on 25 April 1861 on the Dutch mail steamer *Macassar* heading back north through the Moluccas. Just two weeks before in the United States, Fort Sumter was attacked marking the advent of the American Civil War. The news would not reach the archipelago for a few weeks.

The *Macassar* touched at Banda on 29 April and on 1 May at Amboyna. On 3 May, she continued north towards Ternate. The following day, she passed Cayeli (Kayeli) Harbour on the large oval island of Bouru (Buru), about the size of the Isle of Wight, and the third largest island in the Moluccas, directly west of Amboyna. It was such a small Dutch outpost that the steamer did not even drop anchor but

Amboyna to Bouru. "Mr. Wallace's route" map MA1.

fired a gun as a signal to the small Dutch fort. The commandant came out in a native boat to fetch the post and Wallace and his baggage of collecting equipment, wicker chair, books, mosquito net, "bed, blankets, pots, kettles and frying pan, plates, dishes and wash-basin, coffeepots and coffee, tea, sugar and butter, salt, pickles, rice, bread and wine, pepper and curry powder, and half a hundred more odds and ends".[918] Wallace was now travelling in some style, and he could afford it. In addition, he had Ali and an unnamed "other hunter" with him.[919]

The head Dutch official was an "Opzeiner, or overseer, a native of Amboyna—Bouru being too poor a place to deserve even an Assistant Resident". Wallace moved into a hut in the village. "Most of the houses being built of wood frame & gaba gaba (the leaf stem of the sago tree) without a particle of white-wash, & the floors of bare black earth not even raised above the outside level, were very gloomy & damp."[920]

It was too soon for collecting as it was still the wet season and the area around Cayeli was poor at any rate. On the 10th, he sent Ali to a place called Pelah (Pela) to see if it was good collecting ground. Ali reported that it was not. On the 11th, the Rajah of Cayeli took Wallace in a boat with eight rowers up the river to Wayapo. At this point, Wallace's *Journal* ends. If there was another notebook for his remaining travels, it has been lost.

Wayapo was a very muddy and poor place too and on the 13th, he returned to Cayeli. On the 19th, he went south to a place called Waypoti where he moved into "a low hut with a very rotten roof, showing the sky through in several places".[921] He reflected about his ever-changing abodes and how readily he adapted to new collecting stations.

It sometimes amuses me to observe how, a few days after I have taken possession of it, a native hut seems quite a comfortable home. My house at Waypoti was a bare shed, with a large bamboo platform at one side. At one end of this platform, which was elevated about three feet, I fixed up my mosquito curtain, and partly enclosed it with a large Scotch plaid, making a comfortable little sleeping apartment. I put up a rude table on legs buried in the earthen floor, and had my comfortable rattan-chair for a seat. A line across one corner carried my daily-washed cotton clothing, and on a bamboo shelf was arranged my small stock of crockery and hardware. Boxes were ranged against the thatch walls, and hanging shelves, to preserve my collections from ants while drying, were suspended both without and within the house. On my table lay books, penknives, scissors, pliers, and pins, with insect and bird labels, all of which were unsolved mysteries to the native mind.[922]

The collecting was still not very good so he hired two men to clear a patch of forest which was finished on the 24th. The clearing produced some nice long-horned and jewel beetles.

Unfortunately, Wallace had "stupidly left my only pair of strong boots on board the steamer" and his shoes were now falling apart. He had to go barefoot which could lead to "a wound which might lay me up for weeks".[923] On 26 May, Wallace sent a letter to the commandant requesting some shoes. On the 29th, his men returned with nothing. On 1 June, he saw the steamer go past in the distance on her way to Ternate. On 4 June, he recorded twice "Sent letter to Comm. for shoes" and "Sent for shoes to Cayeli". Alas for Wallace, on 6 June he scribbled furiously, "Men returned with letter, *no shoes*!!!" Yet again on 10 June, his patience nearly at an end, Wallace noted: "Sent letter for shoes & letters".[924] The following day Charles Allen joined him, staying until the 21st. On the 12th, Wallace, now understandably exasperated, "Sent men again for letters & shoes". Presumably, this last attempt was successful and the notebook falls silent on the subject of shoes.

Mimicry

Wallace discovered two birds, a honeysucker and an oriole "which I constantly mistook for each other, and which yet belonged to two distinct and somewhat distant families".[925] In *The Malay Archipelago,* he assigned this near identical appearance of unrelated birds to the "mimicry" first proposed by his friend Henry Bates in Amazonian butterflies.

In 1861, Bates argued convincingly that a series of quite unrelated butterflies had come to closely resemble each other in appearance — and that natural selection was the only way to explain it. Some butterflies were not eaten by birds because they had a foul taste or odour. Some unrelated butterflies had come to remarkably resemble them by "numerous small steps of natural variation and selection".[926] Those butterflies which resembled the foul-tasting ones somewhat were often left alone by birds. Over time, this process of selection saw those that "mimicked" the foul-tasting butterflies become more numerous, while their less "mimic"-coloured relatives were destroyed by birds.

Wallace was not the only one to discover new birds on Bouru. Ali also bagged a new bird. Wallace recorded it emphatically in his notebook as "1 Pitta!!!".[927] He consulted his note-encrusted copy of Bonaparte's *Conspectus* under the family Pittidae and concluded it was a new species. It was the Red-bellied Pitta (*Pitta erythrogaster*) [see colour insert], a short-tailed perching bird with a brown head, midnight blue wings and a crimson breast.

After "securing this prize", they returned to Cayeli, packed and left on the steamer *Ambon* as she passed on 2 July 1861. Wallace had sixty-six species of birds, including, he thought, seventeen new species. They stayed for two days at Ternate packing up. The *Ambon* departed again on 7 July and Wallace left the place of his famous discovery, and the Spice Islands, forever. The *Ambon* touched at Menado and Macassar where the fateful tiger beetles had so intrigued him.

Surabaya, Java, 16 July–15 September 1861

Wallace and his servants arrived at the Dutch port of Surabaya on Java on 16 July 1861.[928] Java is the third largest island in the archipelago and was the centre of Dutch population, wealth and power since the 17th century. It was also the major rice producer of the region. It was the fluctuation of Javanese rice prices that had helped spark the Chinese riots in Singapore in 1854. The population of Java was about eleven million. Surabaya, with its large enclosed harbour, was the principal town and port of East Java with a population of about 90,000. Wallace called Java "the Garden of the East, and probably without any exception the finest island in the world".

Once installed in a hotel, possibly Hotel der Nederlanden, Wallace spent two weeks packing up and sending off his collections which were apparently shipped on 20 July and 5 August. Staying in a hotel in a large city was now a very

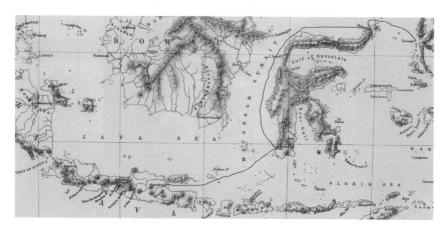

Ternate to Surabaya, Java. Base map from Bickmore, 1869.

unusual mode of living for the wild Wallace of the Eastern Archipelago. He wrote his mother that he was "a little confused arriving in a new place with a great deal to do & living in a noisy hotel".[929]

By around 1 August, he had cleared the decks and set out with Ali into the interior of Java. They headed in the direction of the great volcano Arjuno, 3,339 metres high, over forty miles into the interior. They stopped first at the village of Modjokerto.[930] Wallace had a letter of introduction to an English merchant named Mr. Ball (possibly Henry Ball) who invited Wallace to stay with him. The following day, Ball drove Wallace to Modjo-agong. Along the way, they stopped to examine the ancient ruins of "Modjo-pahit", in particular "two lofty brick masses, apparently the sides of a gateway" consisting of the finest brickwork Wallace had ever seen.[931] This was probably Wringin Lawang, the gate at Trowulan from the 14th-century Hindu Majapahit Empire.

After a week of delay, Wallace settled at Modjo-agong where he was intrigued to observe a gamelan concert. Three days later, a local boy was killed by a tiger. Wallace hoped to secure part of it but the skin was riddled with spear wounds and the skull smashed to remove the teeth which were worn as ornaments.

Wallace spent a month at the villages of Wonosalem and Djapannan. He procured two "magnificent specimens [of Javan peacocks] more than seven feet long" and several other gorgeous birds.[932] "In a month's collecting at Wonosalem and Djapannan I accumulated ninety-eight species of birds, but a most miserable lot of insects."[933] It was time to move on to hopefully richer ground. So on 30 August, Wallace "returned to Sourabaya by water, in a roomy boat which brought myself, servants, and baggage at one-fifth the expense it had cost me to come to Modjo-kerto". On 10 September 1861, Wallace and his servants left Surabaya on the steamer *Batavia*. It was 420 miles to the western end of Java, with a stop halfway at Samarang.

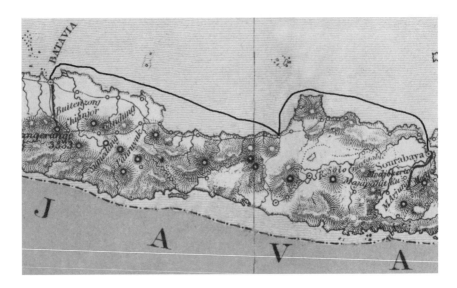

Surabaya to Batavia. "Mr. Wallace's route" map MA1.

Batavia, Java, 18 September–1 November 1861

Wallace was listed in the newspaper *Javabode* as arriving at Batavia, the capital of the Dutch East Indies, on 18 July 1861.[934] As Simon Winchester noted, "The name Batavia had a kind of easy, silky poetry to it."[935] Batavia (after the Roman name for the Netherlands) had been a Dutch base since 1619. The population when Wallace arrived was over 350,000, including 50,000 Chinese and perhaps 5,000 Westerners.

A coach ride two miles up into the tidy suburbs took Wallace to the most luxurious hotel in the East, the famous Hotel des Indes, built in 1829. He found it "very comfortable, each visitor having a sitting-room and bedroom opening on a verandah, where he can take his morning coffee and afternoon tea. In the centre of the quadrangle is a building containing a number of marble baths always ready for use; and there is an excellent *table d'hôte* breakfast at ten, and dinner at six, for all which there is a moderate charge per day."[936]

He met two other Englishmen here, the brothers Walter and Henry Woodbury. The former was co-owner of the Woodbury & Page studio. They made many striking photographs of the people and places of the region. Four of their photographs were later engraved on wood as illustrations for Wallace's *The Malay Archipelago.*[937]

Hotel des Indes in 1910. Tropenmuseum.   "Back view of Gov. Genl's. Palace Buitenzorg" c.1865 by Woodbury.

On 21 September 1861, Wallace and his team travelled "by Diligence" (a coach) to Buitenzorg (Bogor) with its large landscaped botanical gardens and opulent Palladian residence of the Governor-General of the Dutch East Indies. Wallace was not impressed with the ornamental layout of the gardens or the chunky gravel footpaths.

On 24 September, he proceeded on horseback, or walking to enjoy the scenery, with coolies carrying his baggage. The pretty villas and terraced hillsides seen from the road won his admiration. Eventually, reaching the pass over Mount Megamendong, he rented half of the "road-keeper's hut". The following day on the 25th, he commenced work. "One day a boy brought me a butterfly between his fingers, perfectly unhurt....It proved to be the rare and curious Charaxes kadenii, remarkable for having on each hind wing two curved tails like a pair of callipers. It was the only specimen I ever saw, and is still the only representative of its kind in English collections."[938]

In mid-October, Wallace visited places called Tchipannas and Gunung Malam at what is now Mount Gede Pangrango. He reached the highest point he ever visited in the East. "I enjoyed the excursion exceedingly, for it was the first time I had been high enough on a mountain near the Equator to watch the change from a tropical to a temperate flora."[939] He spent 19–25 October at

Pasangrahan resting place for travellers at Nnangteng, Java. c. 1865 Woodbury. Probably similar to the rest houses where Wallace stayed.

Calliper butterfly. MA1:178.

Megamendoeng. He tried another area, "a coffee-plantation some miles to the north", but the collecting was not worth his time and the rains set in so he returned to Batavia on 28 October. Once again, collections were packed for shipment and he prepared to move on. But he was nearing the end of his odyssey.

Sumatra, 8 November–15 January 1862

Having obtained Dutch government permission for his next destinations, Wallace and his servants embarked on the steamer *Prins van Oranje* on 1 November 1861.[940] On board the steamer, he met a German geologist, Baron Ferdinand von Richthofen, the uncle of World War I flying ace the "Red Baron", Manfred von Richthofen. And, remarkably, von Richthofen was reading Darwin's *Origin of species*. Wallace asked if he "was a convert", von Richthofen "smiled & said 'It is very easy for a Geologist.'"[941]

The *Prins van Oranje* steamed 200 miles northwest to the island of Banca (Bangka) lying off the east coast of Sumatra. Banca had been one of the world's richest sources of tin since the 18th century. Here, the steamer stopped at the pretty little village of Muntok on her way to Singapore. Wallace disembarked on 2 November. After two days, he took a passage on "a good-sized open sailing-boat…to the mouth of the Palembang river [Sumatra], where at a fishing village, a rowing-boat was hired to take me up to Palembang, a distance of nearly a hundred miles by water."[942]

The town of Palembang was part of the kingdom of the same name but under Dutch rule. The town was clustered along the river and the houses even

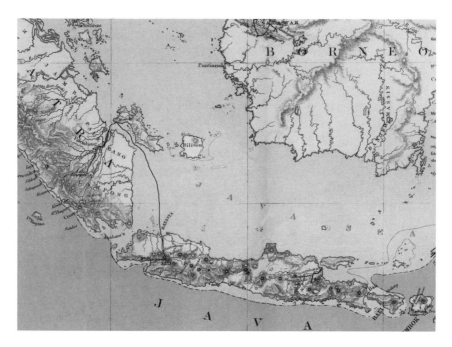

Java to Sumatra. "Mr. Wallace's route" map MA1.

had extensions built on rafts on the river that rose and fell with the tide. These included shops open to the river side so one could go to market in a boat.

Wallace and his men reached their destination on 8 November. They lodged with an unidentified doctor to whom Wallace brought a letter of introduction.

The area around Palembang was not suitable for collecting so Wallace and his team headed along the road south towards Bencoolen, the former British outpost on the opposite coast. On 13 November, they reached Lorok. Here, the collecting was still not profitable so on the 16th, they moved on a farther twenty-five or thirty miles to an area that he was told was still forested. The same disappointment occurred again so they moved deeper into the interior. On the 20th, they reached Moera-dua (Muara Dua), about forty-eight miles southwest of Palembang.

Once again, the area was not profitable so they decamped to Lobo Raman (Lubuk Raman) on 3 December, about sixty miles from Palembang and halfway to Bencoolen. Here, Wallace stayed in one of the post houses the Dutch had built along the road for travellers. The locale was ideal. The "guard-house is situated quite by itself in the forest, nearly a mile from each of three villages". In the branches of the trees over his house, monkeys frolicked and the area still had

"Palembang, at high water." Bickmore, 1869, p. 530.

elephants, tapirs and rhinoceroses. Some time after 22 December, he caught a brief glimpse of a rhinoceros in the forest. He could find no one who had ever seen an orangutan. But there were other apes — lithe little gibbons which swung nimbly through the branches on long arms. Wallace bought one hoping to bring it back to England. It swung about on the verandah on a tether.

Still, the collecting was not that great here either. There were a few birds; the butterflies were better. He found a remarkable butterfly which looked almost exactly like a leaf. "I captured several specimens on the wing, and was able fully to understand the way in which this wonderful resemblance is produced."[943] As he examined their resemblance to a leaf, he understood that natural selection explained how this appearance had gradually come about. Over a vast period of time, countless individual butterflies were born of varying shades and shapes. Those that happened to resemble leaves somewhat were devoured less often by birds. So over time, these butterflies had been honed by natural selection to resemble leaves to an astonishing degree.

The absent revolutionary

On 14 December 1861, Prince Albert died, leaving Queen Victoria in mourning for the rest of her long life. The events now summarised as "the Darwinian revolution" were in full swing. Wallace, still collecting and thereby investing in his financial future, was taking almost no part in the events he had done so much to bring about. He wrote to Silk, "It is an age since you wrote to me last & yet you might have found plenty to write about without touching upon politics. Essays & reviews & 'the Gorilla war'" for example or "have you read 'Great Expectations.'"

Essays and reviews was a recently published collection by seven Church of England divines who summarised and endorsed a series of new findings in science and biblical scholarship that undermined, now from within the church, more conservative interpretations. It caused a bigger sensation than *Origin of species* and sold as much in two years as *Origin* did in twenty. One of the essayists, the Oxford professor of geometry Baden Powell, referred to the recently published *Origin of species* by "a naturalist of the most acknowledged authority" which "substantiates…*the origination of new species by natural causes*: a work which must soon bring about an entire revolution in opinion in favour of the grand principle of the self-evolving powers of nature".[944] Powell was well disposed since he already believed that species changed.

The "gorilla war" referred to the controversy over the claims of the American adventurer Paul du Chaillu who famously brought the first specimens of gorillas to London and made a sensation with wild tales of his adventures hunting them in Africa — much of which was found to have been fabricated. Even his famous frontispiece illustration depicting him about to shoot a raging gorilla walking fully upright towards him, was copied from an earlier French book.

In Sumatra, the heavy tropical rains kept Wallace under a roof for much of the day. As he told Silk, "Bad times for me, but I walk out regularly 3 or 4 hours every day, picking up what I can, and generally getting some little new or rare or beautiful thing to reward me.…Could you see the table at which I am now writing, your hair would stand on end at the reckless confusion it exhibits!"[945]

About this time, Wallace "received information" that there were two live Birds of Paradise for sale in Singapore.[946] How this news reached him in the middle of Sumatra he never said. It is possible that his young friend George Rappa may have brought the news. Wallace had helped Rappa sell some books through Stevens in 1856. Rappa was listed as returning to Singapore with Wallace on the Batavia steamer in January, but Rappa was not listed as leaving Batavia where the steamer began. Wallace had already been offered a free passage back to England by the Zoological Society of London if he could bring back live Birds of Paradise. Hearing of the chance to save so much money, Wallace decided to return to Singapore a month earlier than he had planned.

Wallace and his men took a boat back to Palembang via a tributary of the Palembang River. They stopped at a place called Sungei Rotan on 7 January to repair the boat. While Wallace was breakfasting on his thirty-ninth birthday, Ali and the other hunter came up with a large male Great Hornbill — a bird the size of a swan. Wallace was taken to the tree where the female was holed up inside with a chick. He could see the tip of her long yellow bill protruding from a hole about twenty feet from the ground. The male seals the female inside a hole in a tree with mud, leaving only a small opening for food and air. With great difficulty, Wallace persuaded some local people to climb the tree and bring him the bird and its chick. It was rare to find a hornbill chick. "It was exceedingly plump and soft, and with a semi-transparent skin, so that it looked more like a bag of jelly, with head and feet stuck on, than like a real bird…one of those strange facts in natural history which are 'stranger than fiction.'"[947]

Female hornbill, and young bird. MA1:212.

Wallace returned to Palembang around 9 January and on the 15th, they sailed back out to Muntok to catch the steamer *Macassar* from Batavia to Singapore on the 16th.

"Ho! for England!"

On Saturday 18 January 1862, the *Macassar* arrived at Singapore bearing Wallace, Rappa and Ali.[948] Where Wallace stayed went unrecorded, perhaps at a hotel or with his friends Rappa, Geach or another resident named John Fisher. According to a 1904 obituary, "Wallace was for some time the guest of Mr Fisher in Singapore, and received much substantial assistance from him at a difficult period in his residence in the East".[949]

Charlie Allen

Charles Allen would finish collecting shortly after Wallace's departure for England. Allen later worked in mining in the region and eventually settled in Singapore. He married and raised a large family and became the manager of Fisher's Perseverance estate located in what is now the Geylang district of Singapore. The estate grew and processed lemon grass to make citronella oil which was used in soaps and insect repellent. In 1887, Allen became the owner of the estate.

The poor London boy finally made his fortune in the East. He died in 1892. Allen's daughters married well, one even to the architect of Singapore's famous Raffles Hotel.[950] A 1906 article on Wallace (actually a mistaken obituary!) stated: "The late Mr Charles Allen, whose many friends yet in Singapore will remember, used often to speak of his friend Wallace, of whom Mr Allen's family hold some interesting reminiscences, connected with his residence in Singapore."[951] Sadly, none of these reminiscences were provided, and are probably lost forever. Allen's descendants lived in Singapore until the Japanese invasion in 1942. His great-granddaughter still lives in Fremantle, Western Australia.

Ali

Wallace made his final arrangements to leave the East. His gibbon, a siamang, was a big hit in Singapore where such a creature was unknown. It was sent on a sailing ship via the Cape route, possibly on the German barque *Henry & Oscar* bound for Falmouth on 23 January.[952] But the siamang did not survive the journey.

Wallace had photographs taken of himself with Geach and one of Ali who, for the first time, adopted European dress. Presumably, this had to do with his final payment and gifts from Wallace. "On parting, besides a present in money, I gave him my two double-barrelled guns and whatever ammunition I had, with a lot of surplus stores, tools, and sundries, which made him quite rich. He here, for the first time, adopted European clothes, which did not suit him nearly so well as his native dress, and thus clad a friend took a very good photograph of him."[953] Which of his friends might have been a photographer is unknown. There were commercial photographers in Singapore such as Mr T. Heritage from London whose shop on Queen Street had been open for six months.[954]

Ali later returned to his wife on Ternate. Nothing else is known about his subsequent life or if any of his descendants survive. Remarkably, Ali was recorded one more time in 1907 when the American zoologist Thomas Barbour visited Ternate.

> Here came a real thrill, for I was stopped in the street one day as my wife and I were preparing to climb up to the Crater Lake. With us were Ah Woo with his butterfly net, Indit and Bandoung, our well-trained Javanese collectors, with shotguns, cloth bags, and a vasculum

Geach and Wallace. A version of this photograph, with Geach removed, was published in Marchant, 1916. National Portrait Gallery, London.

for carrying the birds. We were stopped by a wizened old Malay man. I can see him now, with a faded blue fez on his head. He said, "I am Ali Wallace." I knew at once that there stood before me Wallace's faithful companion of many years, the boy who not only helped him collect but nursed him when he was sick. We took his photograph and sent it to Wallace when we got home. He wrote me a delightful letter acknowledging it and reminiscing over the time when Ali had saved his life, nursing him through a terrific attack of malaria. This letter I have managed to lose, to my eternal chagrin.[955]

"My faithful Malay boy—Ali". ML1:382.

Paradise Birds

On 6 February, Wallace paid 400 Singapore dollars to the commission agents Mark Moss and William Waterworth on Serangoon Road and received a receipt for the two "Paradize Birds".[956] That evening, the P&O steamer *Emeu* arrived bringing the Hong Kong mails of 1 February.[957] The next day, Wallace wrote to Sclater, the secretary of the Zoological Society, "They were in the hands of a European merchant who was well aware of their value & asked an exorbitant price. As however they seemed in excellent health, had been in Singapore 3 months & in possession of a Bugis trader a year before that, I determined if possible to obtain them. After protracted negotiations I have purchased them."

Wallace checked with the P&O office to see if they had received word that he and the birds should receive free passage to London. He was "much surprised & disappointed to find that no order for a free passage had been sent out, but merely instructions to *take care* of the birds if sent on board".[958] So Wallace was forced to buy a first-class ticket to Southampton for $552, plus a thirty-three-dollar passage through Egypt. He would later claim the cost of his journey from the Zoological Society. This plus the $400 for the Lesser Birds of Paradise [see colour insert] made a whopping $1,000 for the Zoological Society to bear.[959]

Way Back Home

On 8 February, Wallace's boxes and bird cage were weighed on the wharf. Presumably, Geach, Fisher, Rappa and Ali were there to send him off as he was rowed out in a small sampan to the giant P&O steamer *Emeu* never to return. Wallace left Singapore on the same steamer line on which he had arrived eight years before. He might have had mixed feelings of satisfaction and regret as he watched the red-tile rooftops of Singapore disappear behind the steamer. Many years later, he would look back on this time as "the best part of my life".[960]

The *Emeu* steamed first to Penang where Wallace was given a small collection of insects by James Lamb.[961] The steamer continued southwest for eight days to Galle where Wallace wrote to Stevens.

> I had first to make an arrangement for a place to stand the large cage on deck. A stock of food was required, which consisted chiefly of bananas; but to my surprise I found that they would eat cockroaches greedily, and as these abound on every ship in the tropics, I hoped to be able to obtain a good supply. Every evening I went to the store-room in the fore part of the ship, where I was allowed to brush the cockroaches into a biscuit tin.[962]

The *Emeu* then proceeded to Bombay arriving on 22 February 1862.

> The ship stayed three or four days at Bombay to discharge and take in cargo, coal, etc., and all the passengers went to a hotel, so I brought the birds on shore and stood them in the hotel verandah, where they were a great attraction to visitors. While staying at Bombay a small party of us had the good fortune to visit the celebrated cave-temple of Elephanta on a grand festival day, when it was crowded with thousands of natives—men, women, and children, in ever-changing crowds, kneeling or praying before the images or the altars, making

gifts to the gods or the priests, and outside cooking and eating—a most characteristic and striking scene.[963]

Wallace left Bombay on 25 February on the P&O steamer *Malta*. She arrived at Suez on 12 March. His birds were in good health. He took the new train line which brought him to Cairo in nine hours.[964] The birds now caused him considerable worry: "the night was clear and almost frosty. The railway officials made difficulties, and it was only by representing the rarity and value of the birds that I could have the cage placed in a box-truck."[965]

On 13 March, Wallace left Alexandria on board the *Ellora*. Unfortunately, the P&O ship was so clean Wallace could find no cockroaches to feed the birds. The *Ellora* saw the *Euxine* which had taken Wallace out in 1854 and at 1.40 am on 17 March she entered the harbour of Malta.[966] Wallace alighted to put on a store of insect food for the precious birds from a hotel where "I could get unlimited cockroaches at a bakers close by". The following day, Wallace sent a telegram to Sclater about the birds. The telegraph office apparently had no idea what he was sending because they transcribed his message as "garadisi bards"![967]

Despite recording that "I stayed a fortnight" in Malta, he must have departed on the 25th or 26th of March.[968] The steamer journey to Marseilles took three days. According to recent arrangements, British subjects did not require a passport to travel through France but only "a simple declaration of their nationality".[969] From Marseilles, Wallace took a train via Lyon to Paris. "At Marseilles I again had trouble, but at last succeeded in getting [the birds] placed in a guards van, with permission to enter and feed them *en route*. Passing through France it was a sharp frost, but they did not seem to suffer." The train from Paris (Chantilly) to Boulogne took about six hours. Crossing the channel on a South Eastern Railway Co. ship took only two hours.

So on 31 March 1862, Wallace again stepped ashore on England's green and pleasant land at Folkestone, although he wrote in his *Notebook* 5 that he arrived on 1 April. One last incorrect date from Mr. Wallace! He stayed at the large waterfront Pavilion Hotel.[970] "When we reached London I was glad to transfer [the paradise birds] into the care of Mr. Bartlett, who conveyed them to the Zoological Gardens. Thus ended my Malayan travels."[971]

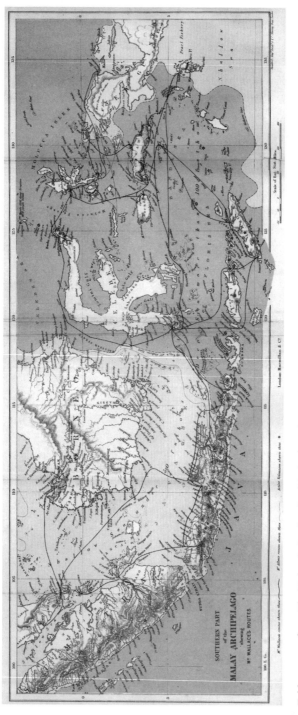

Fold-out map from *The Malay Archipelago* (1869) showing the routes of Wallace and Allen and active volcanoes. Shallow seas are coloured lighter than deeper seas. Engraved by Edward Stanford's Geological Establishment, London.

Chapter 12

COUNTING UP

Despite the romantic appeal of the traditional story that Wallace set out in search of the origin of species, in fact, as he himself wrote, the "main object of all my journeys was to obtain specimens of natural history, both for my private collection and to supply duplicates to museums and amateurs".[972] And in this, as in much else, his eight years in the Eastern Archipelago were "successful, financially, beyond my expectations. Celebes, the Moluccas, the Aru Islands, and New Guinea were, for English museums and private collections, an almost unknown territory. A large proportion of my insects and birds were either wholly new or of extreme rarity in England, and as many of them were of large size and of great beauty, they brought very high prices. My agent had invested the proceeds and a year after my return I found myself in possession of about £300 a year."[973] Of course he was after profit, he had to make his own way in the world.

So his voyage had a happy ending. After all he lived to tell the tale! He had survived depravations, injuries, shipwrecks and worse. He had become a recognised figure in the scientific community whose views were taken seriously. His odyssey completed his journey back to the financial independence of a gentleman. Sadly, it was not to last. As Peter Raby recounts in his biography of Wallace, the pressures of a financially needy family and a series of disastrous investments eventually lost Wallace the capital he had struggled so long to accumulate.

Wallace's other fortune brought back from the Eastern Archipelago was his private collection of natural history specimens. Originally, he had planned a complete catalogue of the beetles of the region, his "'Coleoptera Malayana,' to contain descriptions of the known species of the whole Archipelago, with an essay on their geographical distribution, and an account of the habits of the genera and species from my own observations".[974] But he never wrote his great treatise. Wallace became more interested in other concerns, and so his collection languished and was never utilised for its original purpose.[975] Much of it was eventually sold over the years to help meet rising debts.

Wallace's commercial collections were sold to institutions like the British Museum, the Derby Museum, Liverpool, the Berlin Museum and the University Museum, Oxford. Individual collectors who purchased Wallace specimens included Saunders, Francis Walker, Frederick Smith, Mr. Carlson, J. B. Davis, Edwin Brown, E. W. Janson, F. P. Pascoe and even French collectors Abbe Marseul and Count Minszech. According to Baker, "the British Museum purchased a total of 7,758 insects of all orders, a substantial number but still an insignificant portion, 7%, of the whole". Whereas Wallace's much smaller collection of mammals and reptiles "purchased by the British Museum was, apparently, less than 150".[976]

Almost every writer on Wallace cites his impressive total of 125,660 specimens. He gave a breakdown: 310 mammals, 100 reptiles, 8,050 birds, 7,500 shells, 13,100 moths and butterflies, 83,200 beetles and 13,400 other insects.[977] The collection was remarkable, including the world's biggest bee and rarest cat. P. L. Sclater calculated there were 212 new species of birds. Wallace also bagged 200 new species of ants and an impressive 900 new species of beetles.[978]

Wallace's Help

But it is a mistake, repeated by most writers, to imply that Wallace collected 125,660 specimens himself. He employed full-time collecting assistants throughout his voyage. His records do not allow us to see exactly how many specimens were collected by assistants. Charles Allen alone collected about 40,000.[979] There was (not counting boat crews, porters and cooks) Ali, Fernandez, a shooter in Malacca, Baderoon, Baso and two servants at Maros, Lahagi, Lahi and Jumaat at Dorey, servants at Menado including "Thomas began shooting (25 cent a day)"[980] and Cornelius,[981] two hunters at Lotta, a man at Langowan, two boys with blowpipes at Panghu, two men at Lempias, Petrus Rehatta, Headonus, Mesach and Theodorus Matakena at Amboyna.[982] On Ceram, there was a lad from Awaiya and two hunters, at Bouru another hunter, in Sumatra a hunter and even Geach was employed to add to the collection.[983] Specimens were also purchased from local peoples like the Dyaks in Borneo, the Bird of Paradise hunters on Bessir and dealers and colonials like Mesman and Captain Brooke. Working for Wallace was dangerous too. Ali was struck by fevers, two crewmen were stranded on a deserted island, possibly to die, Baderoon became a slave in Dobbo and Jumaat died at Dorey. Wallace's collecting total was so high not because he was a superhuman collector, but because he paid a small army of assistants to maximise specimen production.

Wallace was far from the only naturalist in the archipelago, though the others are seldom mentioned. He is much to blame for this tradition since he often claimed it was a region "which hardly any naturalist had then properly explored".[984] Ida Pfeiffer's travels may have led Wallace to travel to the archipelago in the first place. Then there were other naturalists who studied the region such as von Rosenberg, Mohnike, Doleschall, Motley, Zollinger, Huguenin, von Richthofen, Bernstein, Cantor and Collingwood. Not to mention the Dutch Scientific Commission and scores of lesser-known figures. Wallace soon had serious competitors on his heels such as the American naturalist Albert Bickmore and the French entomologist P. J. M. Lorquin.[985] Many others followed. So to fully restore Wallace to his original context reveals him to be not a lone discoverer or "heroic pioneer" in the scientific exploration of Southeast Asia, but one of a community of investigators, who has only subsequently become so much more famous as to outshine, and render invisible, the others. It is rather ironic, since Wallace has been called "Darwin's moon", that Wallace has so many moons.

Meeting Eminent Men

Back in London, Wallace began to attend scientific meetings. He was no longer a nobody, but "the great naturalist traveller".[986] Still somewhat shy, he never fully integrated into the elite scientific community. He remained an active, if somewhat peripheral, figure in the world of Victorian science. At one of these scientific meetings, Wallace finally met his hero, Charles Lyell. In the summer of 1863, Wallace had lunch at the Lyell's house. Lady Lyell was not impressed with the gauche Wallace. He was "shy, awkward, and quite unused to good society". The status-conscious Lyell was apparently not particularly impressed either. One day that November, Lyell bumped into Henry Bates at the London Zoo. Lyell "was wriggling about in his usual way, with spy-glass raised by fits and starts to the eye, and began; 'Mr Wallace, I believe—ah—' 'My name's Bates.' 'Oh, I beg pardon, I always confound you two.'"[987] Wallace later came to be a frequent visitor with the patrician Lyell who treated his admirer with affectionate condescension.

Wallace and Darwin finally met between 12 June and 15 July 1862.[988] "Mr. Darwin invited me to come to Down for a night, where I had the great pleasure of seeing him in his quiet home, and in the midst of his family."[989] Wallace gave Darwin a honeycomb from Timor.[990] Since April 1859, they had dropped the more formal address "My dear Sir" in their letters in favour of "My dear Mr Wallace" which, as an etiquette book of the day noted, was "the style adopted between intimate friends".[991] It is a pity we do not have a photograph of the two men strolling quietly down the sandwalk. Though they would later politely disagree about sexual selection, animal colouration and famously on the adequacy of natural selection to account for human evolution, not to mention spiritualism, they remained loyal friends for the rest of their lives.[992]

In the same year, Wallace sought the hand of the daughter of his chess friend, the auctioneer Lewis Leslie. Miss Leslie was an excellent catch for Wallace, coming from a substantially wealthier family. The marriage would have marked the highest rung of his social ladder. Alas one day in the autumn of 1864, Miss Leslie broke off the engagement without warning. Wallace was utterly devastated. Peter Raby perceptively guessed a possible reason why Wallace may have been so unceremoniously dumped.

> The Leslies lived in Kensington, the Wallaces in Notting Hill. Leslie was an auctioneer, but with offices in Mayfair. Wallace, for all his scientific reputation, was far from affluent, and still had no obvious prospects, even two years after his return from the East; he lived above an unsuccessful photographer's shop. The Leslies' neighbours in Campden Hill included the Duke of Argyll, the Earl of Antrim, and the artists Holman Hunt and Augustus Egg. Were Wallace's social credentials a little too fragile? Or was his lack of an official job seen as a disadvantage?[993]

Perhaps part of the sting Wallace felt was from this lifelong disadvantage? It is reminiscent of that other great emotional sting when his parents' efforts at economy exposed him as poor before his schoolfellows.

On the rebound, Wallace became engaged to Annie Mitten, the daughter of his friend the botanist and country pharmacist William Mitten. In April 1866, Wallace, aged forty-three, married Miss Mitten. Wallace recalled that she was "then about eighteen years old". She was in fact twenty. They had three children: Herbert (1867–1874), Violet (1869–1945) and William (1871–1951). His grandsons and great-grandchildren survive him.

The Malay Archipelago

After many delays, finally in 1869, Wallace published his great work, *The Malay Archipelago: the land of the orang-utan, and the bird of paradise. A narrative of travel, with studies of man and nature.* The title included two of the most exotic (and valuable) creatures from the archipelago and ones that represented either side of the Asian and Australian divide. Wallace rearranged the material in his *Journal* into five sections, corresponding to a five-fold "geographical, zoological, and ethnological" division of the archipelago: Indo-Malay islands, Timor group, Celebes group, Moluccan group and Papuan group. Crawfurd had earlier divided the archipelago into five "divisions" though there were differences.[994] Wallace also continued Crawfurd's racial division of the archipelago between Malays and Papuans.

Wallace opened the book with a generous dedication: "To Charles Darwin, *author of 'The origin of species,'* I dedicate this book, not only as a token of personal esteem and friendship but also to express my deep admiration for his genius and his works." The book focused more on the beauty or rarity of the specimens collected than their habits. This was Darwin's only quibble. He wanted more information on animal behaviour. But he thought very highly of the book as he wrote to Wallace, "Of all the impressions which I have rec^d from y^r book the strongest is that y^r perseverance in the cause of science was heroic. Your descriptions of catching the splendid butterflies have made me quite envious, & at the same time have made me feel almost young again, so vividly have they brought before my mind old days when I collected, tho' I never made such captures as yours. Certainly collecting is the best sport in the world."[995]

With the style of his writing, Wallace tried to show he was a scientific insider and also to claim authority within that community. Subtle clues showed that he was a gentleman naturalist. In his Amazon book, he had already referred to plenty of birds giving him "very good sport" — something he had never had the opportunity to enjoy in England.[996] *The Malay Archipelago* is peppered with frequent references to his servants.

Wallace often used a favourite technique, claiming that "the facts beat me" or he was "compelled" to believe something. He was forced to a conclusion and so it was not conditional or open to further assessment. This was the rhetorical power of "nature" and "facts" used by the phrenologists and mechanics' institute literature. He would later use it to demand respect for his belief in spiritualism and other unconventional beliefs. "The facts beat me. They compelled me to accept them, *as facts,* long before I could accept the spiritual explanation of them." "If I have now changed my opinion, it is simply by the force of evidence."[997] *The Malay Archipelago* ends with the sort of social reform commentary, written of as of the highest importance in the world, that would become more and more typical of Wallace's interests. His youthful Owenism began to reappear.

The book is not without a considerable number of mistakes, as we have seen. Pieter Veth, the eminent Dutch scholar of the East Indies, pointed out in his Dutch translation of *The Malay Archipelago* (1870–1) that Wallace erred in his statements on the size of some of the islands of the archipelago, including Borneo, and many other details.[998] Curiously, Wallace did not make use of Veth's corrections in later editions. Apparently, he did not read Veth's important commentary — the most authoritative edition of the book ever published. Because it is in Dutch, Veth's work has remained a great untapped resource.

The Malay Archipelago's wonderful stories and not least Wallace's self-effacing style soon won the book a warm following. Later, naturalists exploring the region were often inspired by the book, such as Wallace's German translator Adolf Meyer, the Russian Nikolai Miklouho-Maclay and the Italians O. Beccari and L. M. d'Albertis. The Scottish naturalist Henry Forbes later opened his own book on the archipelago with the remark "Mr. A. R. Wallace's 'Malay Archipelago' is so accurate and exhaustive an account of the Eastern Isles, that there have been left but few gleanings for those who have followed him to gather."[999] *The Malay Archipelago* remains a much loved book and has never gone out of print. It is still carried by travellers to the archipelago today.

Wallace in 1902. Frontispiece to *My life*.

Journey's End

Wallace applied unsuccessfully for many positions such as Director of Bethnal Green Museum, Assistant Secretary of the Royal Geographical Society and Superintendent of Epping Forest. He appealed to the "eminent men" he knew, such as Lyell, Darwin, Hooker and Huxley, to support his applications.

In the mid-1860s, Wallace became an ardent believer in the new movement of spiritualism with its séances, table turnings, ghost writing and so forth. Thanks to Wallace's spiritualism, we have preserved the only known evidence that he read Ida Pfeiffer's book while writing *The Malay Archipelago*. He mentioned her account of a ghost story in Java in the *Spiritual magazine* in 1868. It was his only mention of her name in print.[1000]

His public support for spiritualism and later anti-vaccination did much to damage his scientific reputation. His earlier caution about revealing radical ideas, as seen in his vague language in the Sarawak law, Aru and permanent varieties papers, was gone. Perhaps the experience of being so unexpectedly vindicated out of the blue with the Ternate essay or his increased status or just his greater maturity emboldened him to reveal unorthodox views more openly. Despite his shy manner in person, in his later life he never shied away from disagreement or stating what he believed.

In 1870, he answered the published challenge of a flat-earth conspiracy theorist to prove the earth is round. Using his old surveying skills, Wallace showed that a six-mile stretch of the old Bedford Canal was indeed not flat. The flat-earther refused to look through the telescope, since he already knew what could be seen because the Earth is actually flat! The referee awarded the massive £500 wager to Wallace. Sadly, the flat-earth madman persecuted Wallace relentlessly for years thereafter in a campaign of libel and defamation that cost far more in court costs and suffering than Wallace ever could have dreamt.[1001] For example, the madman sent this letter to Annie Wallace.

> Madam, if your infernal thief of a husband is brought home some day on a hurdle, with every bone in his head smashed to pulp, you will know the reason. Do tell him from me he is a lying infernal thief, and as sure as his name is Wallace he never dies in his bed. You must be a miserable wretch to be obliged to live with a convicted felon. Do not think or let him think I have done with him.[1002]

Wallace never fully utilised all of his knowledge of the birds and insects of the Eastern Archipelago. Instead, his background formed a respectable platform from which to address other topics that came to interest him. His interests and publications became ever more diffuse, spreading his productive energies widely. By the end of his life, supported by a state pension of £200 that Darwin, Huxley and others secured for him in 1881, Wallace had achieved more fame and respect than he seemed to desire or felt he deserved. In 1868, he was awarded the Royal Medal of the Royal Society. In 1870, he received the Gold Medal of the Société de Geographie of Paris and in 1882, the year of Darwin's death, the University of Dublin awarded Wallace an honorary doctorate. In 1908, he was awarded the Order of Merit.

When Darwin died, Wallace joined the international outpouring of admiration for a man described as the towering scientific figure of the age whose *Origin of species* had effected "the vast revolution…in our conception of nature".[1003] Years later, Wallace explained why Darwin's name was so famous throughout the world.

This lofty place has been given to him by his contemporaries, first, on account of his great discovery of the law of natural selection; next for his thorough study, for twenty years before making it known in all its applications and far-reaching consequences; and, lastly, for the unexampled mass of facts and experiments he had accumulated, and the extreme fairness and great logical power he had devoted to its exposition and demonstration. Such an amount of carefully-examined facts and sound reasoning, characteristic of the whole series of his works, has probably never before been adduced in support of a grand and newly-discovered theory.[1004]

From 1886–7, Wallace made a successful lecture tour of the United States. The fundamentalist movement that has turned America into the creationist capital of the Western world had not emerged and Wallace was greeted as a great discoverer. Had he travelled there in the 1920s, around the time of the notorious Scopes' monkey trial, his tour would have been rather different. Wallace's lectures formed the basis of his book *Darwinism* (1889), probably the finest book on evolution since *Origin of species*, and to this day a superb overview of the evidence for evolution.

But it also ended with a slide into anti-materialism that seems out of place with the rest of the book. One reviewer, the biologist and Darwin acolyte George John Romanes, said in contrast to the naturalist Wallace "we all agree in admiring", the last chapter was "the Wallace of spiritualism and astrology, the Wallace of vaccination and the land question, the Wallace of incapacity and absurdity". This sparked a bitter and unflattering feud.[1005] Elsewhere, Wallace allowed himself to quarrel in endless newspaper and periodical debates with figures far beneath him in stature and intelligence. It was a significant difference with the genteel Darwin that is seldom noted.

Wallace struggled for years against great hardships to elevate himself to a higher socio-economic and scientific status which he achieved so brilliantly with his voyage in the Eastern Archipelago. Outliving almost all of his contemporaries, he found himself in the surprising position of the grand old man of Victorian science. The anthropologist George Bettany wrote in 1889, "The intelligent minds which honour the name of Darwin, will not forget to honour that of his fellow-discoverer, Alfred Russel Wallace."[1006] On 7 November 1913, Wallace died peacefully in his sleep at the ripe old age of ninety.

Reflections

Historians of science often conclude their books with reflections on how their research reveals something new about how "science" works. It's part of the genre. Sometimes it's a good way to justify funding. But unless one is writing about quite recent science, it seems the lessons could be rather dubious. What is meant by "science", like the rest of culture, changes too much over time. As the anthropologist Marshall Sahlins quipped, "Such is the flux one can never step in the same culture twice."[1007] The more closely science is examined, the more fully this is appreciated. Conversely, the less someone has studied this, the simpler and "all the same" it seems. The naturalists of Victorian times are not 21st-century field biologists. It seems to me that caveats about specificity of time and place cannot be stressed strongly enough. That being said, what can be gleaned from the story of Wallace, Darwin and the origins of evolutionary theory?

The great struggle for the history of science since its emergence as a field of professional scholarship in the mid-20th century has been to escape the old-fashioned sort of armchair history associated with retired scientists writing about their heroes. According to legend at least, this was supposedly a history where lone geniuses had great ideas and "science" was something totally detached from society. Instead, science was pure, neutral and untainted by any values or external factors. Historian of science John Henry noted, "A striving for an ever richer contextualization can be seen, then, as the driving force in current historiography of science."[1008] It has proved enormously powerful and enriching in many ways. But now, to echo Winston Churchill, the battle over putting science in context is over. The battle to appreciate that context is not always socio-political has only begun. Much of the literature on Darwin and Wallace is particularly stuck in this mid-20th-century Marxist conception of context.

As the famous historian of Tudor England G. R. Elton wrote in 1967 as part of his relentless critique of the Marxist history of E. H. Carr, "Historians naturally praise in their fellows those enterprises and distinctions which come closest to their own mode of thought and work."[1009] This remains true today. The sociologist Stanley Lieberson has well observed that "in the absence of a rigorous study, the choice among competing explanations rests much more on the rhetorical skills of the proponents and the dispositions of the recipients than on the likely truth of any. If there is a disposition to assume that a change in tastes or fashion must have *social meaning* (in the sense that some minor or major social change must be responsible), then a plausible account will surely be found, particularly if it is not investigated systematically."[1010]

Anything in the socio-political context that seems similar to scientific ideas is confidently attributed to that source. Historians of science in a seminar or conference will nod with assent when the science of so-and-so is said to be derived from the social or political context. But, as historian Robert Richards pointed out, we "must be careful not to mistake analogies for homologies, not to assume that because one set of ideas is similar to another it must have descended from that other", and other work must be done to "establish the firmer ground of probability for real genealogical relations".[1011]

The mystery of mysteries has always been — how did Darwin and Wallace formulate "the same" theory of evolution? It cannot be a coincidence. And of course it isn't. They were Victorian Englishmen from the same time and culture. Clearly, there were common influences on both. Their shared contexts give confidence to those who claim that that it was the social and political context of Victorian capitalism and competitive individualism that inspired their evolutionary theories. Was it? Consider Moore's explanation of Wallace's 1858 "Malthusian moment". Moore claimed that "the geographic, economic, and social common context" of surveying in 1840s Britain returned to Wallace on Gilolo. Memories of the "Welsh farmers; of himself, the Saxon surveyor; of the tithes, the poor rates, the riots" led Wallace to remember Malthus and "by a deft pulled focus, Wallace arrives at natural selection".

I think this is an appeal to the wrong lineage. The connections are distant and tenuous, the similarities with the content of the 1858 Ternate essay no more specific and plausible than a dozen other possibilities. We should make our choices not on what sort of context *must* determine the details of science, but based on the actual details of each case.[1012]

Since Darwin and Wallace are from the same broad context, does it follow which part of their context was the common influence? Of course not. Their biological views were not derived from the social relations of Victorian Britain. Neither was interested nor involved in politics. It is odd that given the widespread acceptance amongst scholars today that readers make their own meaning out of a text that "reading Malthus must mean social or economic theory" persists.[1013] Malthus for them was instead, as Wallace put it, "philosophical biology".[1014] There was a great deal about other species in Malthus. Darwin and Wallace took no notice of the politics.

Darwin and Wallace both loved natural science and immersed themselves in the same scientific community. Within that community, they had similar experiences collecting and organising species and varieties. Their theoretical views were rearrangements of many existing elements circulating in the scientific community and literature on the very same topics: the history of life and living species. Most importantly, they read and were influenced by the same authors: Humboldt, Kirby and Spence, Loudon, Prichard, Lawrence, Lindley, *Vestiges*, Whewell, von Buch, Strickland, Somerville, Sclater, Boisduval, Ramsay, Lesson, Powell, Owen, Murchison, Milne-Edwards, Forbes, Blyth, Pictet, Malthus and a thousand others. And most crucially for both: Charles Lyell.

Similar Steps Towards Natural Selection

Darwin and Wallace went through a series of considerations about species that are, despite all the differences of detail, still strikingly similar. Both began with a fascination for the history of the world and the changes that living things had undergone. Both came to appreciate the evidence for extinctions and a succession of species types in the geological record. Both then accepted that species were derived from earlier species via genealogical descent or a process of unbroken biological generations rather than special creations. Next, they postulated that if one species could change into a new daughter species, it could also give rise to two or more daughter species, and thus form a branching family tree. Both considered for a time that geological subsidence created isolated pieces of land, which allowed stranded species to evolve away from ancestral populations. Both men went through a phase of rejecting that organisms were perfectly adapted since this smacked of the natural theology they both came to reject.

Thereafter, Darwin and Wallace envisioned the process of new species formation as a sudden saltation, i.e., "at one blow". For Darwin, this was a new but very similar species like the rhea. For Wallace, this was a new variety born fully formed like an albino. Wallace never had time for his theorising to move beyond this first rendition before he read Darwin. Finally, they came to realise that naturally occurring variants would be sifted by a struggle for existence which resulted from Malthusian superfecundity and thus adapted to environments and ways of life. For Darwin, the spark was Galápagos mockingbirds and South American fossils; for Wallace, the tiger beetles.

But why these two men? In addition to all the particular influences we have seen, both men, in their different ways, were somewhat intellectual outsiders. They did not conform too slavishly to the views of the time. Each was rather quirky in his way. They thought a little outside the usual orthodoxy, noticed things unappreciated by others and were open to new ideas more so than the average.

Of course there were many differences. Besides those usually mentioned, we should recall that Darwin was a deist and Wallace a freethinker. Darwin suffered from ill health for most of his life; Wallace usually enjoyed robust health. And Darwin could not draw to save his life, whereas Wallace was a fine draughtsman.

Like most professional historians of science, I do not see anything mysterious about Wallace's reputation today. Although I too would like to increase awareness of Wallace, I do not agree that anything unexpected or sinister has happened to his reputation. It is just as one would predict with the passage of time. A curtain of forgetfulness sweeps generations of thinkers from us; *extremely* few remain in front of it. Figures far more famous in Wallace's lifetime such as Cuvier, Faraday, Whewell, Lyell and Owen are equally unknown today. Darwin is a rare anomaly.

Wallace's story is a cross-section of the evolution of the life sciences in the first half of the 19th century. His early knowledge of the natural world was extremely vague and unstructured. But his successive scientific interests, connections and influences mirrored the developments in the scientific world. He became enthralled with the scientific classification of living things in a nested taxonomy. He came to accept the new authoritative philosophies that natural laws, rather than the finger of an invisible creator, make nature work the way it does. He was moved by the vast age of the earth and the exciting discoveries of the progressive history of life during that deep time. He then came to realise that just as the earth had slowly developed naturally, so had life. Successive types had branched off from one another over countless eons. And finally, newly arising types were sifted by a struggle for existence that adapted them to their new environments. Wallace will remain an inspiring example of what can be achieved as a self-driven adventurer, enthusiast and independent thinker.

THE PARADISE-BIRDS IN THE ZOOLOGICAL SOCIETY'S GARDENS.

Mr. A. R. WALLACE, the well-known traveller and naturalist, who has been engaged these last eight years in exploring the little-known islands of the Indian Archipelago, has for some time held a commission to obtain living birds of paradise for the Zoological Society of London; but, although he visited in person the islands inhabited by several species of this magnificent group of birds, he failed in his efforts to preserve these birds alive when captured, and had given up all hopes of being successful in his object. It was only a short time before Christmas last, when in the interior of Sumatra, that Mr. Wallace received information of two specimens of the lesser bird of paradise (*Paradisea papuana*) being alive in captivity at Singapore. Mr. Wallace immediately proceeded to that place, purchased the birds, which were then in the hands of a European merchant, and left by the following mail for England, arriving in safety in London with his valuable burden on the 1st of this month.

The two paradise-birds have been lodged in the upper part of the Zoological Society's old museum, a room having been fitted up for their reception with a large cage of galvanised iron wire 20ft. long by 11ft. in width. As they are both males it has been found necessary to keep them apart, the sight of one another, or even of a paradise-bird's plume waved near them in the air, producing in them great excitement. The cage has therefore been divided by a screen, which excludes the light, and the two birds placed in the separate compartments. The remarkable side plumes which ornament the males of the true *Paradisea* when in full dress are as yet but partially developed in these specimens; but in a few weeks, if the birds continue to thrive, will probably attain their full dimensions.

The paradise-birds in the Zoological Society's gardens are fed on rice, bread, vegetables, and fruit, but require also insect food, and seem particularly partial to mealworms and cockroaches. They are very tame, readily taking a tidbit from the hand of the attendant, and, considering the confinement they have been subject to during their long journey from Singapore, in wonderful health and condition.

BIRDS OF PARADISE IN THE ZOOLOGICAL SOCIETY'S GARDENS, REGENT'S PARK.

The best for last. Wallace's most valuable specimens of all. Two male Lesser Birds of Paradise. *The Illustrated London News* Suppl. (Apr. 1862): 375.

NOTES

Books and articles are cited in author, short title and date form, and a full listing of these is in the References. Some standard sources are abbreviated as short titles, e.g., MA, ML, etc. The full reference is given below. These are also listed alphabetically in the References, once under the abbreviated title and again under the author's or editor's name. Letters quoted from the original manuscript are cited with their institutional call number. Wallace's publications are sometimes listed, for brevity, by the "S" numbers from the bibliographical list by Charles H. Smith. These are all given in full in the References. In a few instances, I cite Darwin's notebooks as C, D, TAN, etc. These references can be used to look up the passages cited in the standard edition by Barrett *et al.* (1987).

Abbreviations

BL: British Library, London.

Calendar: Burkhardt, F. and Smith, S. eds. 1985. *A calendar of the correspondence of Charles Darwin, 1821–1882*. New York & London: Garland (superseded by the online database http://www.darwinproject. ac.uk/).

CCD: Burkhardt, F. *et al.* eds. 1985-. *The correspondence of Charles Darwin*. 19 vols. Cambridge: U. Press.

CUL: Cambridge University Library.

DAR: Darwin Archive at Cambridge University Library.

Journal 1: Linnean Society of London MS178a.

Journal 2:	Linnean Society of London MS178b.
Journal 3:	Linnean Society of London MS178c.
Journal 4:	Linnean Society of London MS178d.
Lyell:	Lyell, C. 1835. *Principles of geology: being an inquiry how far the former changes of the earth's surface are referable to causes now in operation.* 4th edn. 4 vols. London: Murray.
MA:	Wallace, A.R. 1869. *The Malay Archipelago: The land of the orang-utan, and the bird of paradise. A narrative of travel, with studies of man and nature.* 2 vols. London: Macmillan & Co.
ML:	Wallace, A.R. 1905. *My life: A record of events and opinions.* 2 vols. London: Chapman & Hall.
NHM:	Natural History Museum (London), Wallace Collection.
Notebook 1:	Linnean Society of London MS179.
Notebook 2/3:	Natural History Museum (London) Z MSS 89 O WAL.
Notebook 4:	Linnean Society of London MS180.
Notebook 5:	Natural History Museum (London) Z MSS 89 O WAL.
RGS:	Royal Geographical Society, London.
S[+ no.]	Smith, Charles H. ed. 1998–. *The Alfred Russel Wallace page* [bibliography].
ZSL:	Zoological Society of London.

1. Darwin to Lyell 18 [June 1858] CCD7:107. Darwin exaggerated somewhat. Wallace used "struggle for existence" which was the title of chapter 5 of Darwin's book-in-progress.
2. F. Darwin, *Life and letters of Charles Darwin*, 1887, 2:197. This and all of the other publications of Darwin, Wallace and many others cited in this book are freely available in full in John van Wyhe, *The complete work of Charles Darwin online* (http://darwin-online.org.uk/) and *Wallace online* (http://wallace-online.org/).
3. Browne, *Charles Darwin*, 2002, p. 39; Moody, The reading of the Darwin and Wallace papers, 1971.
4. For example, the botanist and anthropologist G.T. Bettany, who published one of the first biographies of Darwin, wrote a biographical introduction on Wallace in 1889, which shows how the Darwin–Wallace affair was formerly described. There was also no conspiracy or unfairness to Wallace in the biographies by George, *Biologist philosopher*, 1964 or Williams-Ellis, *Darwin's moon*, 1966.
5. Camerini, *Wallace reader*, 2002, p. 12. The phrase was repeated by J. Rosen, *New Yorker*, 12 Feb. 2007.
6. Ruse, *Darwinian revolution*, 1979; Glick & Kohn, *On evolution*, 1996, p. 152 and Wilson, *Forgotten naturalist*, 2000 claim that Wallace *did* send the Ternate essay for publication.
7. Also in Rose & Rose, *Alas, poor Darwin,* 2000, p. 11.
8. Wilson, *Forgotten naturalist*, 2000.
9. See for example Beccaloni & Smith, Celebrations for Darwin downplay Wallace's role, 2008 and Benton, Race, sex and the 'earthly paradise', 2009.
10. Wallaceism: R. Uhlig, 2010. *Genius of Britain*. London: Collins. The same claim is made by several other writers, e.g., Davies, *Darwin conspiracy*, 2008, p. 161 and McCalman, *Darwin's armada*, 2010, p. 324.
11. The Ternate letter stamps are reproduced in several publications, including McKinney, *Wallace*, 1972; Wilson, *Forgotten naturalist*, 2000; Davies, *Darwin conspiracy*, 2008; Lloyd, Wimpenny & Venables, Wallace deserves better, 2010.
12. Hallmark, Charles Darwin 'facing accusations of plagiarism' (press release), 2008.
13. Quoted in Brackman, *A delicate arrangement*, 1980, p. 348.
14. See Moore, Wallace's Malthusian moment, 1997.
15. Sulloway, Darwin and his finches, 1982; Sulloway, *Beagle* collections of Darwin's finches, 1982; and John van Wyhe, Where do Darwin's finches come from?, 2012.
16. In general, I will use the Victorian place names to be consistent with the quotations and contemporary sources used. Current names are given in brackets at first occurrence. The Southeast Asian region was referred to by various names in Wallace's day. In order of frequency these were: East Indies, Indian Archipelago, Eastern Archipelago, Malay Archipelago and East Indian Archipelago. "Eastern

Archipelago" is used in this book to avoid confusion with the title of Wallace's book *The Malay Archipelago* (1869).

17. Wallace to Darwin 2 Jan. 1864 CCD12:5.
18. Anon., *The youthful travellers*, 1823, p. 113.
19. For example Bryant, *Naturalist in the river*, 2003, p. 8, Wallace "was no gentleman"; Davies, VPRO documentary, 2010; Shermer, *In Darwin's shadow*, 2002, pp. 14, 30, 35. It has occasionally been pointed out that Wallace was not working-class, e.g., Berry, *Infinite tropics*, 2002.
20. There is a vast scholarly literature on social class in Victorian Britain, for example Cannadine, *Rise and fall of class in Britain*, 1999. On the origins of language of a "middle class", see Wahrman, *Imagining the middle class*, 1995.
21. Baptism Register, NL/Pa 13, p. 15, Gwent Record Office.
22. ML1:7.
23. ML1:29.
24. Smith, *Alfred Russel Wallace page*, 1998–.
25. van Wyhe, *Darwin in Cambridge*, 2009.
26. *Hertford trade directory*, 1832, p. 741.
27. ML1:110 and Turnor, *Hertford*, 1830, pp. 332–335.
28. Phillips & Wetherell, The great reform act of 1832, 1995.
29. ML1:13.
30. ML1:58.
31. See Raby, *Wallace*, 2001, p. 13 and Hertfordshire Archives, D/Z113 A1/5.
32. Desmond & Moore, *Darwin*, 1992, p. 466 and Desmond, *Huxley*, 1997, p. 244 mistakenly claim that Wallace was apprenticed. See ML1:79.
33. ML1:61.
34. ML1:104.
35. Secord, *Victorian sensation*, 2000, p. 333. However see Claeys, Wallace, women, and eugenics, 2008. Thanks to Charles Smith.
36. Owen, *Lecture on consistency*, 1840.
37. ML1:87.
38. Hume, *Essays and treatises on several subjects*, 1793, 4:177.
39. ML1:89.
40. Lamarck's transmutation views were detailed in his *Recherches sur l'organisation des corps vivants*, 1802; *Philosophie zoologique*, 1809 and *Histoire naturelle des animaux sans vertèbres*, 1815–1822.
41. Rudwick, *Bursting the limits of time*, 2005.
42. Sulloway, Darwin and his finches, 1982.
43. Barlow, Darwin's ornithological notes, 1963, p. 262.
44. Herbert, The red notebook, 1980, pp. 63, 66.
45. Barlow, *Autobiography of Charles Darwin*, 1958, p. 120.

46. Browne, *Voyaging*, 1995, p. 385.

47. Malthus, *An essay on the principle of population*, 1826. There is a large literature on Malthus for Wallace and Darwin. See Winch, *Malthus*, 1987; Browne, *Secular ark*, 1983 and Jones, Wallace, Robert Owen and the theory of natural selection, 2002. For an account of "Malthusianism" at this time, see Winch, Darwin fallen among political economists, 2001. For refutations of the persistent assertions that influence from Malthus' book entails borrowing from economics in natural history, see Kohn, Theories to work by, 1980, p. 145; Hull, Deconstructing Darwin, 2005; Gordon, Darwin and political economy, 1989; Gale, Darwin and the concept of a struggle for existence, 1972 and Bowler, Malthus, Darwin, and the concept of struggle, 1976.

48. *Notebook E*, p. 26e. Barrett *et al.*, *Darwin's notebooks*, 1987, p. 404.

49. *Notebook E*, p. 58. Barrett *et al.*, *Darwin's notebooks*, 1987, p. 412.

50. *Notebook E*, pp. 115e; 137; 71.

51. Barlow, *Autobiography of Charles Darwin*, 1958, p. 120.

52. Darwin to W. D. Fox [15 June 1838] CCD2:92. See also Darwin to Lyell [14] Sept. [1838] CCD2:107.

53. ML1:137.

54. Morrell, *Science, culture and politics*, 1997, pp. 194, 196 and [Luxford], Popular works on natural history. *Westminster Review*, 1845, 44:103.

55. Wallace, Kington Mechanic's Institution, 1845, pp. 67–68.

56. ML1:196.

57. As noted by C. H. Smith, a slightly different chronology from ML1:192 is given in Wallace [to T. D. A. Cockerell], 1903.

58. ML1:194. Loudon, *Encyclopedia of plants*, 1836.

59. ML1:195.

60. Rudwick, *Worlds before Adam*, 2008.

61. Secord, Introduction to Lyell's *Principles*, 1997.

62. Barlow, *Autobiography of Charles Darwin*, 1958, p. 101.

63. Wallace, *Trichius fasciatus*, 1847.

64. van Wyhe, *Phrenology*, 2004.

65. van Wyhe, The authority of human nature, 2002.

66. van Wyhe, The diffusion of phrenology, 2007.

67. Darwin to W. D. Fox [3 Jan. 1830] CCD1:97.

68. Winter, *Mesmerized*, 1998.

69. Cooter, *Phrenology in the British Isles*, 1989, p. 157. For Bates, see Clodd, *Memories*, 1870, pp. 65–66.

70. Engels, *Dialectics of nature*, 1954, pp. 69–70.

71. ML1:236.

72. Engels, *Dialectics of nature*, 1954, pp. 71–72. Engels refers to Wallace's interest in spiritualism from the mid-1860s. See Kottler, Wallace, 1974.

73. Wallace mistakenly wrote that his brother died in 1846, ML1:14; 239. This period is well described in Raby, *Wallace*, 2001, pp. 23–24.

74. Hughes, Wallace: some notes on the Welsh connection, 1989.

75. Moore, Wallace in Wonderland, 2008, p. 356.

76. The original delineations survive in NHM WP18/38 and WP18/39. I am grateful to Judith Magee for sending me copies. Wallace discussed these in *Wonderful century*, 1898, p. 174 and ML1:258ff.

77. Wallace, *Wonderful century*, 1898, p. 174.

78. On *Vestiges* and steam printing, see Secord, *Victorian sensation*, 2000.

79. Wallace, *Wonderful century*, 1898, p. 138.

80. ML1:234.

81. van Wyhe, George Combe's Constitution of man, 2003.

82. For use of this language in Wallace writings see for example: [Discussion.] 1864; Origin of human races, 1864; Anthropology address, 1867; [Discussion.] 1868; *Contributions*, 1870, p. 319. See Durant, Scientific naturalism, 1979.

83. Rockell, The last of the great Victorians, 1912, p. 663.

84. On Darwin see Barrett *et al.*, *Darwin's notebooks*, 1987.

85. Under the stairs myth: J. Doherty, In Darwin's garden, BBC 2009, episode 1; Bryson, *A short history of nearly everything*, 2003; Stott, *Darwin's ghosts*, 2012; Moore, British Council Darwin Now podcast, 2009; Reeve, *Darwin's home*, p. 8: "Darwin stored the first pencil sketch of his species theory completed in 1842, in the cupboard under the stairs". This myth arises from misunderstanding that the 1842 sketch was found under the stairs after Emma Darwin's death in 1896. This was not the 1844 essay and is no evidence that Darwin kept either draft there before *Origin* was published. See F. Darwin, *Foundations*, 1909, p. xvii.

86. CCD3:43.

87. van Wyhe, Darwin's 'Journal', f22.

88. Darwin, *A monograph of the sub-class Cirripedia*, 2 vols. 1851, 1854.

89. DAR 113. This was later published in F. Darwin, *Foundations*, 1909.

90. Across his life Darwin consistently referred to his early essay as not aimed at publication. Six years after writing it, he referred to it as "my rude species = sketch", Darwin to Hooker 13 June [1850] CCD4:343. When surprised by Wallace in 1858, Darwin had an urgent need to publish an overview of his theory, but he still did not publish this essay. Instead he spent a gruelling thirteen months writing the *Origin of species*. The essay was "never for an instant intended for publication", Darwin to Wallace 25 Jan. [1859] CCD7:241.

91. Darwin to Jenyns 25 [Nov. 1844] CCD3:84–85.
92. Darwin to Jenyns 14 Feb. [1845] CCD3:142–143.
93. Gérard, 1844. De l'espèce dans les corps organisés. Extract from Orbigny, *Dictionnaire universel d'histoire naturelle*, 5:428–452. Hooker to Darwin [4–9 Sept. 1845] CCD3:250.
94. Darwin to Hooker [10 Sept. 1845] CCD3:252–253.
95. Darwin to Hooker [18 Sept. 1845] CCD3:255.
96. Darwin to Hooker [5 or 12 Nov. 1845] CCD3:263–264.
97. On Stevens, see Allingham, *A romance of the rostrum*, 1924 and Baker, Wallace's record of his consignments, 2001. On Wallace's relationship with Stevens see Camerini, Wallace in the field, 1996.
98. Douglas, *World of insects*, 1856, p. 203.
99. Bastin, Introduction to *Malay Archipelago*, 1986 notes: "These were the commission rates charged by Stevens when he was acting as agent to Wallace and Bates when they were in the Amazon Valley during 1848–52 and it is reasonable to assume that they were the same…during 1854–62."
100. ML1:197.
101. Perhaps the first to claim a break was Williams-Ellis, *Darwin's moon*, 1966. Brackman, *A delicate arrangement*, 1980 claimed there was no evidence for dissension. However see Raby, *Wallace*, 2001, pp. 44–45 and Slotten, *Heretic in Darwin's court*, 2004, p. 54. On Wallace in the Amazon see the excellent Knapp, *Footsteps in the forest*, 1999.
102. Raby, *Wallace*, 2001 is very good on Wallace's time in London after the Amazon. There is a vast literature on the history of collecting. See for example Jardine, Secord & Spary, *Cultures of natural history*, 1997; Knight, *Ordering the world*, 1981; Camerini, Wallace in the field, 1996. There is also a large and growing literature on popular science in the period, for example Fyfe, *Science and salvation*, 2004 and Fyfe & Lightman, *Science in the marketplace*, 2007.
103. Baker, Discovery of the Hymenoptera, 1995.
104. St. John, *Life in the forests*, 1862, 1:162.
105. See *Transactions of the Entomological Society of London*, 1855, 3:80, 87.
106. Baker, Discovery of the Hymenoptera, 1995.
107. Runciman, *The white Rajah*, 2010.
108. Brooke to Wallace 1 Apr. [1853] BL Add 46441.
109. *Literary Gazette*, 16 July 1853, p. 701.
110. Wallace, My relations with Darwin, 1903. If the meeting took place, there are several possible dates when Darwin was in London. See CCD5 Appendix I.
111. "At the request of *the Royal geographical society* in London, permission was granted to Mr. A.R. Wallace, who wanted to explore the Dutch East-Indian possessions

in the interest of geography and natural history, to visit the Moluccan islands, Celebes and Timor." Staten Generaal van Nederland, *Verslag*, 1853, p. 112. Translation by Kees Rookmaaker to whom I am indebted for this reference.

112. ML1:340.

113. Rookmaaker & van Wyhe, Charles Allen, 2012.

114. Wallace, *Wonderful century*, 1898, p. 3.

115. Maunder, *The scientific and literary treasury*, 1853, p. 725.

116. MA2:135.

117. [Thackeray], *Journey from Cornhill to grand Cairo*, 1846, p. vii.

118. *Accounts and papers of the House of Commons.* vol. xcv.

119. Anon., *Popular overland guide*, 1861, p. 1.

120. Anon., Ocean routes, 1854, p. 28.

121. In comparison those of the USA covered 730,000. Homans, *Cyclopedia of commerce*, 1860, p. 1754.

122. *The Times*, 6 Mar. 1854, p. 9. Wallace was listed as a passenger in *Allen's Indian Mail*, 20 Mar. 1854, p. 147.

123. *The Times*, 6 Mar. 1854, p. 9.

124. FitzRoy, *Narrative*, 1839, 1:18, and Thomson, *HMS Beagle*, 1995, p. 129. With thanks to Gordon Chancellor.

125. ML1:326.

126. Stocqueler, *Overland companion*, 1850, p. 10.

127. Wallace to Silk 19 Mar. 1854 NHM WP1/3/27. The new transcriptions of Wallace's eastern voyage correspondence in this book are taken from the forthcoming new edition transcribed and edited by van Wyhe and Rookmaaker, *Alfred Russel Wallace: Letters from the Malay Archipelago*, 2013.

128. Anon., *Popular overland guide*, 1861, p. 21.

129. Anon., *Popular overland guide*, 1861, p. 22.

130. Eames, *Another budget*, 1855, p. 4.

131. Stocqueler, *Hand-book of British India*, 1854, pp. 88–89.

132. Eames, *Another budget*, 1855, p. 4.

133. Anon., *Popular overland guide*, 1861, p. 5.

134. There was more than one publication with the title *How to make money*; Wallace's cabin mate was probably reading Freedley, *How to make money*, 1853.

135. Eames, *Another budget*, 1855, p. 21.

136. Anon., *Diary of a field officer*, 1853, p. 4.

137. Anon., *Popular overland guide*, 1861, p. 59.

138. Wallace to Silk 26 Mar. [1854] NHM WP1/3/27.

139. [Thackeray], *Journey from Cornhill to grand Cairo*, 1846, p. 82.

140. Stocqueler, *Hand-book of British India*, 1854, p. 86.

141. Stocqueler, *Hand-book of British India*, 1854, p. 86.

142. ML1:335–336.

143. ML1:335. For more details on Galle at the time see Stocqueler, *Hand-book of British India*, 1854, p. 383.

144. [Neale], *The old arm-chair*, 1854, p. 141.

145. ML1:336.

146. ML1:335–336.

147. She was expected c. 15 April 1854, *The Straits Times*, 18 Apr. 1854, p. 3. *NewspaperSG* (http://newspapers.nl.sg/) by the National Library of Singapore is an outstanding resource that enables many of the newspapers of the period to be consulted and partially searched (the OCRed text is not corrected) for free.

148. Jagor, *Reiseskizzen*, 1866 includes wonderful drawings of the kinds of boats seen in Singapore Harbour. See also the drawings by Charles Dyce reproduced in Liu, *Singapore a pictorial history*, 2001, p. 29.

149. The monthly totals were printed in *The Straits Times*, c. 33,000 per month in 1855.

150. *The Straits Times*, 18 Apr. 1854, p. 4.

151. Creswell, From Dudley colliery to Borneo, 1878.

152. *The Straits Times*, 25 Apr. 1854, p. 8.

153. Useful histories include: Buckley, *An anecdotal history of old times in Singapore*, 1902 and Turnbull, *History of Singapore*, 2009. Liu, *Singapore a pictorial history*, 2001 provides a wealth of illustrations of Wallace's Singapore.

154. *The Straits Times*, 14 Feb. 1854, p. 7.

155. *The Straits Times*, 14 Feb. 1854, p. 7, *Macphail's Edinburgh ecclesiastical journal*, 1855, p. 261.

156. Thomson, *Some glimpses*, 1864, p. 15.

157. Holmberg, A community of prestige (thesis), 2009.

158. Anon., *Singapore directory for 1854*.

159. Anon., *Catalogue of books in the Singapore library*, 1860. The holdings fluctuated very quickly as new books arrived on the bi-monthly steamers and older works were discarded. No traces of user records are known to survive in the National Museum of Singapore or the National Archives of Singapore.

160. *The Straits Times*, 18 Apr. 1854, p. 4.

161. Nolan, *History of the British empire*, 1859, 1:201.

162. Straits Settlement, Annual report 1860–61, p. 43.

163. *The Straits Times*, 14 Feb. 1854, p. 5.

164. Wallace to *Literary Gazette*, Sept. 1854 NHM WP1/8/38.

165. Wallace, Letters from the Eastern Archipelago, 1854.

166. *The Straits Times*, 1 Apr. 1856, p. 5.

167. *The Straits Times*, 25 Sept. 1855, p. 4.

168. NHM WP1/362.

169. *Singapore: A country study*, 1991, p. 22.

170. *Southern literary messenger*, 1854, 20:220.

171. *British and foreign medico-chirurgical review* vol. 5.

172. Wilkes, *Narrative of the United States Exploring expedition*, 1845, 5:382.

173. *Notebook 4*, p. 65.

174. MA2:216.

175. Pickering & Hall, *The races of man*, 1854, p. 181. The OED claims it is disparaging. This is certainly incorrect at the time of Wallace. See Veth, *Insulinde*, 1870, 1:44 note 1.

176. MA1:31.

177. ML1:348.

178. McDougall, *Letters from Sarawak*, 1854, p. 10.

179. McDougall, *Letters from Sarawak*, 1854, p. 10.

180. Fallows, *Progressive dictionary of the English language*, 1835. The references in the OED only date to 1857–8.

181. *Chambers's journal*, 20 July 1861, p. 35.

182. *Transactions of the medical and physical society of Bombay*, 1862, 7:140.

183. Anon., *Popular overland guide*, 1861.

184. MA1:52. See however his charming description of Singapore, some of which is still fitting today, in MA1:31–34. The first, *Journal 1*, commences in June 1856. The gap in coverage from his arrival in April 1854 to March 1855 (the first diary entries in *Notebook 4*) is about the same as covered in the extant notebooks and may have been covered in the notebook Wallace said was lost. A section of *Notebook 4* contains diary entries from March–June 1855. If the notebooks were all labelled later this would explain why what Wallace labelled *Journal 1* may actually have been his second.

185. Barrett *et al.*, *Darwin's notebooks*, 1987; Chancellor & van Wyhe, *Darwin's notebooks from the voyage of the Beagle*, 2009 and Kohn, Theories to work by, 1980.

186. See the introduction to Chancellor & van Wyhe, *Darwin's notebooks from the voyage of the Beagle*, 2009.

187. Wouters, *Informalization*, 2007, pp. 27–28. On the importance of letters of introduction to the travelling collector see Camerini, Wallace in the field, 1996.

188. Anon., *Etiquette*, 1854, p. 73. Wallace must have had many letters of introduction during his travels but as usual with such ephemeral materials, none survive.

189. Thomson, *Some glimpses*, 1864. Wallace to Silk 15 Oct. 1854 NHM WP1/3/33.

190. *The Calcutta Review* 37, 1861, p. 43.

191. *The Straits Times*, 3 Jan. 1854, p. 2. Skittles evolved into modern bowling.

192. *Singapore Free Press*, 11 Apr. 1839, p. 1.

193. *Singapore Free Press*, 9 Mar. 1843, p. 1.

194. *Singapore Free Press*, 20 Mar. 1845, p. 2.

195. Quoted in Anon. The French in the South Seas, 1853.

196. *The Straits Times*, 29 Apr. 1851, p. 3.

197. *The Straits Times*, 29 Apr. 1851, p. 5.

198. Train, *An American merchant in Europe, Asia and Australia*, 1857, pp. 66–67.

199. His estate was declared insolvent in Sept. 1857. *The Straits Times*, 21 Apr. 1858, p. 2.

200. [Kinloch], *De Zieke Reiziger* [the sick traveller], 1853, p. 11.

201. Wallace, [Letter to *Zoologist*], 1854, S14.

202. Wallace, [Letter to *Zoologist*], 1854, S14.

203. I am grateful to Richard Corlett for this estimate.

204. Wallace to M. A. Wallace 30 Apr. 1854 NHM WP1/3/28. Another possibility is the Frenchmen for China on the *Euxine*. Wallace mistakenly called them Jesuits. MA1:34–35, 43 & ML1:337, 348.

205. Liew, The Roman Catholic Church of Singapore (thesis), 1993.

206. MA1:35–36.

207. Pfeiffer, *A woman's journey round the world*, [1852], p. 123.

208. MA1:36.

209. See Ng, Corlett & Tan, *Singapore biodiversity*, 2011, pp. 50–51. I am grateful to Richard Corlett for informing me of this.

210. Wallace to M. A. Wallace 30 Apr. 1854 NHM WP1/3/28.

211. Wallace, Localities given in Longicornia Malayana, 1869, p. 691.

212. MA1:37–38.

213. Wallace to M. A. Wallace 28 May 1854 NHM WP1/3/29.

214. Wallace, The entomology of Malacca, 1855, p. 4637.

215. Wallace, Letter, 1855, p. 4804.

216. Wallace, On the ornithology of Malacca, 1855, p. 97. Now called the Oriental Magpie Robin (*Copsychus saularis*). Richard Corlett informs me that the Magpie Robin was almost extinct in Singapore by the 1980s, "but has made a comeback as a result of deliberate releases of imported birds".

217. Wallace, *Zoologist* 12, 1854.

218. *Notebook 1*, p. 3. Jukes, *A sketch of the physical structure of Australia*, 1850.

219. *Notebook 1*, p. 5. Fry, On the relation of the Edentata to the reptiles, 1846.

220. Fry, On the relation of the Edentata, 1846, p. 284, quoting Strickland, Description of a chart of the natural affinities, 1844.

221. Rudwick, *The meaning of fossils*, 1985, p. 153. See Pietsch, *Trees of life*, 2012, pp. 68–69.

222. Wallace, *Zoologist* 12, 1854, p. 4397.

223. MA1:37.

224. *Singapore Chronicle*, 8 Sept. 1831, p. 3.

225. "We find that in stating the number of people annually killed in Singapore by tigers at 150, we were much below the mark. We are assured by a person who has the best means of knowing the amount, that more than three hundred natives are every year carried off. This gives nearly one in every hundred of the population of this island to the tigers annually." *Asiatic journal and monthly register* ns, 1843, 40:46. Thanks to Richard Corlett.

226. *Singapore Free Press*, 16 Nov. 1843, p. 3.

227. See the lengthy discussion of tigers in Singapore in Cameron, *Our tropical possessions*, 1865. He repeated the death-per-day claim on p. 91.

228. Pfeiffer, *A lady's second journey round the world*, 1855, 1:51.

229. *Southern literary messenger* 20:220.

230. *Illustrated magazine of art* 3:278. Cameron, *Our tropical possessions*, 1865. See Boomgaard, *Frontiers of fear*, 2001. With thanks to Tim Barnard.

231. Mereweather, *Diary of a working clergyman*, 1859, p. 330.

232. McDougall, *Letters from Sarawak*, 1854, p. 12.

233. Anon., *Catalogue of books in the Singapore library*, 1860, p. 19.

234. *The Straits Times*, 31 Jan. 1854, p. 4.

235. *The Straits Times*, 14 Aug. 1855, p. 5.

236. *The Straits Times*, 4 Sept. 1855, p. 4; *The Straits Times*, 18 Dec. 1855, p. 4.

237. Collingwood, *Rambles of a naturalist*, 1868, p. 254.

238. *Punch*, 27 Oct. 1855, p. 170.

239. *The Straits Times*, 1865, pp. 102–103.

240. *The Straits Times*, 14 Nov. 1863, p. 2. See also *The Straits Times*, 7 Nov. 1863, p. 4. Also reporting four deaths from tigers.

241. Buckley, *An anecdotal history of old times in Singapore*, 1902, 1:221.

242. *The Straits Times*, 13 Aug. 1902, p. 4. See also *Singapore Free Press*, 13 Aug. 1902.

243. Original story and photo: *Malayan Saturday Post*, 8 Nov. 1930, p. 38.

244. *The Straits Times*, 27 Mar. 1935, p. 11.

245. Wallace, Letters from the Eastern Archipelago, 1854.

246. *The Straits Times*, 30 May 1854, p. 4.

247. *The Straits Times*, 9 May 1854, p. 4, The riots are described in great detail in *The Straits Times*, 16 May 1854, p. 4.

248. *The Straits Times*, 16 May 1854, p. 4.

249. *Allen's Indian Mail*, 18 July 1854, p. 378.

250. Wallace to Shaw 1 Nov. 1854 RGS Letter book 1854 (JMS 8/17); Wallace, Introduction, 1873, p. 285 and Wallace, Localities given in Longicornia Malayana,

1869, p. 691. Pfeiffer, *A woman's journey round the world*, 1852 gives a detailed account of a visit to Pulau Ubin.

251. MA2:141.

252. Rookmaaker & van Wyhe, Direction for Collecting in the Tropics by A.R. Wallace. http://wallace-online.org/content/frameset?pageseq=1&itemID=CUL-DAR 270.1.2&viewtype=text

253. *The Straits Times*, 30 May 1854, p. 8.

254. *Notebook 1*, p. 130.

255. Wallace, Introduction, 1873, p. 288.

256. *The Straits Times*, 18 July 1854, p. 8.

257. ML1:341.

258. *The Straits Times*, 13 July 1854. Wallace to M. A. Wallace [c. 24] July 1854 NHM WP1/3/30.

259. MA1:39.

260. Decroix, *History of the Church and churches in Malaysia and Singapore*, 2005, pp. 225–226.

261. [Kinloch], *De Zieke Reiziger*, 1853, pp. 122–123.

262. Thomson, *Some glimpses into life in the Far East*, 1864, p. 25.

263. MA1:118.

264. Wallace, On the bamboo and durian of Borneo, 1857.

265. MA1:43.

266. Wallace to M. A. Wallace July 1854 NHM WP1/3/30.

267. *The Straits Times*, 1 Aug. 1854, p. 4. Thanks to Jerry Drawhorn.

268. Wallace to M. A. Wallace 30 Sept. 1854 NHM WP1/3/32.

269. Wallace, Extracts of a letter, 1855.

270. Wallace, Ornithoptera brookiana, 1855.

271. Wallace, Extracts of a letter, 1855, stated there were five servants, but six in MA1:45.

272. Wallace, Ornithoptera brookiana, 1855.

273. MA1:49.

274. Anon., A visit to Kew Gardens, 1852, p. 216.

275. Wallace to Shaw 1 Nov. 1854 RGS Letterbook. The mountain is now recorded as 4,186 feet.

276. van Wyhe, Darwin's 'Journal', f32. Darwin, *A monograph of the sub-class Cirripedia*, 2 vols. 1851, 1854.

277. Barlow, *Autobiography of Charles Darwin*, 1958, p. 118.

278. Darwin to W. D. Fox 27 Mar. [1855] CCD5:293.

279. Wallace to Shaw 1 Nov. 1854. RGS Letterbook.

280. *The Straits Times*, 26 Sept. 1854, p. 8.

281. *The Straits Times*, 21 Nov. 1854, p. 6.

282. [Simonides], *The Borneo question*, 1854.

283. Wallace to Silk 15 Oct. 1854 NHM WP1/3/33.

284. Wallace to Silk 15 Oct. 1854 NHM WP1/3/33.

285. Baker, Wallace's record of his consignments, 2001, p. 260.

286. *The Straits Times*, 17 Oct. 1854, p. 4.

287. Baker, Wallace's record of his consignments, 2001.

288. *The Straits Times*, 31 Oct. 1854, p. 8. Although in Wallace's *Notebook 1*: "sent by the 'Royal Alice', leaving Singapore 17 Oct. 1854".

289. Wallace to M. A. Wallace 30 Sept. 1854 NHM WP1/3/32.

290. Wallace to M. A. Wallace 30 Sept. 1854 NHM WP1/3/32.

291. Wallace to Shaw 1 Nov. 1854. RGS Letterbook.

292. Wallace's travels are complex and involve voyages to many islands unfamiliar to most readers. To make following Wallace's journey more intelligible I have used selection's from Wallace's fold-out map in *The Malay Archipelago* (1869). However, as this map was recycled from earlier publication(s) by the publisher, it is extremely cluttered with place names that are not mentioned by Wallace. There are even comments referring to another expedition, for example in the middle of Borneo "Dutch steamer reached here in March". This makes the map difficult to use. I have removed most of the superfluous names from the map sections reproduced in this book to make them easier to follow. Some of the lines have also been redrawn or place names added for clarity.

293. *The Straits Times*, 17 Oct. 1854, p. 8 and *The Straits Times*, 24 Oct. 1854, p. 8.

294. Apparently he based this date on his note recording the start of his collecting from 1 Nov. in *Notebook 1*, p. 120. In a letter to N. Shaw dated 1 Nov. 1854, Wallace wrote "I am only three days here". The maps of Wallace's journeys in Sarawak and Borneo in Wilson, *Forgotten naturalist*, 2000 are helpful.

295. *The Straits Times*, 16 Dec. 1851, p. 7.

296. Pfeiffer, *A woman's journey round the world*, [1852], pp. 59–61, 63.

297. Cox & Metcalfe, The Borneo Company Limited, 1998.

298. Marchant, *Wallace letters*, 1916, 1:38.

299. St. John, *Life in the forests*, 1862, pp. 89–90.

300. Bunyon, *Memoirs of McDougall*, 1889. I am grateful to Kees Rookmaaker for bringing this work to my attention.

301. St. John, *Life of Sir James Brooke*, 1879, p. 316.

302. *The Straits Times*, 12 Dec. 1854, p. 8.

303. St. John, *Life of Sir James Brooke*, 1879, p. 273.

304. Wallace, The entomology of Malacca, 1855.

305. Wallace to F. Sims 20 Feb. 1856 NHM WP1/3/37.

306. St. John, *Life of Sir James Brooke*, 1879, p. 274.

307. Wallace later noted in his copy of Darwin's *Origin of species* that orangutan's and humans were derived from a common, extinct, ancestor: "So with the orangutan & man." See Beddall, Wallace's annotated copy of Darwin's *Origin of species*, 1988, p. 283. Moore, Wallace's Malthusian moment, 1997, p. 299 was even misled to connect Wallace's supposed thoughts on orangutans as evolutionary ancestors, not of humans in general as St. John wrote, but of the Dyaks, to Wallace's Sarawak law, which was written before Wallace had seen an orangutan.

308. Bunyon, *Memoirs of McDougall*, 1889, p. 133.

309. *Notebook 1*, p. 122. The shells were probably purchased by Saunders. For Wallace's shells, see Wallace, List of the land shells, 1865.

310. MA1:54.

311. ML1:354.

312. Cited in Beccaloni, Homes sweet homes, 2008.

313. St. John, *Life of Sir James Brooke*, 1879, p. 14.

314. ML1:382.

315. Thomson, *Some glimpses*, 1865, p. 32. Some recent authors assume the term "boy" referred to Ali's age.

316. ML1:354.

317. Wallace's notes occupy a single page of *Notebook 1* (p. 17) in ink between other notes in pencil. The notes are given here with Wallace's responses in italics.

From, <u>Pictet's Paleontology</u>
Laws of Ecological development.

1. Species, limited Geolog. duration.	*Undisputed.*
2. Contemporaneous spec⁵· generally appearᵈ· or disapᵈ together?	*generally. unimportant.*
3. Older, fauna, gtᵗ· difference from recent	*True, with specific exceptions*
4. Recent faunas more diversified than ancᵗ·	*Doubtful, but does not affect develt. theory*
5. Most perfect Animals recent.	*True where a whole series can be traced often apparently false*
6. Order of appearance, like order of development.	*Important.*
7. From birth to death of species genus, or family, no interruption.	*Very important, undisputed.*
8. Faunas shew temp. has varied ?	
9. Ancient species more widely distributed?	*Because more sea & therefore more equal temp.*
10. Fossil animals same plan as living.	*Undisputed. Important.*

Wallace apparently read Pictet vol. 1, but it is not clear if he used the first or second edition.

318. Darwin, *Journal of researches*, 1845, pp. 173, 377.
319. Darwin, *Journal and remarks,* 1839, pp. 209–210.
320. There are at least two likely Forbes' articles that Wallace may have read, but he did not record the precise reference and textual clues are inconclusive: Forbes, Anniversary address of the president, 1854 and Forbes, On the manifestation of polarity, 1854. On Forbes see Browne, *Secular ark*, 1983, pp. 144–154, for polarity see p. 149.
321. Forbes, Anniversary address of the president, 1854, p. lxxix.
322. Wallace to H. Bates 4 Jan. 1858 NHM WP1/3/41.
323. ML1:354. On *Vestiges* influence here, see Schwartz, Darwin, Wallace, and Huxley, 1990 and McKinney, *Wallace*, 1972. See Lyell 3:99–100.
324. Hodge, Darwin and the laws of the animate part of the terrestrial system, 1983, pp. 30–31.
325. Wallace, On the law which has regulated the introduction of new species, 1855, pp. 190–191.
326. Wallace never used 'antitype' in his private notes and only twice more in a publication, once in the Ternate essay, and in: On the zoological geography of the Malay Archipelago, 1860. He often used the word "type" in *Notebook 4*.
327. Brackman, *A delicate arrangement*, 1980, p. 27. See also Lloyd, Wimpenny & Venables, Wallace deserves better, 2010.
328. Strickland, On the true method of discovering the natural system, 1840, p. 190. See Ospovat, The development of Darwin's theory, 1995 for a summary of the meanings of "branching" in earlier authors such as Pallas, Oken, Lamarck, Geoffroy, von Baer, Owen, and Milne Edwards. See also Chambers, *Vestiges*, 1844, pp. 191ff and Bulmer, Theory of natural selection of Wallace, 2005. Wallace objected to Swainson's circles in: Attempts at a natural arrangement of birds, 1856, p. 212.
329. Wallace, On the law which has regulated the introduction of new species, 1855, p. 191. For further discussion see Fagan, Theory and practice in the field, 2008.
330. The finger bones of manatees and whales were well illustrated in his copy of the *English Cyclopedia*.
331. "Large air cellules in the thigh bones of *Ostrich* (Proof that it is not to render the body *lighter* for flight.)" *Notebook 4*, p. 112.
332. [Chambers], *Vestiges*, 1844, p. 121. That rudimentary organs were not evolutionary vestiges was noticed by Fichman, *Wallace*, 1981, p. 43 and Raby, *Wallace*, 2001, pp. 102–103. Wallace had encountered similar suggestions about sudden appearances of varieties in Lawrence's Lectures on Man. See also Martin Rudwick's

summary of Geoffroy's theories of the sudden appearance of monstrosities: *Worlds before Adam*, 2008, p. 239.

333. Wallace, On the habits of the Orang-utan of Borneo, 1856, pp. 30–31. Wallace also discussed rudimentary organs in this context in *Notebook 4*, pp. 97–100.

334. *Notebook 1*, p. 121. *The Straits Times*, 13 Mar. 1855, p. 8.

335. Evenhuis, Publication and dating, 2003.

336. Severin, *Spice Islands*, 1997; Kohn, On the origin of the principle of diversity, 1981, p. 1106; Wilson, *Forgotten naturalist*, 2000, p. xii; Fichman, *An elusive Victorian*, 2004; Davies, *Darwin conspiracy*, 2008; Desmond & Moore, *Darwin's sacred cause*, 2009, p. 297; Lloyd, Wimpenny & Venables, Wallace deserves better, 2010 and Stott, *Darwin's ghosts*, 2012.

337. Severin, *Spice Islands*, 1997. Wallace himself later wrote: "My paper written at Sarawak rendered it certain to my mind that the change had taken place by natural succession and descent — one species becoming changed either slowly or rapidly into another." ML1:360.

338. This was noted by Bulmer, Theory of natural selection of Wallace, 2005.

339. "(Long before Owen published I had in M. S. worked out the succession of types in Old World.—) as I remember telling Sedgwick, who of course disbelieved it.)" Darwin to Lyell 27 [Dec. 1859] CCD7:456.

340. *Notebook 4*, p. 122; Wallace to Darwin [27 Sept. 1857] "my views on the order of succession of species" and Wallace to H. Bates 4 Jan. 1858 "my paper on the succession of species".

341. Wallace, The dawn of a great discovery, 1903.

342. Hamilton, Anniversary address of the president, 1856, p. cxviii.

343. Wilson, *Lyell's scientific journals*, 1970. But see ibid. p. 55.

344. Darwin, *On the origin of species*, 1859, p. 355.

345. CCD5:522. Some writers have accused Darwin of ignoring or concealing Wallace's Sarawak law paper in his historical sketch in later editions of *Origin of species*. However, the probable reason that Darwin did not list it was that the Sarawak paper did explicitly mention or declare evolution.

346. Darwin to Wallace 22 Dec. 1857 CCD6:514.

347. Darwin, *Origin*, 1859, p. 355. Darwin told Wallace he would say this, in Darwin to Wallace 6 Apr. 1859 CCD7:279.

348. Brooke to Wallace 4 July 1856 BL Add 46441 ff.2–5.

349. ML1:355.

350. Hamilton, Anniversary address of the president, 1856, p. cxviii.

351. Darwin to Wallace 1 May 1857 CCD6:387.

352. van Wyhe, Mind the gap, 2007.

353. "Wallace — Creationist or not — was receiving the nicest kind of trespass notice" Desmond & Moore, *Darwin*, 1992, p. 454 and Desmond & Moore, *Darwin's sacred cause*, 2009, pp. 253–254.

354. H. Bates to Wallace 19 Nov. 1856 NHM Catkey-418383.

355. Marchant, *Wallace letters*, 1916, 1:38–39.

356. St. John, *Life of Sir James Brooke*, 1879, p. 319.

357. Creswell, From Dudley colliery to Borneo, 1878.

358. Two years after Wallace's visit, an English miner named Marshall Creswell travelled to the Si Munjon mines to work under Coulson on a three-year contract. Creswell kept a journal that was later published in the *Newcastle Courant*; see Creswell, From Dudley colliery to Borneo, 1878. In ML1:341, Wallace said Coulson was a Yorkshireman; he was actually from Durham.

359. Wallace to Stevens 8 Apr. 1855 (*Zoologist* 13, 1855).

360. Wallace to Stevens 8 Apr. 1855 (*Zoologist* 13, 1855).

361. In his book on his travels in the Amazon, Wallace was never more specific than to mention his "insect-net". In a list of equipment in *Journal 2* he noted: "2 bag nets". On the history and context of Victorian collecting equipment, see Larsen, Equipment for the field, 1996.

362. Stevens, *Directions for collecting*, [1850?].

363. Bates gave particularly detailed instructions for skinning birds: Bates, Hints on the collection of objects of natural history, 1864, p. 311. For instructions to commercial collectors by Wallace himself see "Direction for Collecting in the Tropics by A.R. Wallace", p. 238.

364. Baker, Discovery of the Hymenoptera, 1995, discusses Wallace's labelling practices in detail. For the insect pins used, see Glaubrecht & Kotrba, Wallace's discovery of "curious horned flies", 2004. For Darwin's labels, see Rosen & Darrell, Darwin's specimen collections, 2011.

365. *Journal 3*, Oct. 1858 Kaióa. See also MA2:31.

366. Herbert, "A Universal Collector", 2010 and Fagan, Wallace, Darwin, and the practice of natural history. 2007.

367. Chancellor & van Wyhe, *Darwin's notebooks from the voyage of the Beagle*, 2009, p. 115. The collecting methods of Wallace and Darwin are compared and contrasted in Fagan, Wallace, Darwin, and the practice of natural history. 2007. For an overview of the large literature on the history of field work, see the special issue of *Osiris*, 1996.

368. Keynes, *Darwin's zoology notes & specimen lists*, 2000, p. 151.

369. Darwin, *Birds Part 3 of The zoology of the voyage of H.M.S. Beagle*, 1841, p. 64.

370. Desmond & Moore, *Darwin*, 1992, p. 171: Darwin "had tagged his specimens in a desultory manner and had rarely bothered to label by island"; Glick & Kohn,

On evolution, 1996, p. xv: "he had not even bothered to label his specimens by island".

371. Baker, Wallace's record of his consignments, 2001.
372. Baker, Wallace's record of his consignments, 2001, p. 255.
373. van Wyhe, http://darwin-online.org.uk/specimens.html
374. Anon., Huber on ants, 1812, p. 144.
375. [Burton], Stephens's British insects, 1828. The same was repeated in Burton, *Lectures on entomology*, 1837, p. 47. Thomson, *HMS Beagle*, 1995, pp. 143–144 also makes this point.
376. *Notebook 4*, p. 6.
377. Borneo Bay Cat, *Catopuma badia* (Gray, 1874). See Sunquist & Sunquist, *Wild cats of the world*, p. 49. I am grateful to Kees Rookmaaker for calling this to my attention.
378. *Notebook 4*, p. 7. Somerville, *Physical geography*, 1849, 2:216. On Somerville, see Secord, *Collected works of Mary Somerville*, 2004, vol. 1.
379. *Notebook 4*, p. 8. Knighton, *Tropical sketches*, 1855, p. 195.
380. Larkum, *A natural calling*, 2009. I am grateful to Tony Larkum for calling this to my attention.
381. Raby, *Wallace*, 2001, p. 171.
382. Wallace, Borneo, *Literary Gazette*, 1855.
383. Crawfurd, *Malay-English dictionary*, 1852, p. 122. Wallace's copy is in the library of the Linnean Society. MA1:62.
384. *Notebook 4*, p. 5.
385. Jones, The orangutan in captivity, 1982.
386. *Notebook 4*, p. 9. Wallace loosely quoted this passage in his paper "On the habits of the Orang-utan of Borneo", 1856, p. 29.
387. *Notebook 4*, p. 13.
388. *Notebook 4*, p. 16. We should not read too much symbolic meaning into this language of "the monster". It may simply mean that the animal struck Wallace as huge. Orangutans are the largest arboreal animals. A fly fisherman commonly calls a large trout a monster. Wallace also called a large snake on Bouru a "monster". MA2:133.
389. *Notebook 4*, p. 19.
390. Wallace, *Narrative of travels on the Amazon and Rio Negro*, 1853, p. 42.
391. St. John, *Life in the forests of the Far East*, 1862, 1:22.
392. Browne, A bigger picture of apes, 2006.
393. This woodcut (also used on the front cover vignette of MA) by artist Josef Wolf (who also drew the frontispiece orangutan) was drawn from a photograph by Walter Woodbury who was based in Java. The orangutan might therefore have

been a Sumatran rather than a Bornean animal. I am grateful to Ian Singleton for confirming that the animal might be the Sumatran species although it is difficult to tell from a young animal. Several other experts kindly consulted on my behalf by Richard Corlett felt it was impossible to determine from the woodcut. On Joseph Wolf, see Palmer, *Life of Joseph Wolf*, 1895.

394. Wallace, A new kind of baby, 1856.

395. *Notebook 4*, p. 27.

396. MA1:71–72.

397. In Wallace to H. Bates 30 Apr. 1856 NHM WP1/3/39, Wallace stated that he shot 15. But 16 in his paper on the Orang-utan or Mias, 1856. His notebook records show that 11 were shot by Wallace, the provenance of the others are not recorded. Some were purchased and some shot by others. See *Notebook 4*, pp. 10–11 and *Notebook 2/3*, pp. 9–10, 12.

398. Brooke to Wallace 31 Oct. 1857 NHM WP1/9/32.

399. St. John, *Life in the forests of the Far East*, 1862, 1:22. Wallace, On the Orang-utan or Mias, 1856, himself said as much in his article, "They were moreover all obtained in a very limited tract of country."

400. I am grateful to Jerry Drawhorn for this information.

401. Mrs Harris was the fictitious friend of Mrs Gamp's in Charles Dickens' *Martin Chuzzlewit* (1843–4). *Punch* used Gamp and Harris to lampoon the *Standard* and *Morning Herald*.

402. Wallace to F. Sims 25 June 1855 NHM WP1/3/34.

403. MA1:59.

404. MA1:81. However, he stated "three months" in Journey up the Sadong River, 1857. In *Notebook 2/3*, p. 71 he noted "July. All month in house with sore ancle."

405. *Notebook 4*, p. 3.

406. Knight, *English cyclopaedia*, 1854, 2:55.

407. *Notebook 4*, p. 31.

408. *Notebook 4*, p. 32.

409. McKinney, *Wallace*, 1972 and Raby, *Wallace*, 2001 treated this as the title of Wallace's future species book but this is based only on this heading for a single paragraph of notes. It was not indicated as a title of his book. When Wallace *did* write a title for a projected book he put it in quotation marks as with his "Coleoptera Malayana". *Notebook 4*, p. 133 and Wallace to F. Bates 2 Mar. 1858.

410. *Notebook 4*, pp. 35–36.

411. *Notebook 4*, p. 37. Wallace refers to Lyell 2:433: Lyell declared that "varieties will differ in some cases more decidedly than some species, if we…assume…that there are fixed limits, beyond which descendants from common parents can never deviate from a certain type".

412. *Notebook 4*, p. 39.

413. *Notebook 4*, pp. 41–42. Lyell 2:437ff.

414. MA1:101. Wallace stated that he left the mines with Bujon, but in S29 that he met Bujon at the village of Jahi.

415. Wallace to M. A. Wallace 25 Dec. 1855 NHM WP1/3/36.

416. MA1:107.

417. *Catalogue of books, in the Singapore library*, 1860, p. 4.

418. *Notebook 4*, p. 56.

419. MA1:132. St. John (1863, 1:166) stated that "Peninjau" meant "look out". In MA1:128 & 131 Wallace referred to "Peninjauh" as the name of the mountain rather than the bungalow — an error repeated by modern writers.

420. MA1:133.

421. MA1:132.

422. See Secord, Nature's Fancy, 1981.

423. Memorandum [Dec. 1855] CCD5:510. There has been endless confusion about who wrote first, Darwin or Wallace. That the letter came via Stevens (also in Darwin's list of recipients) see Wallace to A. Newton of 3 Dec. 1887, S459a.

424. Darwin to E. L. Layard 9 Dec. 1855; G. H. K. Thwaites 10 Dec. 1855 and C. A. Murray 24 Dec. 1855. The quotation is from the latter. CCD5:530.

425. *Notebook 2/3*, facing p. 1, records nightly insect captures up to the 18th. Part of this list was reproduced in MA1:134–135.

426. Wallace to F. Sims 20 Feb. 1856 NHM WP1/3/37.

427. J. Brooke to Wallace 5 Nov. 1856 BL Add 46441.

428. *Notebook 2/3*, p. 20.

429. Wallace to F. Bates 2 Mar. 1858 NHM WP1/3/42.

430. On the Sarawak collections see Cranbrook *et al.*, Wallace, collector, 2005.

431. In a 20 Feb. letter Wallace wrote, "The day I arrived here a vessel sailed for Macassar". The brig *Cocyra* sailed for Macassar on 16 Feb. (*The Straits Times*, 19 Feb. 1856, p. 8). Since Wallace was unable to get to Macassar for at least another two months, he must have left Sarawak on the *Santubong*. No ships bound for Macassar were listed in March or April 1856. In addition, Wallace wrote to Stevens 10 Mar. 1856, "I have been here already a month."

432. Wallace to F. Sims 20 Feb. 1856.

433. Wallace to T. Sims [20 Feb. 1856]. Presumably enclosed with the letter to F. Sims 20 Feb. 1856. On T. Sims see Ashton, Memoirs of a photographic pioneer, 1930. Thanks to Kees Rookmaaker for the latter reference.

434. *Notebook 4*, p. 56.

435. Thomson, *Account of the Horsburgh Light-house*, 1852. On Thomson see Hall-Jones & Hooi, *An early surveyor in Singapore*, 1979.

436. See S. Jayakumar & T. Koh, *Pedra Branca*, 2009.

437. These notes were probably written in Singapore, but as they are undated, they could also have been written at Sarawak. Owen, Description of the skull of a large species of Dicynodon, 1856.

438. *Notebook 4*, p. 59. Lindley, *An introduction to botany*, 1832, p. 521.

439. *Notebook 4*, pp. 60–61. Darwin, *Journal of researches*, 1845, p. 52. McKinney, *Wallace*, 1972 dates these notes to Lombok July 1856, presumably because of the next dated note, on p. 64. But the Darwin notes follow more closely to Singapore dates. As Wallace copied out the entire paragraph from Darwin and never noted reading him again, it was apparently not a book in his possession. It was likely available in Singapore. Several writers have claimed on the basis of this paragraph that Wallace carried a copy of Darwin's book with him, e.g., Kottler, Darwin and Wallace, 1985, p. 369, Moore, Wallace's Malthusian moment, 1997.

440. *Notebook 4*, p. 62. McKinney, *Wallace*, 1972 dated this to c. July 1856 but see note above. Blyth, An attempt to classify the "varieties" of animals, 1835.

441. *Notebook 4*, pp. 57–58 Huc, *L'Empire chinois*, 1854. I am grateful to Jerry Drawhorn for calling this reference to my attention and discussing Wallace's use of the Singapore Library. Wallace mentioned reading Huc in an April 1856 letter to his sister.

442. Wallace, Attempts at a natural arrangement of birds, 1856. Also written at this time was Wallace, Observations on the zoology of Borneo, 1856.

443. Wallace, On the habits of the Orang-utan of Borneo, 1856. See also Wallace, The philosophy of birds' nests, 1867. Lyell 2:415 attributed habits preceding structure to Lamarck. Jones, Wallace, Robert Owen and the theory of natural selection, 2002, suggests Wallace's aversion to instinct stems from the influence of Robert Owen. A similar rejection of the attribution of structure to habit appears in Wallace, *Narrative of travels on the Amazon and Rio Negro*, 1853, p. 85. Darwin objected to this and another passage in his unpublished species book. See Stauffer, *Natural selection*, 1975, pp. 379–380. Even in the Ternate essay Wallace mentioned: "Organization and resulting habits".

444. Wallace, On the habits of the Orang-utan of Borneo, 1856, p. 31. Whewell, *Plurality of worlds*, 1853.

445. MA1:77. Also the account he recorded in *Notebook 4*, p. 9: "If the boa attacks a mias he seizes and bites it in two."

446. Wallace to Stevens 12 May 1856 CUL Add. 7339/233.

447. *Notebook 4*, p. 55.

448. *Notebook 1*, p. 18.

449. MA1:210.

450. *The Straits Times*, 6 May 1856, p. 4. I am grateful to Erik Holmberg for identifying Buko Kang district.

451. MA1:37.

452. *The Straits Times*, 13 May 1856, p. 5.

453. A live lord. *The Straits Times*, 13 May 1856, p. 5. Wallace maintained that the tiger pits in Singapore were made by the Chinese. See Wallace, Popular natural history, 1880, p. 234 note.

454. The previous quotations are all from Wallace to Stevens 12 May 1856 CUL Add. 7339/233.

455. van Wyhe, Darwin's 'Journal'. http://darwin-online.org.uk/content/frameset?view type=side&itemID=CUL-DAR158.1-76&pageseq=3 f34

456. Darwin's name for the unpublished work, "big book", is constantly quoted, but it is hard to find any citation of the source. There are even varying renditions such as "Big Species Book" in Glick & Kohn, *On evolution*, 1996, p. xvi. There seem to be two sources where Darwin used the term. "I find to my sorrow it will run to quite a big Book." Darwin to W. D. Fox 3 Oct. [1856] CCD6:238 and "I am working very steadily at my big Book" Darwin to Lyell 10 Nov. [1856] CCD6:256.

457. Darwin's estimate on the first page of *Origin* is that it would take "two or three more years to complete it" i.e., by 1861–2. If we subtract the thirteen months spent writing *Origin* this would mean he could have published around 1860. Darwin's completion estimates before receiving Wallace's Ternate essay are consistent with these estimates.

458. *The Straits Times*, 3 June 1856, p. 8.

459. Wallace to Stevens 12 May 1856 CUL Add. 7339/233.

460. S25.

461. *The Straits Times*, 27 May 1856, p. 8.

462. *Journal 1*, 2.

463. *Journal 1*, 1. Registered in *Notebook 2/3*, p. 18 as "sea side, dark volcanic sand."

464. *Journal 1*, 3.

465. Wallace to Stevens 21 Aug. 1856 CUL Add 7339/234.

466. [Nordhoff], *The merchant vessel: a sailor boy's voyages to see the world*, 1856.

467. *Journal 1*, 3.

468. *Notebook 4*, p. 64.

469. *Notebook 2/3*, p. 18.

470. Zollinger reported that "Labuan Tring" (bamboo harbour) was the name of the harbour in the bay of Lombock. Zollinger, 1851, *Journal of the Indian Archipelago and eastern Asia* 5: 328.

471. MA1:242.

472. Daws & Fujita, *Archipelago: The islands of Indonesia*, 1999, p. 79.

473. Camerini, *Wallace reader*, 2002, p. 12.

474. Zollinger, *Tijdschr. v. N. I.*, 1847, IX, 2, bl. 198–205; Zollinger, The island of Lombok. *Journal of the Indian Archipelago and eastern Asia*, 1851, 5:335–336. Wallace mistakenly described Zollinger as Dutch in Wallace, A list of the birds, 1863, p. 481 and MA1:317. See also Camerini, An early history of Wallace's Line, 1993, p. 710.

475. Wallace to Stevens 21 Aug. 1856 CUL Add 7339/234.

476. Sclater, On the general geographical distribution of the members of the class Aves, 1858. Curiously, this claim was made by Bickmore, *Travels in the East Indian Archipelago*, 1869.

477. Wallace to H. Bates 4 Jan. 1858 NHM WP1/3/41. Fichman, Biology and politics, 1997, pp. 11–14, 28, 46–47 discusses Wallace's penchant for boundary-marking.

478. Wallace, On the zoological geography of the Malay Archipelago, 1860, pp. 172–184. When he received this Darwin replied that it was reminiscent of the earlier findings of Earl, On the physical structure and arrangement of the islands in the Indian Archipelago, 1845. Darwin to Wallace 9 Aug. 1859 CCD7:323.

479. Huxley, On the classification and distribution of the Alectoromorphae and Heteromorphae. *Proceedings of the Zoological Society of London*, 1868, p. 313. See also Camerini, An early history of Wallace's Line, 1993. Wallace changed the exact path of the line in later publications.

480. Wallace to H. Bates 4 Jan. 1858 NHM WP1/3/41.

481. Earl, On the physical structure and arrangement of the islands in the Indian Archipelago, 1845. It is not known when Wallace read Earl. Wallace first cited Earl in: On the physical geography of the Malay Archipelago, 1863.

482. Darwin eventually discussed the colours of the skins of Lombok "Penguin drakes", the bizarre upright Indian runner ducks, in *The variation of animals and plants under domestication* in 1868, although he noted they were sent by Sir James. *Variation* 1:280. This was apparently the only time Darwin discussed any species from Lombok.

483. *Journal 1*, 21.

484. *Journal 1*, 22.

485. Hantus = spirits. Hantus was not found in Wallace's Malay dictionary: Crawfurd, *Malay-English dictionary*, 1852. See MA2:166. *Journal 1* and MA1:251–252.

486. *Journal 1*, 10.

487. See P. Armstrong, *Under the blue vault of heaven*, 1991. The "missionary" was in fact a Dutch trader named J.P. Freyss. Identified by Veth, *Insulinde*, 1870, 1:310. See Freijss, Reizen naar Manggarai en Lombok in 1854–1856. *Tijdschrift voor Indische Taal-, Land- en Volkenkunde*, 1860, 9:443–530.

488. Cameron, *Our tropical possessions*, p. 12.

489. MA1:272. For more reports of harsh punishments in Lombok see Zollinger, The island of Lombok. *Journal of the Indian Archipelago and eastern Asia*, 1851, 5:459–469.

490. *Journal 1*, 29.

491. *Journal 1*, 30.

492. Wallace to H. Bates 30 Apr. 1856 NHM WP1/3/39.

493. I. S. Homans, *An historical and statistical account of the foreign commerce of the United* ... 1857, p. 112. See Vickers, *A history of modern Indonesia*, 2005.

494. *Chambers's encyclopaedia*, 1874, vol. 6.

495. Baker, Wallace's record of his consignments, 2001, p. 307.

496. MA1:332.

497. Berthoud, Davis & Reid, *Joseph Conrad*, 2008, p. 525. Thanks to Shannon Bohle for sending me a copy of this work. On Mesman see Bosma & Raben, *Being "Dutch" in the Indies*, 2008, p. 145.

498. *Journal 1*, 34.

499. MA1:334.

500. Pfeiffer, *A lady's second journey round the world*, 1855, 2:21–22.

501. *Journal 1*, 35. Phaenicophaus = Phaenicophaeus. The species name "callirhynchus" was taken from Bonaparte 1850, p. 98, but this was a mistake by Bonaparte for *Phoenicophaeus calyorhynchus* (Temminck, 1825) which is the Yellow-billed Malkoha. With thanks to Kees Rookmaaker.

502. CCD6:387–388.

503. Wallace did not give this date or place name in MA or his *Journal* but it can be found in *Notebook 4*, pp. 40–41.

504. *Journal 1*, 39. Bickmore, *Travels in the East Indian Archipelago*, 1869, p. 36 also reported that water buffaloes reacted strangely to Westerners.

505. Baker, Wallace's record of his consignments, 2001, p. 268.

506. Wallace to J. & M. Wallace 6 Dec. 1856.

507. *Journal 1*, 45.

508. Wallace to F. Sims 10 Dec. 1856.

509. *Journal 2*, p. 102 and Veth, *Insulinde*, 1871, 2:237. Wallace, Narrative of search after Birds of Paradise, 1862, p. 153.

510. Chancellor & van Wyhe, *Darwin's notebooks from the voyage of the Beagle*, 2009, p. 82.

511. Wallace, On the natural history of the Aru Islands, 1857, mistakenly stated that he took two boys, p. 474.

512. MA2:164.

513. MA2:195.

514. *Journal 1*, 48.

515. *Journal 1*, 48.

516. 31 Dec. according to *Journal 1*, 49 and MA2:174, but "January 1st, 1857" in S38, p. 473.

517. MA2:174–175.

518. *Journal 1*, 51.

519. Wallace to Stevens 10 Mar. 1857. S35. [*Cyphogastra calepyga*]

520. *Journal 1*, 55.

521. MA2:193.

522. Wallace, On the Arru Islands, 1858, p. 165.

523. *Journal 2*, 102.

524. MA2:199–200.

525. MA2:201–202.

526. *Notebook 2/3*, p. 127.

527. Commentary following Wallace, On the Arru Islands, 1858, p. 170.

528. O'Connor, S., M. Spriggs, P.M. Veth, The archaeology of the Aru Islands, Eastern Indonesia. *Terra Australis*, 2005, 22:90. On changes in Aru by 1872 see J.T. Cockerell, *The Brisbane Courier*, 21 Jan. 1874, p. 5. Thanks to Kees Rookmaaker for the latter reference.

529. *Notebook 4*, p. 46 gives 16 Feb., 5 Feb. is noted without change of location on p. 44.

530. Enclosed with a cover letter from J.T. Cockerell BL Add. 46442 ff. 92–93. This letter was mentioned by George, *Biologist philosopher*, 1964, pp. 40–41, who incorrectly stated that Wallace was invited to return to Aru as governor. Jan Van Der Putten kindly looked at the letter for me and provided a more accurate translation. The letter was a token of respect to Wallace who was still remembered.

531. *Journal 2*, 75.

532. von Rosenberg, *Der malayische Archipel*, 1878, p. 332, my translation.

533. In *Notebook 4*, p. 46b he stated 10 May but he noted in his *Journal* that he had lost track of the days while away from Dobbo.

534. Wallace to Stevens 10 Mar. 1857. S35.

535. In the *Journal* and MA he stated a "month" but in: On the great Bird of Paradise, 1857, three weeks. He did not indicate which ship or date his consignment left Macassar. The voyage to Singapore was about ten days; see *The Straits Times*, 28 July 1857, p. 4.

536. *The Straits Times*, 8 Sept. 1857, p. 4.

537. Collecting notes at Maros start on 11 Sept. 1857. *Notebook 4*, p. 48. However, a note on the top of *Notebook 4*, p. 49 reads "from. Aug. 16th. /57" which may suggest he left Macassar then. If so, it leaves unexplained why his collecting records begin on 11 Sept.

538. Beddall, Wallace, Darwin, and the theory of natural selection, 1968 and Slotten, *Heretic in Darwin's court*, 2004, p. 144.

539. Knight, Cetacea. *English Cyclopedia*, p. 869–915.
540. *Notebook 4*, p. 90. This passage was quoted by Wallace in a letter to Darwin [Dec? 1860] CCD8:504. McKinney, *Wallace*, 1972 also dated these notes, on unstated grounds, to "late in 1857". Von Buch, *Description physique des Iles Canaries*, 1836, which is not listed in the 1860 Singapore library catalogue.
541. *Notebook 4*, p. 40. This pencil insertion is in darker pencil. It sounds more like Wallace's theoretical views from the second Macassar visit, and refers to the same publication as the 1857 Macassar notes, but this is a speculative interpretation.
542. Lyell 2:447.
543. Ospovat, The development of Darwin's theory, 1995, p. 58, 246 note 62.
544. Wallace, On the natural history of the Aru Islands, 1857.
545. Wallace, On the natural history of the Aru Islands, 1857. *Notebook 4*, p. 50, the quotation is not verbatim from Lyell 3:154. Wallace's italics.
546. Wallace, On the theory of permanent and geographical varieties, 1858. Wallace had made notes on Blyth's categories of varieties paper in *Notebook 4*, p. 62, apparently in Singapore in March–May 1856. In the final paragraph, Wallace used letters to represent species which is reminiscent of Strickland, On the true method, 1840, p. 189. The permanent varieties paper also drew on notes from *Notebook 4*, p. 45.
547. Knox's articles appeared between 1855 and 1857. On Knox, see Desmond, *Politics of evolution*, 1989.
548. Wollaston, *On the variation of species*, 1856. Wollaston's book was reviewed in *Entomologist's weekly intelligencer*, 1856, 1:78–79. See Browne, *Voyaging*, 1995, p.539. However, Wallace to F. Bates 2 Mar. 1858 suggests Wallace had not seen Wollaston's book but only a report of it.
549. The phrase was also used in Lyell 2:131, 183.
550. Prichard, *The natural history of man*, 1843, p. 11.
551. There is a much clearer declaration of Wallace's views on this subject in a paper he published in *Ibis* in 1860, after his evolutionary Ternate essay and Darwin's *Origin of species* had appeared: "Now, I believe in all these cases, where the difference is constant, we must call them distinct species. A 'permanent local variety' is an absurdity and a contradiction; and, if we once admit it, we make species a matter of pure opinion, and shut the door to all uniformity of nomenclature". Note that Baden Powell, *Essays on the spirit of the inductive philosophy*, 1855, p. 379, made almost the identical point. I am grateful to Jon Hodge for pointing out this passage in Baden Powell.
552. See *Notebook 4*, pp. 48b–49b for "Amasanga" collecting records dated 11 Sept.– 8 Nov. 1857. Wallace (*Zoologist* 16, 1858) wrote "I arrived in August". *Notebook 2/3*, p. 132ff also covers this time. This is a note on insect habits, colours and environment.

553. Bosma & Raben, *Being "Dutch" in the Indies*, 2008, p. 147.

554. Mentioned in Wallace, On the Great Bird of Paradise, 1857.

555. MA2:364. In *Notebook 4* he called the area Amasanga, but did not use this name elsewhere.

556. MA2:365.

557. *Journal 2*, 108. See *Notebook 4*, pp. 48b–49b. On 26 Sept. he wrote up his *Journal* entry on the excursion. The date may give the incorrect impression that he was at the falls on the 26th.

558. Wallace to Darwin [27 Sept. 1857] CCD6:457–458.

559. *Journal 2*, 111.

560. Wallace to F. Bates 2 Mar. 1858 NHM WP1/3/42.

561. *Notebook 2/3*, p. 141. It is not clear when or where these were collected. Two collected in Maros in *Notebook 4*, p. 51b seem possible, except for the reference to salt water- which is apparently absent in Maros far inland.

562. *The Straits Times*, 23 Jan. 1858, p. 2.

563. Campo, *Engines of empire*, 2002, pp. 38–42.

564. *The Straits Times*, 26 Jan. 1861, p. 2; *The Straits Times*, 29 June 1861, p. 4.

565. Bastin, Introduction to *The Malay Archipelago*, 1986, gives this date, but copied it from George, *Biologist philosopher*, 1964 where no source is given. The 1860 Dutch mail schedule reproduced in Brooks, *Just before the origin*, 1984, gives "5–6 days" Macassar to Timor, so 25 Nov. is plausible.

566. In MA1:288 Wallace stated he stayed there a day. Strangely, he wrote in his *Journal* that Coupang was on the "NE end of the large island of Timor"; it is the southwest.

567. MA1:450. Wallace was mistaken, the Portuguese built no fort on Banda. He apparently referred to Fort Belgica built by the Dutch in 1611 which still stands.

568. MA2:450. See Veth, *Insulinde*, 1870, vol. 1, note 2 for Banda.

569. Pfeiffer, *A lady's second journey round the world*, 1855, 1:366.

570. See Veth, *Insulinde*, 1870, vol. 1, note 5.

571. Mohnike, *Blicke auf das Pflanzen- und Thierreich*, 1883. I am grateful to Anna Mayer for calling this work and other information to my attention.

572. Wallace to Stevens 20 Dec. 1857. S44.

573. Wallace to H. Bates 4 Jan. 1858 NHM WP1/3/41.

574. See Stagl, Carl Ludwig Doleschall, 1999 and Glaubrecht & Kotrba, Wallace's discovery of "curious horned flies", 2004. I am grateful to Anna Mayer for calling these works to my attention. See also Veth, *Insulinde*, 1870, vol. 1, notes to Chapter 20.

575. MA2:463.

576. *Notebook 4*, pp. 52b–53b.

577. MA2:465.

578. MA2:466.

579. MA2:467–468.

580. Pfeiffer, *A lady's second journey round the world*, 1855, 1:364.

581. *Journal 2*, 119.

582. *Journal 2*, 120.

583. Darwin to Wallace 1 May 1857 CCD6:387.

584. Wallace to H. Bates 4 Jan. 1858 NHM WP1/3/41.

585. Marchant, *Wallace letters*, 1916, 1:36.

586. Pfeiffer, *A lady's second journey round the world*, 1855, 1:401.

587. Bleeker, *Reis door de Minahassa en den Molukschen archipel*, 1856, 1:163. I am grateful to Kees Rookmaaker for calling this work to my attention.

588. Veth, *Insulinde*, 1871, vol. 2, chapter 21, note 10.

589. Wallace misspelled the name as "Duivenboden". 'van Renesse' was added to his name in 1860. Wiersma, Duijvenbode, 1885. Thanks to Kees Rookmaaker. See Bosma & Raben, *Being "Dutch" in the Indies*, 2008, p. 164 and Dickenson, Indian archipelago, 1838.

590. Apparently in the area called Daolasi on Akehude beach. Heij, *Biographical notes of Antonie Augustus Bruijn*, 2010, p. 38.

591. Bleeker, *Reis door de Minahassa en den Molukschen archipel*, 1856, 1:172.

592. MA2:2. Bickmore, *Travels in the East Indian Archipelago*, 1869, p. 302 gives instead "Prince of the Moluccas".

593. Several writers mistakenly claim that van Duivenbode owned the house.

594. *Journal 2*, 123.

595. MA2:3–4.

596. Wallace to H. Bates 4 Jan. 1858 NHM WP1/3/41.

597. *Notebook 4*, p. 109.

598. Wallace to A.B. Meyer 22 Nov. 1869 *Nature*, 1895, 52:415; Wallace to A. Newton 3 Dec. 1887, in F. Darwin, *Charles Darwin*, 1892, p. 189–190; Wallace, *Natural selection*, 1891, p. 20–21; *Wonderful century*, 1898, p. 140; Wallace, The dawn of a great discovery, 1903; ML1:360–363; Anon., *Darwin–Wallace celebration*, 1908, p. 6–7, 117–118.

599. Bastin, Introduction, 1986; George, *Biologist philosopher*, 1964, Raby, *Wallace*, 2001; Slotten, *Heretic in Darwin's court*, 2004, p. 144; Beccaloni and also J. Mallet in Smith & Beccaloni, *Natural selection and beyond*, 2008, p. 27; Severin, *Spice Islands*, 1997; Moore, Wallace's Malthusian moment, 1997; Larson, *Evolution's workshop*, 2001; Fagan, Wallace, Darwin, and the practice of natural history, 2007; Benton, Race, sex and the 'earthly paradise', 2009; Desmond & Moore, *Darwin's sacred cause*, 2009, p. 299 and Stott, *Darwin's ghosts*, 2012. Fichman, *Wallace*, 1981, p. 166 note 48 seems to be one of the few authors to doubt this.

600. Moore, Wallace's Malthusian moment, 1997. Beccaloni in Smith & Beccaloni, *Natural selection and beyond*, 2008, p. 27. Raby, *Wallace*, 2001, p. 133.

601. ML1:338; Wallace to F. Sims 10 Oct. 1861 NHM WP1/3/51. At least ten publications by Wallace from the archipelago were signed from locations without a post office including Batchian, Ampanam, Lombock, Awaiya and Ceram. Letters to his cousin C.A. Wilson were signed "Gilolo". *South Australian Register* 21 Sept. 1859. Thanks to Jerry Drawhorne.

602. *Notebook 4*, p. 54b.

603. MA2:16. Many writers have assumed that this illness is the famous one during which he thought of natural selection.

604. Oxford University Museum of Natural History Library. The document, mostly in another handwriting, was apparently for writing his life or identifying his collections years later. The durations of the stays and some dates were added in Wallace's handwriting. The second estimate for Sept. "2 weeks?" is accurate. The document is mentioned in Smith, *A history of the Hope entomological collections*, 1986.

605. Wallace, Notes on the localities given in Longicornia Malayana, 1869, p. 695. The second stay was two weeks (14 Sept.–1 Oct. 1858). The few contemporary fragments we have support this. *Notebook 2/3* lists only sixteen Coleoptera collected at Ternate before Dodinga, Gilolo. But Dodinga has only fourteen further species — a paltry number even for a two week stay. *Notebook 2/3*, pp. 144–147.

606. This is not unprecedented. Wallace misdated a surviving letter to his sister with the wrong year, Wallace to F. Sims 20 Feb. 1855 [1856].

607. Winchester, *Krakatoa*, 2003, p. 105.

608. Dana, Thoughts on species, 1857 seems a tantalising candidate. But Evenhuis, Publication and dating, 2003 indicates that this issue of *Annals* was not published until 1 Jan. 1858, if so then it could not have arrived at Ternate in time. Seventy-five days was the minimum transit time.

609. Wallace used this expression in his recollections: Wallace to A. Newton 3 Dec. 1887, in F. Darwin, *Charles Darwin*, 1892, p. 189–190; Wallace, *Natural selection*, 1891, p. 20–21; Wallace, My relations with Darwin, 1903, p. 78; ML1:360–363.

610. Curiously, Brooks, *Just before the origin*, 1984, pp. 174–199 claimed it was his idea. Young, *Darwin's metaphor*, 1985, p. 44; Moore, Wallace's Malthusian moment, 1997 and van Wyhe, Alfred Russel Wallace, 2013.

611. McKinney, *Wallace*, 1972.

612. *Notebook 4*, p. 40. Written apparently in Macassar, July–19 Nov. 1857. McKinney, *Wallace*, 1972 was the first to show the similarities between the Ternate essay and passages in *Notebook 4*.

613. Wallace to F. Bates 2 Mar. 1858 NHM WP1/3/42.

614. ML1:359.

615. *Notebook 4*, p. 59, probably in March–May 1856 in Singapore. Wallace quoted Lindley, *An introduction to botany*, 1832, Chapter 2 Irregular Metamorphosis, p. 521.

616. *Notebook 4*, p. 150. My italics. Wallace was responding to Lyell 3:21.

617. Browne, *Secular ark*, 1983 rightly describes influence of Malthus on Wallace as one of population theory, not economic theory. See also the important discussions in Gale, Darwin and the concept of a struggle for existence, 1972; Bowler, Malthus, Darwin, and the concept of struggle, 1976; Winch, Darwin fallen among political economists, 2001.

618. Wallace, The disguises of insects, 1867. Thanks to George Beccaloni for calling this to my attention.

619. [Wallace], Mimicry, 1867, p. 8. The term "camouflage" came into use during the First World War, and was soon applied to species. See for example Howes, *Insect behavior*, 1919, pp. 156, 168.

620. Lyell 2:453: "new peculiarities…do not attest any tendency to departure to an indefinite extent from the original type of the species". Also, "the descendants of common parents may deviate indefinitely from their original type." Lyell 2:406 and "departure from a common type", Lyell 2:438. See *Notebook 4*, p. 45: "Changes which we bring about artificially in short periods may have a tendency to revert to the parent stock though this in animals is not proved." McKinney, *Wallace*, 1972 first noticed the overwhelming influence of Lyell in Wallace's notes and that the Ternate essay was primarily a reaction to Lyell. See also Wallace's notes in *Notebook 4*, p. 90 on von Buch, *Description physique*, 1836, p. 147–148. Wallace wrote: "brought back to the primitive type". This passage was quoted by Wallace in a letter to Darwin [Dec? 1860] CCD8:504. Also *Notebook 4*, p. 149–150, quoting Lyell 3:21. It is unclear when Wallace wrote this in his notebook. He later quoted this passage in *Darwinism*, 1889, p. 4.

621. Lyell 2:445: "there is so decided a tendency in the seedlings to revert to the original type, that our utmost skill is sometimes baffled in attempting to recover the desired variety". Wallace contested this in *Notebook 4*, p. 45 in July 1855. See also Lyell 2:438: "Domestic animals in South America have reverted to their original character" and Lyell 2:465: "no indefinite capacity of varying from the original type". Also Lyell 2:450: "But, before we can infer that there are no limits to the deviation from an original type which may be brought about in the course of an indefinite number of generations we ought to have some proof that, in each successive generation, individuals may go on acquiring an equal amount of new peculiarities, under the influence of equal changes of circumstances."

622. Lyell discussed this: "in the universal struggle for existence, the right of the strongest eventually prevails; and the strength and durability of a race depends mainly on

its prolificness, in which hybrids are acknowledged to be deficient". Lyell 3:9, 140, 162. This sense of struggle stressed that some survive over others. So, like Blyth, Lyell said the struggle for existence kept hybrids in check and thus species limits preserved.

Darwin's *Journal of researches*, 1845, p. 175, contains a reminiscent passage: "We do not steadily bear in mind, how profoundly ignorant we are of the conditions of existence of every animal; nor do we always remember, that some check is constantly preventing the too rapid increase of every organised being left in a state of nature. The supply of food, on an average, remains constant; yet the tendency in every animal to increase by propagation is geometrical."

623. Bowler, Wallace's concepts of variation, 1976 and Bulmer, Theory of natural selection of Wallace, 2005.

624. *Notebook 4*, p. 40. This page is dated July 1855 however this line is an insertion in darker pencil. It sounds more like his theoretical views from Macassar in 1857. *Vestiges*, Owen and Huxley also proposed, in their different ways, abrupt appearances of new forms. See Rupke, *Owen*, 2009 and Desmond, *Huxley*, 1997. For Wallace's earlier and different views on species and varieties see his letter to H. Bates 9 Nov. 1847 in Marchant, *Wallace letters*, 1916, 1:91–92. See the useful discussion in Bulmer, Theory of natural selection of Wallace, 2005, p. 132 quoting an 1896 letter from Wallace claiming that "My 'varieties' therefore included 'individual variations'". I do not see that Wallace's recollections decades later are applicable to what the Ternate essay originally meant in 1858.

625. Bowler, Wallace's concepts of variation, 1976: variety = variant population not geographically isolated.
Brooks, *Just before the origin*, 1984: variety = variant population but geologically isolated.
Kottler, Wallace, the origin of man, 1985: varieties = variant individuals in some cases and races in others. Kottler doubted that varieties did not arise from a process of natural selection.
Browne, *Charles Darwin*, 2002, p. 62: "Wallace had spoken mostly about the replacement of species by other species, groups by groups, rather than the individual changes that preoccupied Darwin".
Bulmer, Theory of natural selection of Wallace, 2005: variety = variant population; variation "usually meant a heritable character…or an individual possessing such a character".
Fagan, Theory and practice in the field, 2008, p. 86 argued again for individuals, also citing, p. 90: "George Beccaloni (pers. commun.) argues, following Kottler (1985), that by 'varieties' Wallace meant not subspecies or 'permanent varieties,' but individuals differing from the norm."

626. Wallace, Note on the theory of permanent and geographical varieties, 1858, p. 5888.

627. Wallace, *Narrative of travels on the Amazon and Rio Negro*, 1853, p. 361; Letters from the Eastern Archipelago, 1854, p. 260; On the Orang-utan or Mias of Borneo, 1856, p. 475 and *Zoologist* 16, 1858, p. 6122.

628. Beddall, Wallace's annotated copy of Darwin's *Origin of species*, 1988, p. 269. A strikingly similar consideration was noted by Darwin in his *Notebook C*, p. 85. In 1847, Wallace had read of varieties arising as he wrote to H. Bates: "Lawrence's 'Lectures on Man.'...The great object of these 'Lectures' is to illustrate the different races of mankind, and the manner in which they probably originated, and he arrives at the conclusion (as also does Prichard in his work on the 'Physical History of Man') that the varieties of the human race have not been produced by any external causes, but are due to the development of certain distinctive peculiarities in some individuals which have thereafter become propagated through an entire race." Marchant, *Wallace letters*, 1916, 1:91; ML1:254–255. It is by no means clear how long Wallace remained convinced of this manner of the origination of races.

629. See for example: S13:258; S27:230; S35:92; S33:5655–5656; S36:272; S47; "that variety of mankind" meaning the 'race' of South American Indians. S10.

630. Wallace changed his 1858 views by 1867: "Perhaps no principle has ever been announced so fertile in results as that which Mr. Darwin so earnestly impresses upon us, and which is indeed a necessary deduction from the theory of Natural Selection, namely—that none of the definite facts of organic nature, no special organ, no characteristic form or marking, no peculiarities of instinct or of habit, no relations between species or between groups of species—can exist, but which must now be or once have been *useful* to the individuals or the races which possess them." Wallace, Mimicry, 1867, pp. 2–3.

631. Divergence tends to receive disproportionate attention in the writings of pro-Wallace commentators such as Brooks, *Just before the origin*, 1984.

632. Kohn, On the origin of the principle of diversity, 1981. See also Kottler, Darwin and Wallace, 1985 and Beddall, Darwin and divergence, 1988.

633. Kohn, On the origin of the principle of diversity, 1981, p. 1105.

634. "Seeing this gradation and diversity of structure in one small, intimately related group of birds, one might really fancy that from an original paucity of birds in this archipelago, one species had been taken and modified for different ends." Darwin, *Journal of researches*, 1845, p. 380. Again, Darwin, *A monograph on the fossil Lepadidæ*, 1851, p. 48: "This, the most ancient genus of the Lepadidæ, seems also to be the stem of the genealogical tree; for Pollicipes leads, with hardly a break, by some of its species into *Scalpellum villosum*..."

635. Kohn, On the origin of the principle of diversity, 1981, p. 1105.

636. Brackman, *A delicate arrangement*, 1980, p. 17 and Davies, *Darwin conspiracy*, 2008, p. 146. Darwin to Hooker 8 [June 1858] CCD7:102.

637. See Kohn, On the origin of the principle of diversity, 1981, p. 1107 and Beddall, Wallace's annotated copy of Darwin's *Origin of species*, 1988, p. 9.

638. Browne, Darwin's botanical arithmetic, 1980. See also Beddall, Darwin and divergence, 1988; Ospovat, The development of Darwin's theory, 1995; Kohn, Internal dialogue, 1985; Pearce, Darwin and the economy of nature, 2010 and Raby, *Wallace*, 2001, p. 287–289.

639. Kohn, On the origin of the principle of diversity, 1981, p. 1106.

640. Wallace made this point in July 1855 in *Notebook 4*, p. 44: "Many of Lamarck's views are quite untenable & it is easy to controvert them but not so the simple question of a species being produced in time from a closely allied distinct species which however may of course continue to exist as long or longer than its offshoot."

641. Passages like this unambiguously apply natural selection to groups rather than individuals because "varieties" and "races" are here used synonymously.

642. E.g., Desmond, *Huxley*, p. 245; Fagan, Theory and practice in the field, 2008, p. 86 and Smith in Smith & Beccaloni, *Natural selection and beyond,* 2008, p. 347 and Smith, Wallace's unfinished business, 2004; Cronin, *The ant and the peacock*, 1993, p. 19; Desmond & Moore, *Darwin*, 1992, p. 468 and Smith, Wallace, past and future, 2005.

643. This is the only quotation in the essay. The other lines in quotation marks are emphasised remarks. Wallace refers to Owen, Description of the skull of a large species of Dicynodon, 1856. It is actually not a quotation from Owen, Wallace copied it from *Notebook 4*, p. 54 which was probably written in March 1856 in Singapore.

644. ML1:360.

645. Wallace to Hooker 6 Oct. 1858; Wallace to Stevens 29 Oct. 1858 CUL Add. 7339/235; *Notebook 4*, p. 122.

646. F. Darwin, *Charles Darwin*, 1892, p. 190. Italics in the original.

647. Wallace to Darwin [27 Sept. 1857] CCD6:457. Wallace to H. Bates 4 Jan. 1858 NHM WP/1/3/41.

648. Smith, *Wallace; an anthology*, 1991.

649. de Clercq, *Ternate*, 1890, claims that Sidangoli was not a village but an inlet and that the attribution of the house to the Sultan of Tidore rather than the Sultan of Ternate, whose territory this was, an "absurdity", p. 26 of English translation.

650. *Journal 2*, 125.

651. *Notebook 4*, p. 109.

652. *Notebook 4*, p. 109.

653. de Clercq, *Ternate*, 1890, translation p. 37. De Clercq was ever ready to spot mistakes it seems, even correcting Wallace for stating that he "unloaded his luggage on the 'beach;' he meant on the bank of the river or creek.", p. 36.

654. According to de Clercq, *Ternate*, 1890, p. 32: "Wallace restricts the dwelling place of the Alfurus to the east coast and the interior…but this is due to his ignorance of the real situation. He is also confused by the fact that Moslems are to be found in all the coastal villages."

655. *Journal 2*, 127. Several writers have assumed that this passage was related to Wallace's inspiration to think of Malthus and the Ternate essay.

656. Wallace to Silk 30 Nov. 1858 NHM WP1/3/45. A point earlier observed by Vetter, Wallace's other line, 2006.

657. Bickmore, *Travels in the East Indian Archipelago*, 1869, p. 311.

658. See for example Bronowski, The ascent of man (BBC 1973) episode 8; Wilson, *Forgotten naturalist*, 2000; Lowrey, Wallace as ancestor figure, 2010; Larson, *Evolution*, 2004; Benton, Darwin and Wallace as environmental philosophers, 2009 and Benton, Race, sex and the 'earthly paradise', 2009.

659. Desmond & Moore, *Darwin*, 1992, p. 467.

660. Equating different non-European races is reminiscent of the debate between Marshall Sahlins and Gananath Obeyesekere regarding the peoples of Hawaii. See Sahlins, *How "natives" think: about Captain Cook, for example*, 1996.

661. An important exception to the usual descriptions of Wallace's attitudes to other races is Vetter, The unmaking of an anthropologist Wallace returns from the field, 2010. Vetter's article deals with Wallace's overlooked anthropological ideas and seeks to put them in context of Dutch paternalistic colonialism and racial progress. In fact, Wallace used the word "savages" to describe non-European races more than thirty times in *The Malay Archipelago*. Darwin used the term nineteen times in his *Journal* and seventeen times in the 1845 edition.

662. *Journal*, 208, 71, 182, 146.

663. Wallace, The native problem in South Africa, 1906, p. 176.

664. Pearson, Wallace's Malay Archipelago journals and notebook, 2005, p. 67.

665. *Notebook 2/3*, p. 140. CCD6:514.

666. Some writers have speculated that Wallace would have sent his essay to Darwin via the more expensive route "via Marseille" (which diverged from the P&O at Malta) rather than "via Southampton" and the letter would have arrived four days earlier. This is apparent in the difference between Wallace's departure date from Southampton, 4 March 1854, and the date of the most recent mails to arrive with him at Singapore — 8 March.

667. Darwin to Wallace 22 Dec. 1857 CCD6:514.

668. ML1:363. The same information can be gleaned from Darwin's letters and Lyell, *Antiquity of man*, 1863, p. 408: recalled the letter asked that "it might be shown to me if thought sufficiently novel and interesting."

669. Browne, *Secular ark*, 1983, p. 182 and Raby, *Wallace*, 2001, p. 133 suspected that Darwin's letter arrived in Ternate as early as February or as late as March. The February steamer arrived only c. forty-nine days after Darwin sent his letter — far too early for it to arrive. However, 9 March was seventy-seven days later, exactly when it should be expected. In John van Wyhe & Kees Rookmaaker, a new theory to explain the receipt of Wallace's Ternate Essay by Darwin in 1858, 2012, I mistakenly claimed that Raby "assumed that Darwin's letter arrived in February". Raby, *Wallace*, 2001, p. 133 called attention to the fact that Wallace must have written in reply to the Darwin letter. Davies, *Darwin conspiracy*, 2008, p. 142 confirmed the arrival of the letter on 9 March with a complete itinerary from London to Ternate. He deserves credit for showing clearly when Darwin's letter arrived in Ternate.

670. Davies, *Darwin conspiracy*, 2008, p. 142.

671. Anon. *Verslag van het beheer en den staat der Nederlandsche bezittingen*, 1858, p. xiii, 521. See Campo, *Engines of empire*, 2002, p. 41. I am grateful to Kees Rookmaaker for finding these references.

672. J. Motley to Wallace 22 May 1858 Banjermassin [Borneo] BL Add 46435.

673. Brooks, *Just before the origin*, 1984, p. xviii.

674. Others who accept early dates are Daniels, *Usk: origin of a thinker*, p. 24 and Gardiner, The joint essay of Darwin and Wallace, 1995, p. 21.

675. I am very grateful to Judith Magee for kindly sending copies of letters bearing postmarks in the NHM Wallace Collection.

676. The editors of the *Correspondence of Charles Darwin* pointed out: "The correspondence between mid-May and mid-June 1858 provides some circumstantial evidence in favour of the 18 June date of receipt. Topics discussed in letters written in this interval are consistent with the normal tenor of Darwin's work, and he shows no sign of anxiety. He says in a letter to Syms Covington, 18 May [1858], that he expects the publication of his species theory to be still some time away. On 16 May [1858], he arranged a meeting with Hooker to discuss his manuscript on large and small genera, stating, I am in no sort of hurry, more especially as I know full well you will be dreadfully severe.—' On 18 [May 1858], he again tells Hooker: 'There is not least hurry in world about my M.S.' In his letter to Hooker of 8 June [1858], he indicates that this topic is still foremost in his mind: I will try to leave out all allusion to genera coming in & out in this part, till when I discuss the 'principle of Divergence', which with 'Natural Selection' is the key-stone of my Book & I have very great confidence it is sound.' This does not fit the mood

of someone who is distressed, as Darwin clearly was in his letter to Lyell, at the prospect of losing priority for his life's work." CCD7:xviii.

677. van Wyhe, Darwin's 'Journal'. http://darwin-online.org.uk/content/frameset?-pageseq=70&itemID=CUL-DAR158.1-76&viewtype=side. The start date of that line would have been written at the commencement of the section of the book, on 14 June.

678. McKinney, *Wallace*, 1972, Brackman, *A delicate arrangement*, 1980, p. 17; Brooks, *Just before the origin*, 1984; Camerini, An early history of Wallace's Line, 1993; Davies, *Darwin conspiracy*, 2008; Quammen, *Song of the dodo*, 1997; Moore, Wallace's Malthusian moment, 1997; Severin, *Spice Islands*, 1997; Shermer, *In Darwin's shadow*, 2002; Smith & Beccaloni, *Natural selection and beyond*, 2008, pp. x, 26–27, 100; Quammen, The man who wasn't Darwin, 2008 and McCalman, *Darwin's armada*, 2010.

679. Shermer, *In Darwin's shadow*, 2002, p. 133.

680. Wallace to A.B. Meyer 22 Nov. 1869 *Nature*, 1895, p. 415; Wallace to A. Newton 3 Dec. 1887, in F. Darwin, *Charles Darwin*, 1892: 189–190; Wallace, *Natural selection*, 1891, pp. 20–21; Wallace, The dawn of a great discovery, 1903; ML1:360–363; Anon., *Darwin–Wallace celebration*, 1908, pp. 6–7, 117–118.

681. Brooks, *Just before the origin*, 1984, p. 181.

682. E.g., Brackman, *A delicate arrangement*, 1980, p. 205, Brooks, *Just before the origin*, 1984, p. 182; Davies, *Darwin conspiracy*, 2008, p. 133; Raby, *Wallace*, 2001, p. 132 and Berry, *Infinite tropics*, 2002, p. 52.

683. CCD7:xviii.

684. *Javasche Courant*, 1 May 1858. I am very grateful to Kees Rookmaaker for assisting me with Dutch sources.

685. *Javasche Courant*, 28 Apr. 1858.

686. *Singapore Free Press*, 6 May 1858, p. 3: "The Dutch mail steamer *Banda*, Captain Bosch, arrived here on the 30th ultimo [Apr. 1858], from Batavia, the 26th. She returned to Batavia on the 1st current, with the Europe mails of the 26th of March." Also *Javasche Courant*, 1 May 1858.

687. *The Straits Times*, 8 May 1858, p. 4. *Sydney Morning Herald*, 9 June 1858, p. 3.

688. The post office records reproduced in Brooks, *Just before the origin*, 1984, p. 254, show that the *Nemesis* arrived in Suez on 3 June 1858 at 00:30 h. The arrival in Suez was also mentioned in *The Times*, 9 June 1858, p. 12.

689. Sidebottom, *The overland mail*, 1948, pp. 158–161.

690. The mail from the *Nemesis* arrived in Alexandria on 4 June at 11:30 h, Brooks, *Just before the origin*, 1984. It was loaded at 17:15 h on the *Colombo*, which departed the next day, 5 June 1858, at 05:00 h, *The Times*, 17 June 1858, p. 8.

691. This solution was first published together with my colleague Kees Rookmaaker in: John van Wyhe & Kees Rookmaaker, A new theory to explain the receipt of Wallace's Ternate Essay by Darwin in 1858, 2012.

692. Davies, How Charles Darwin received Wallace's Ternate paper 15 days earlier, 2012 challenged this reconstruction by asserting, without any evidence, that the mail from the Moluccas and Celebes could only be taken on the *Koningin der Nederlanden* which left Batavia on the 12th of each month. Although the mail company contracts do not say this, Davies was apparently confused by the fact that the mail steamer from Surabaya normally arrived after the *Koningin der Nederlanden* had departed for Singapore. Hence normally, the mail from the Moluccas had to wait for the service on the 12th of the next month. But in April 1858, the Surabaya steamer arrived in Batavia *before* the departure of the 26 April steamer for Singapore. The owner Cores de Vries himself was on the steamer when it arrived in Batavia — this might be why she arrived earlier than usual. *The Straits Times*, 8 May 1858, p. 2, has a report from the *Sourabaya Cour* of 17 April 1858. This newspaper was probably on the *Banda* with the Ternate essay. This seems to confirm that the Moluccas mail was indeed taken on to Singapore on 26 April and arrived on 30 April.

693. This view is also adopted by Berry & Browne, The other beetle-hunter, 2008.

694. This point was made by Browne, *Charles Darwin*, 2002, p. 16.

695. Wallace to Sims 25 Apr. 1859 NHM WP1/3/46. ML1:367.

696. Shermer, *In Darwin's shadow*, 2002, pp. 119, 129.

697. In passing it may be noted that despite frequent assertions in the literature that Darwin sent Gray an abstract of his theory as a form of, as Brackman, *A delicate arrangement*, 1980, p. 54 put it "an escape hatch for the moment at which he would proclaim priority in the field" and Davies, *Darwin conspiracy*, 2008, p. 126 called an "insurance policy", Gray had in fact asked Darwin to explain his views. See Gray to Darwin 7 July 1857 CCD6:423.

698. Wallace, The dawn of a great discovery, 1903.

699. Haughton, Presidential address, 1857.

700. Barrett *et al.*, *Darwin's notebooks*, 1987. Transcriptions and facsimiles of these notebooks are also published in John van Wyhe, *The complete work of Charles Darwin online* (http://darwin-online.org.uk/).

701. F. Darwin, *Foundations*, 1909.

702. These papers are now in the Darwin Archive at CUL and are published in John van Wyhe, *The complete work of Charles Darwin online*, http://darwin-online.org.uk/manuscripts.html

703. Kohn, On the origin of the principle of diversity, 1981, p. 1108.

704. Darwin's estimate on the first page of *Origin* is that it would take "two or three more years to complete it"; that is, by 1861–2. If we subtract the thirteen months

spent writing the *Origin* this would mean he could have published it in about 1860. There are other estimates by Darwin which confirm this, such as Darwin to Wallace 22 Dec. 1857: "I do not suppose I shall publish under a couple of years." CCD6:515. See also Stauffer, *Natural selection*, 1975, p. 10.

705. Wallace gave the spelling "*Hesther Helena*" in the *Journal* and MA. I have followed the spelling given by the Dutch Resident of Ternate in Report of C. Bosscher 1859 in Overweel, *Topics relating to Netherlands New Guinea in Ternate Residency*, 1995, p. 23. The *Esther Helena* was named after van Duivenbode's daughter-in-law Esther Helena Hartman (1834–1913).

706. Haenen & Huizinga, *The general and political reports of Ternate residency (1824–1889)*, 2001, p. 141. I am grateful to Kees Rookmaaker for calling this work to my attention.

707. MA2:299.

708. *Notebook 2/3*, p. 146.

709. Aritonang & Steenbrink, *A history of Christianity in Indonesia*, 2008, pp. 350–351.

710. MA2:302.

711. *Journal 3*, 131. MA2:305.

712. *Notebook 4*, p. 55b.

713. These are now in the British Museum. See the catalogue entries in the manuscript catalogue in John van Wyhe, *Wallace online* (http://wallace-online.org/). The chicken is mentioned in *Notebook 4*, p. 118.

714. Anon., *Nieuw Guinea*, 1862, p. 65. I am grateful to Kees Rookmaaker for calling this work to my attention.

715. MA2:313, 319.

716. Anon., *Nieuw Guinea*, 1862, p. 67.

717. Anon., *Nieuw Guinea*, 1862, p. 78.

718. Report of C. Bosscher 1859 in Overweel, *Topics relating to Netherlands New Guinea in Ternate Residency*, 1995, p. 20.

719. Lesson, *Histoire naturelle des oiseaux de paradis et des épimaques*, 1835.

720. *Notebook 4*, pp. 126–130.

721. MA2:314. On the flies and their subsequent fate see Glaubrecht & Kotrba, Wallace's discovery of "curious horned flies", 2004.

722. *Notebook 4*, p. 56. This presumes that Wallace here got the date right.

723. Anon., *Darwin–Wallace celebration*, 1908, p. 6.

724. Darwin to Lyell 18 [June 1858] CCD7:107.

725. Barlow, *Autobiography of Charles Darwin*, 1958, p. 121.

726. Bowler, Wallace and Darwinism, 1984, p. 278 reminiscent of Kohn, On the origin of the principle of diversity, 1981. For further analysis of the differences see Bowler, *Evolution history of an idea*, 1984; Ruse, *Monad to man*, 1996; Kottler, Darwin and Wallace, 1985.

727. See for example Boyer, Cultural transmission and the biology in history, 1994; Loritz, *How the brain evolved language*, 1999 and Turner, *Brains, practices, relativism*, 2002.

728. Lyell, *The geological evidences of the antiquity of man*, 1863, p. 408.

729. In John van Wyhe, *Darwin's shorter publications*, 2009, pp. 296–297. Darwin to W. E. Darwin [20 June 1858] CCD7:113.

730. CCD7:118.

731. Darwin to Lyell 26 [June 1858] CCD7:117.

732. See Darwin to Hooker 13 [July 1858] CCD7:129.

733. Darwin to Hooker [29 June 1858] CCD7:121.

734. The fair copy (DAR 113) is annotated by Hooker and Darwin.

735. *Notebook 4*, p. 57.

736. MA2:322.

737. Bunbury diary 1866, Bunbury. *Life of Sir Charles J.F. Bunbury*, 1906, pp. 237–238.

738. *Times*, 1 July 1858, p. 1.

739. Charter 1802.

740. *Journal of the proceedings of the Linnean Society: Zoology*, 1859, 3:liii.

741. Many commentators have mistakenly claimed that the meeting was specially called in order to read the Darwin and Wallace papers, e.g., Kohn, *A reason for everything*, 2004; Wilson, *Forgotten naturalist*, p. xii; Stott, *Darwin's ghosts*, 2012. This is presumably derived from the perfectly correct description of the meeting as "special".

742. Anon., *Darwin–Wallace celebration*, 1908, p. 85. Virtually identical to that published in *Proceedings of the Linnean Society*, 1857–8, pp. liv–lv. The date of "Oct. 1857" was a mistake corrected in the printed version of the paper to "*September 5th*, 1857". Similarly the sketch was written in 1842 not 1839, as corrected for the published version.

743. F. Darwin, *Life and letters of Charles Darwin*, 1887, 2:125–126.

744. The full minutes are in Anon., *Darwin–Wallace celebration*, 1908, available in John van Wyhe, *The complete work of Charles Darwin online*.

745. See England, Natural selection before the *Origin*, 1997.

746. Barlow, *Autobiography of Charles Darwin*, 1958, p. 122 referring to *Journal of the Geological Society of Dublin*, 1857–1860, 8:152.

747. Wallace, *Wonderful century*, 1898, p. 141.

748. Browne, *Charles Darwin*. 2002, p. 42. Bell, Presidential address, 1859. Bell said almost the same thing in his address of 1856.

749. Darwin to Gray 4 July 1858 CCD7:125–126. It is endlessly repeated that Darwin may have sent the enclosure to Gray as a sort of insurance policy in case he was

forestalled. If this theory were true, one would think Darwin would have bothered to write the date on his own copy rather than needing to write to Gray in America for it.

750. Darwin to Hooker 13 [July 1858] CCD7:129.

751. England, Natural selection before the *Origin*, 1997 and Moody, The reading of the Darwin and Wallace papers, 1971.

752. *Edinburgh new philosophical journal*, 11:148.

753. Quoted in Harrisson, Wallace and a century of evolution in Borneo, 1958, p. 38. Thanks to Kees Rookmaaker.

754. Rupke, *Richard Owen*, 2009, p. 159.

755. Newton, The early days of Darwinism, 1888.

756. Brackman, *A delicate arrangement*, 1980, p. xi. Browne, Making Darwin, 2010. See also Browne, *Charles Darwin*, 2002, p. 16 where she writes that Darwin felt obliged to "acknowledge to Lyell that Wallace had got there first". Rebecca Stott claims in *Darwin's ghosts*, 2012 that the Linnean Society actually officially decided in Darwin's favour on a priority question. This is presumably literary licence. There was no such adjudication.

757. Smith, Wallace (DNB), 2004. Similarly in Smith, *Wallace; an anthology*, 1991, p. 3: "without obtaining Wallace's permission first", an opinion shared by Brooks, *Just before the origin*, 1984 and Lloyd, Wimpenny & Venables, Wallace deserves better, 2010, p. 344: "Protocol should have dictated that Wallace's paper be read first but Lyell and Hooker arranged to have it presented after that of Darwin." Quammen, *Song of the Dodo*, 1997. Quammen's biased version is copied by other writers, see for example Daws & Fujita, *Archipelago*, 1999, p. 124 and Heij, *Antonie Augustus Bruijn*, 2010, p. 47.
Bryson, *At home*, 2010, p. 468.

758. In 1851–2, there was a "'Note on the occurrence of an Eatable *Nostoc* in the Arctic Regions and in the Mountains of Central Asia.' By J.D. Hooker, M.D., F.R.S., F.L.S. &c. Accompanied by a communication from the Rev. M.J. Berkeley, F.L.S., on the same subject."
"Mr. Adam White, F.L.S., exhibited numerous insects belonging to Mr. S. Stevens, F.L.S., collected by Mr. Bates in South America, and others belonging to Mr. Frederick Smith…and read extracts from Mr. Bates's letters to Mr. Stevens".
"'On the various forms of *Salicornia*.' By Joseph Woods, Esq., F.L.S.: with some additional remarks by Richard Kippist, Esq., Libr. L.S."
"Mr. Hogg, F.L.S., communicated a letter 'On the Artificial introduction of a breed of Salmon into the river Swale, and a tributary stream in Yorkshire,' which appeared in the 'Durham Advertiser' for April 16th in the present year, under the signature of Isaac Fisher, together with an unpublished letter from the same gentleman in answer to a request from Mr. Hogg for further information."

759. Matthew, Nature's law of selection, 1860, in which Matthew claimed to have preceded Darwin as the discoverer of natural selection in the appendix to Matthew, *On naval timber and arboriculture*, 1831. Darwin later included Matthew among his predecessors in the 'Historical sketch' published in the third and later editions of *Origin*. See CCD8:156. See also Johnson, The preface to Darwin's *Origin of species*, 2007.

760. Wallace to Hooker 6 Oct. 1858 CCD6:166. Nicholas Rupke has written about Richard Owen's ideas of scientific priority which further complicate these issues. "A related reason for Owen's refusal to acknowledge Maclise may well have been Owen's belief that credit belongs to the person who is the first to execute a piece of work properly rather than to someone who has the notion but fails to perfect it. In this case then, priority belonged to whoever gave the archetype its correct and definitive anatomical outline and definition." Rupke, *Richard Owen*, 2009, p. 119.

761. Wallace to Darwin 29 May [1864] CCD12:221. There is yet another anti-Darwin claim floating about since Brooks, *Just before the origin*, 1984 — that Darwin did not acknowledge Wallace sufficiently in *Origin of species*. This is echoed in Lloyd, Wimpenny & Venables, Wallace deserves better, 2010: "Wallace got not so much as a mention in the first two editions of the *Origin of species*." Sufficient acknowledgement is also subjective. But some facts might help clarify the issue. Wallace was mentioned by name increasingly across the ever lengthening editions of the *Origin*: 1st edn: 4 times, 2d edn: 5 times, 3d edn: 8 times, 4th edn: 14 times, 5th edn: 16 times and 6th edn: 20 times. In contrast, Wallace mentioned his travelling companion Bates by name four times in his Amazon narrative whereas Bates mentioned Wallace eleven times in his own book. The number of citations therefore seems to prove nothing of interest.

762. Marchant, A man of the time, 1905, p. 546.

763. Beddall, Wallace, Darwin, and the theory of natural selection, 1968.

764. For example Brackman, *A delicate arrangement*, 1980, pp. 65–66; Beddall, Darwin and divergence, 1988, p. 51; Wilson, *Forgotten naturalist*, 2000, p. 190 and Beccaloni in Smith & Beccaloni, *Natural selection and beyond*, 2008, p. 26. See Shermer, *In Darwin's shadow*, 2002, p. 120.

765. Darwin, [Extracts from letters addressed to Professor Henslow], [1835].

766. Wallace to Sclater [11–15] & 20 Sept. 1861 ZSL GB 0814 BADW.

767. Beccaloni in Smith & Beccaloni, *Natural selection and beyond*, 2008, p. 100. My italics. One of Wallace's few surviving article manuscripts at the RGS shows that it was edited down before publication — also without consulting him.

768. Wallace, The dawn of a great discovery, 1903.

769. Anon., *Darwin–Wallace celebration*, 1908, p. 7.

770. Wallace, *Natural selection*, 1891, p. 21.

771. ML1:240.

772. Rookmaaker & van Wyhe. Litchfield, H., [On plagiarism and scientific jealousy]. (*Darwin Online*, http://darwin-online.org.uk/content/frameset?pageseq=1&itemID=CUL-DAR262.23.3&viewtype=text)

The book referred to is Carneri, *Sittlichkeit und Darwinismus*, 1871.

773. Quoted in Brackman, *A delicate arrangement*, 1980, p. 348.

774. For example, Quammen, *Song of the dodo*, 1997, p. 113; Davies, *Darwin conspiracy*, 2008, p. 101 and Lloyd, Wimpenny & Venables, Wallace deserves better, 2010. See Raby, *Wallace*, 2001, p. 287.

775. Kohn, On the origin of the principle of diversity, 1981, p. 1107.

776. Shermer, *In Darwin's shadow*, 2002, p. 140.

777. I am grateful to the late Fred Burkhardt as well as Jon Hodge, Martin Rudwick, Frank Sulloway, Robert Richards and the more than sixty other colleagues who have written or told me that they fully agree with my argument that Darwin did not put off or avoid publishing in John van Wyhe, Mind the gap, 2007. This has been great comfort in the face of the shameful campaign of defamation and persecution (an open secret in part of the history of science community) of my few, but very tenacious, critics after I published my article despite an email threat not to do so or suffer the consequences to my reputation and career of losing the support of senior Darwin scholars.

778. See for example: Irvine, *Apes Angels & Victorians*, 1956; Pike, *The true book about Charles Darwin*, 1962; Huxley & Kettlewell, *Darwin and his world*, 1965; Olby, *Darwin*, 1967; Moorehead, *Darwin and the Beagle*, 1969; Gruber, *Darwin on man*, 1974; Gould, *Ever Since Darwin*, 1977; Colp, *To Be an Invalid*, 1977; Moore, *Post-Darwinian controversies*, 1979; Eiseley, *Darwin and the mysterious Mr X*, 1979; Mayr, *Growth of biological thought*, 1982; Bowler, *Eclipse of Darwinism*, 1983; Young, *Discovery of evolution*, 1992; Gribbin, *Darwin in 90 minutes*, 1997; Hands, *Darwin*, 2001; Desmond, Browne & Moore, Darwin (DNB), 2004. This chapter, and a few other sections of this book, are revised versions of John van Wyhe, Mind the gap, 2007.

779. Desmond & Moore, Transgressing boundaries, 1998, p. 157.

780. Desmond & Moore, *Darwin*, 1992, pp. xvi; 239. Many other writers have said Darwin kept his theory secret: Greene, *The death of Adam*, 1959; de Beer G., *Charles Darwin*, 1963; Olby, *Charles Darwin*, 1967; Hopkins, *Darwin's South America*, 1969; Bowler, *Charles Darwin*, 1990; Bowlby, *Charles Darwin*, 1990 and Stott, *Darwin and the barnacle*, 2003. Although doubting Darwin's delay in some passages, Browne refers to Darwin's theory as "secret" in *Darwin's Origin of species*, 2006, p. 45.

781. Desmond & Moore, *Darwin*, 1992, pp. xv, xvi, 236, 273, 228, 231, 232, 657.

782. Desmond & Moore, *Darwin*, 1992, p. 292.

783. In Wallace, The dawn of a great discovery, 1903, the section heading "Darwin's Secret" appears. However, as Wallace mentions nothing of secrecy in the text, the section heading, as the others in the article and the first part of the article title, appears to have been added by the editor of the newspaper. Wallace also writes in a way inconsistent with secrecy in his other accounts of Darwin: Debt of science to Darwin, 1883, p. 425; [Acceptance speech], 1908, p. 6 and in Marchant, A man of the time, 1909, p. 546. Bettany, *Life of Charles Darwin*, 1887, p. 65, wrote "Darwin was meditating in secret", but there is otherwise no hint of a delay or reason given for widely separate dates for discovery and publication.

784. CCD2:xvi.

785. I am very grateful to Duncan Porter for supplying me with several further names to add to this list from his important article: On the road to the Origin with Darwin, Hooker, and Gray, 1993, which announced "publication of Darwin's complete correspondence...once and for all dispels the myth that no one knew what Darwin was up to prior to publication of the *Origin*". Frank Sulloway also kindly pointed me to his own discussion in *Born to rebel*, 1996, p. 246: "Far from being a 'closet evolutionist,' as Desmond & Moore claim, Darwin told a dozen of his closest friends about his evolutionary ideas. His twenty-year 'delay' in announcing his theory of natural selection was not really a delay. Darwin used this time advantageously to bolster his argument for evolution, and especially to resolve some of its weakest links."

786. I am grateful to Jon Hodge for calling my attention to this overlooked fact. Barlow, Darwin's ornithological notes, 1963, p. 262.

787. Correspondence references: Waterhouse (1843) CCD2:375–376; Hooker (1844) CCD3:1–3; Jenyns (1844) CCD3:67–68; Herbert (1845) CCD3:261; Dieffenbach (1846) CCD3:310 and (1847) CCD4:12–13; Cresy (1848) CCD4:135–136; Huxley (c. 1851) F. Darwin, *Life and letters of Charles Darwin*, 2:96, 1887; Layard (1855) CCD5:524–525; Murray 1855 CCD5:530–531 and CCD2–7; Bunbury (1856) CCD6:36; Wollaston (1856) CCD6:91 n. 7; Dana (1856) CCD6:180–181; Gray (1857) CCD6:431–433, Gray & Falconer (1857) CCD6:445–449; J. Wedgwood DAR 139.12.17. There are further discussions of transmutation in CCD2–7. See the introduction to CCD3. On Butler and Craik (1858) see Dixon & Radick, *Darwin in Ilkley*, 2009, p. 23 and CCD7:84–85, 249. I am grateful to Greg Radick for bringing these additional names to my attention.

Manuscript notes indicating conversation on transmutation include: with Strickland DAR 205.9.149 (1842); Forbes DAR 45.58 (1844) and possibly in: DAR 205.9.185–186; presumably DAR 205.9.188 (1844), DAR 205.5.103 and DAR 205.5.53; Falconer (1844) DAR 205.9.187; DAR 205.5.114 (1845); Waterhouse DAR 205.5.114 (1845), TAN55; H. Wedgwood, E144; TAN51 and DAR

205.5.60; DAR 205.5.60 (1843); E. Wedgwood E144; (1840) DAR 205.5.30; Lonsdale and Bunbury (1842) DAR 205.9.146; Lonsdale C175–177; Wollaston DAR197.2 (1856). See Bunbury's diary 23 Nov. 1845 recording lunch at the Horner's where Darwin talked of his belief in transmutation: Bunbury, *Life of Sir Charles J.F. Bunbury*, 1:213, 1890–91; CCD3:xiv, 237 n. 5 and CCD6:91 n. 7.

788. CCD3:xiv.

789. Barrett *et al.*, *Darwin's notebooks*, 1987, C177.

790. Darwin, *Origin of species*, 6th edn., 1872, p. 424.

791. Barlow, *Autobiography of Charles Darwin*, 1958, p. 123.

792. Rudwick, *Worlds before Adam*, 2008, p. 288.

793. Herbert and Kohn refer to them as "Private notebooks" in Introduction to Barrett *et al.*, *Darwin's notebooks*, 1987, p. 12. See also p. 517.

794. Desmond & Moore, *Darwin's sacred cause*, 2009, pp. 253–254. Thomson, *The young Charles Darwin*, 2009 also claims that Darwin used vague language to conceal his real meaning.

795. Darwin to W. D. Fox [25 Jan. 1841] CCD2:279.

796. See Desmond & Moore, *Darwin*, 1992, p. 473.

797. Darwin to Hooker [2 Oct. 1846] CCD3:346. See also Darwin to W. D. Fox [15 June 1838] CCD2:92 and Darwin to Lyell [14] Sept. [1838] CCD2:107.

798. ML1:257.

799. Wallace to H. Bates 4 Jan. 1858. NHM WP1/3/41.

800. DAR 113. See Sydney Smith 'Historical preface' in Barrett *et al.*, *Darwin's notebooks*, 1987, p. 1. On Norman see CCD7:507.

801. Secord, *Victorian sensation*, 2000.

802. Darwin to Lyell [10 Dec. 1859] CCD7:422.

803. Darwin to J.L.R. Agassiz 22 Oct. 1848 CCD4:178.

804. [24 Apr. 1845] CCD3:181.

805. Moore in British Council 'Darwin Now' podcast (2009); Kohn 'In Darwin's garden' BBC documentary (2009).

806. Jones, *Free associations*, 1959, pp. 203–204. Also misquoted as "like committing a murder" in Olby, *Charles Darwin*, 1967. Ruse, *The Darwinian revolution*, 1979, p. 185 put it thus: "Darwin confessed that it was like admitting to a murder. It was a murder—the purported murder of Christianity, and Darwin was not keen to be cast in this role. Hence the *Essay* went unpublished."

807. Darwin to Hooker [11 Jan. 1844] CCD3:2.

808. Desmond & Moore, *Darwin*, 1992, p. xviii, 313. See also Desmond & Moore, *Sacred cause*, 2009, p. 229. Bryson, *A short history of nearly everything*, 2003, p. 87. Darwin *did not* refer to himself as a devil's chaplain. See Darwin to Hooker 13 July [1856] CCD6:178.

809. Hodge, Review of Desmond & Moore, *Darwin*, 1992, p. 341. I am grateful to Jon Hodge for pointing out this agreement.

810. Darwin to Hooker 10 Feb. [1875], *Calendar* 9850.

811. Darwin to Wallace 22 Dec. 1857 CCD6:514.

812. Darwin to Hooker 24 Nov. 1873, *Calendar* 9158; to H.E. Litchfield 4 Oct. [1877], *Calendar* 11167; to F.M. Balfour 28 Jan. 1881, *Calendar* 13030; to W.T. Thiselton Dyer 16 Nov. 1881, *Calendar* 13487; to O. Salvin 12 Oct. [1871] CCD19.

813. Anon., On the origin of species, 1859.

814. Anon., *Westminster review*, 1860. See also Bunbury's diary entry for 14 July 1858 in Bunbury, *Life of Sir Charles J.F. Bunbury*, 1906, 2:129 and [Robertson], Variation, 1868.

815. Lyell, *Antiquity of man*, 1863, p. 408. Further statements to the fact that Darwin was known to be working on evolution during the gap years are found in Tyndall, Address, 1874, p. 38; Fish, Charles Darwin, 1882; Wallace, Debt of science to Darwin, 1883; Marchant, *Wallace letters*, 1916, 1:104 *et al.*

816. Two letters in which he tells his story as one of working on the theory for a long time are: To ? 23 Oct. 1880 *Calendar* 12771 and to F. Powell [after 3 Dec. 1881] *Calendar* 13529.

817. Darwin to Gray 20 July [1857] CCD6:432.

818. Quoted in Richards, Will the real Charles Darwin please stand up, 1983, p. 52 and Richards, Huxley and woman's place in science, 1989.

819. Darwin, *Origin of species*, 1859, p. 236.

820. Ghiselin, *Triumph of the Darwinian method*, 1969 and Stott, *Darwin and the barnacle*, 2003.

821. See also Sloan, Darwin's invertebrate program, 1985, and Love, Darwin and Cirripedia, 2002.

822. In both his *Journal* and MA, Wallace stated that they departed on the 27th *and* the 29th with no explanation. However in *Notebook 4*, p. 58b he wrote "28th on board schooner 29th 5 am. Sailed."

823. *Journal 3*, 144. Wallace wrote in his *Journal* and MA2:322: "On July 6th the steamer returned". The official narrative reports that the *Etna* returned on 7 July and left on the 9th.

824. *Notebook 4*, p. 54b. Wallace originally wrote "14" but later overwrote it with 15, apparently he began collecting only on the 15th. Also he wrote "Oct. 1 left at 6 am. 2nd. 7 am. at Ternate".

825. See an advertisement by Foxcroft for a similar expedition to Scotland in 1857 in *The Substitute; or, entomological exchange facilitator* 12, 1857.

826. Wallace to Stevens 29 Oct. 1858: "When I mentioned subscribing to Foxcroft, I did not intend subscribing for Lepidoptera if separate, as I cannot afford it, &

besides care little for any but the Papilios & Pieridae. Withdraw my name there-fore for Lepidoptera after the first year if you cannot at once. In Coleoptera am willing to subscribe 2 years & even a third if he changes his locality, always sup-posing however that the collections are divided fairly, that is that each subscriber get the same number of species, - the only advantage the first on the list have, being the best of the unique ones, & quite enough advantage too, in fact too much, for when 20 people subscribe equally to anything, it is absurd that there sh.^d be any advantage attached to being first on the list." CUL Add. 7339/235.

827. Reported in *Athenaeum* (no. 1630) 22 Jan. 1859, p. 119.

828. DAR 270.1.2. Darwin used identical paper see CCD5:152.

829. NHM WP7/10. In fact none of Wallace's letters or surviving manuscripts (except for the *Journal*) seem to be written on thin foreign paper.

830. This letter is usually quoted from ML1:71 or Marchant, *Wallace letters*, 1916, 1:365. However the contemporary copy transcribed here (NHM WP/1/3/44) has deletions consistent with my interpretation, "greatest" and "and assistance" were omitted from the published versions. Note "~~greatest~~" was first considered and "eminent" is used twice.

831. Wallace to Hooker 6 Oct. 1858 CUL Add 7339.237.

832. "First, it is polite and thoughtful, but not excessively so by the standards of the day, where how one expressed oneself was nearly as important as what was being communicated." Shermer, *In Darwin's shadow*, 2002, p. 139.

833. The original underlining was omitted from previously published transcriptions of this letter. Wallace to Silk 30 Nov. 1858 NHM WP/1/3/45.

834. *Notebook 4*, p. [14b]. It is not included in CCD, see CCD7:130. There are three extra chapters and Chapter XI is different from the table of contents in *Natural selection*. Although we can never know why this Darwin letter is lost, it is possible it was sent to someone else and hence Wallace made a copy to keep or carry with him. With thanks to Gordon Chancellor for helpful discussion. The table of con-tents was first noticed and published by Brooks, *Just before the origin*, 1984.

835. MA2:23–24.

836. MA2:32.

837. *Notebook 4*, p. 51b.

838. In his *Journal* and MA2:35 he gave 21 Oct., but in his 29 Oct. 1858 letter to Ste-vens, Wallace stated, "Here I have been also yet only 5 days." CUL Add. 7339/235.

839. MA2:35.

840. Wallace to M. A. Wallace 6 Oct. 1858 NHM WP1/3/44.

841. MA2:37.

842. *Notebook 4*, p. 60b. He entered his birds captured in *Notebook 5*, from p. 1.

843. MA2:40–41. The date 24 Oct. is recorded in *Notebook 5*, p. 1.

844. Quoted here from the original, not from the published version. Wallace to Stevens 29 Oct. 1858 CUL Add. 7339/235.

845. Gray, [Notes on a new bird-of-paradise], 1859.

846. Sclater, Note on Wallace's Standard-wing, 1860.

847. A point noted by Camerini, Wallace in the field, 1996.

848. Wallace to Stevens 29 Oct. 1858 CUL-Add. 7339/235. Richard Kippist (1812–1882), a botanist, was Librarian of the Linnean Society (1842–1881). See Freeman, *Companion*, 2010.

849. Raby, *Wallace*, 2001, p. 142.

850. Wallace to Silk 30 Nov. 1858.

851. *Notebook 4*, p. 61b. Agassiz is noted on pages 140–147 and unnumbered page near back cover. Huguenin is also mentioned in *Notebook 5*, p. 3.

852. Smith, Catalogue of hymenopterous insects collected by Mr. A.R. Wallace, 1860, p. 133. I am grateful to John S. Ascher for calling the bee to my attention.

853. Wallace, Introduction to Smith, 1873, p. 287. MA2:68.

854. Messer, *Chalicodoma pluto*: The world's largest bee, 1984.

855. MA2:51. This passage was first drafted in *Notebook 4*, p. 61.

856. *Notebook 4*, pp. 137–140.

857. *Notebook 4*, p. 61; for the box sent from Batchian see *Notebook 1*, p. 105.

858. *Journal 3*, 148.

859. The letter was published in Wallace, Letter concerning the geographical distribution of birds, 1859. Wallace referred to Sclater, On the general geographical distribution of the members of the class Aves, 1858.

860. See Darwin to Wallace 22 Dec. 1857 CCD6:514. Darwin later told Wallace, "Mr W. Earl published several years ago the view of distribution of animals in Malay Archipelago in relation to the depth of the sea between the islands." Darwin to Wallace 9 Aug. 1859 CCD7:323. See also Darwin, *Origin*, 1859, p. 395.

861. MA2:71–72.

862. Wallace to T. Sims 25 Apr. 1859 NHM WP1/3/46.

863. Darwin to Wallace 25 Jan. [1859] CCD7:241.

864. Wallace, On the zoological geography of the Malay archipelago, 1860.

865. *Journal 3*, 175. Wallace entered the notes from "Drysdale" [sic] in *Notebook 4*, p. 104.

866. Later described somewhat romantically in Wallace, On the progress of civilization in Northern Celebes, 1865.

867. MA1:378. I am grateful to Kees Rookmaaker for identifying Neys.

868. Not 6 July as he wrote in *Journal*; see *Notebook 4*, p. 62. Collecting for this time is noted in *Notebook 4*, pp. 62–63. Other notes in *Notebook 5*, pp. 14ff.

869. *Journal 3*, 184.

870. *Journal 3*, 187.

871. Wallace, The ornithology of Northern Celebes, 1860, p. 141.

872. Wallace, The ornithology of Northern Celebes, 1860, p. 144.

873. Wallace, The ornithology of Northern Celebes, 1860, p. 145.

874. Berry, "Ardent beetle-hunters", 2008, p. 49.

875. MA1:445.

876. *Journal 3*, 193.

877. See MA2:75–76.

878. *Allen's Indian Mail*, 12 Jan. 1861, p. 33.

879. MA1:476.

880. Wallace to Stevens 26 Nov. 1859. S59.

881. S154, p. 694.

882. *Journal 4*, 198.

883. ML1:370.

884. See Pfeiffer, *A woman's journey round the world*, [1852], 1:371.

885. When H.O. Forbes visited Paso in the late 1870s, the Rajah's son remembered Wallace who purportedly stayed in his house. Forbes, *A naturalist's wanderings in the Eastern Archipelago*, 1885, p. 290.

886. Darwin, *Origin of species,* 1859, pp. 4–5, 62. On the publication of *Origin* see Browne, *Darwin's Origin of species*, 2006.

887. Watson to Darwin 21 Nov. [1859] CCD7:385.

888. MA1:475; MA2:86.

889. See CCD8:556. Wallace's annotated copy of *Origin* survives at CUL. See Beddall, Wallace's annotated copy of Darwin's *Origin of species*, 1988.

890. NHM WP7/9/21. My transcription. A transcription was first published by Beccaloni in Smith & Beccaloni, *Natural selection and beyond*, 2008, pp. 96–97.

891. Darwin to Wallace 18 May 1860 CCD8:119–221.

892. Wallace to Silk 1 Sept. 1860 NHM WP1/3/48 and ML1:372.

893. Wallace to H. Bates 24 Dec. 1860 NHM WP1/3/49.

894. In MA1:478 and the Hope Oxford note, Wallace gives the date 20 Feb., but the 24th in MA2:86. The 20th is given in his *Journal* as the date he departed from Allen and the town of Amboyna to return to Passo.

895. *Journal 4*, 210 and MA2:108. In ML1:370 he mistakenly stated that he purchased the prau on Goram.

896. MA2:108.

897. MA2:116.

898. MA2:123.

899. MA2:337.

900. See James, An 'open clash between science and the church'?, 2005.

901. von Rosenburg, 1878, p. 381 pointed out that Wallace always misspelled Umka as Muka.

902. MA2:349.

903. MA2:359.

904. MA2:360–361.

905. Dufour, *Histoire de la Prostitution,* 1851. See *Notebook 4,* p. 91; Wallace to Silk 1 Sept. 1860 and Dawson, *Darwin, literature and Victorian respectability,* 2007.

906. Wallace to Silk 1 Sept. 1860 NHM WP1/3/48.

907. *Notebook 2/3,* p. 88r. But in *Journal* and MA2:368 Wallace stated that he left on 29 Sept.

908. *Journal 4,* 234.

909. Wallace to Stevens 7 Dec. [1860] in *Ibis* 3 no. 10 (Apr. 1861).

910. Wallace to H. Bates 24 Dec. 1860; 2 Jan. 1861 NHM WP1/3/49; WP1/3/63.

911. Wallace to Darwin 30 Nov. 1861 CCD9:357.

912. But in *Journal* and MA, Wallace stated: "12th. [Jan.] Arrived at Delli". Bickmore, *Travels in the East Indian Archipelago,* 1869 provides an excellent near contemporary description of Delli.

913. On Hart see CCD15:285 "He is what you may call a *speculative* man: he reads a good deal, knows a little and wants to know more, and is fond of speculating on the most abstruse and unattainable points of science and philosophy" (Marchant, *Wallace letters,* 1916, 1:79). No reply from Hart has been found in the Darwin Archive-CUL, and he is not cited in Darwin, *Expression,* 1872. On the collecting at this time see *Notebook 5,* p. 34.

914. CCD15:555 state that Geach worked for the Portuguese government in Timor. However, Wallace indicates that Geach was employed by a Portuguese merchant at Singapore. See MA1:301; 303, possibly the prominent Joze d'Almeida listed in the *Singapore directory.* In MA1:296 Wallace stated he stayed with Geach rather than Hart.

915. ML1:375.

916. Wallace to Darwin 30 Nov. 1861 CCD9:357.

917. Wallace to T. Sims 15 Mar. 1861 BL Add.39168 ff.2–27

918. Wallace to M. A. Wallace 20 July 1861 NHM WP1/3/50.

919. MA2:136.

920. *Journal 4,* 244. Wallace misspelled Opziener.

921. MA2:131.

922. MA2:133–134.

923. MA2:131.

924. *Notebook 5,* p. 48.

925. MA2:151.

926. Bates, Contributions to an insect fauna of the Amazon Valley, 1862, p. 513. See Browne, *Darwin*, 2002, pp. 224–225.

927. *Notebook 5*, p. 49. Wallace's copy of Bonaparte, *Conspectus generum avium*, 1850, is in the library of the Linnean Society of London. Wallace, Descriptions of three new species of Pitta from the Moluccas, 1862 gives the following: 1. Pitta rubrinucha, new species, from Bouru. This is now recognised as a subspecies of the Red-bellied Pitta, *Pitta erythrogaster* Temminck, 1823, and the combination is *Pitta erythrogaster rubrinucha*. Thanks to Kees Rookmaaker.

928. *Javabode*, 24 July 1861. Wallace mistakenly stated 18 July in MA1:148.

929. Wallace to M.A. Wallace 20 July 1861. NHM WP1/3/50.

930. *Notebook 5*, p. [53].

931. MA1:158.

932. MA1:169.

933. MA1:172.

934. *Javabode*, 21 Sept. 1861 named as "Walace". I am grateful to Kees Rookmaaker for this reference. MA1:172 gives no date.

935. Winchester, *Krakatoa*, 2003.

936. MA1:173.

937. MA1:xiii. Also one photograph of "a forest stream, West Java" in Wallace, *World of life*, 1910, facing p. 75.

938. MA1:176–178.

939. MA1:181.

940. "The Englishman, Mr. A.R. Wallace, who travelled for several years through the Moluccas to study natural history, and who was mentioned in the Report of 1853, p. 85, has been granted permission for the same purpose to visit the residencies of Banka and Palembang." Staten Generaal van Nederland, *Verslag*, 1861, p. 1498. Translation by Kees Rookmaaker to whom I am indebted for finding this reference.

941. Wallace to Darwin 30 Nov. [1861] CUL DAR 181.6.

942. MA1:190.

943. MA1:205. Moore, Wallace's Malthusian moment, 1997, claimed that "Not once in a thousand pages does the [*Malay Archipelago*] mention Malthus or natural selection". This is incorrect, see MA1:141, 207, 419 and MA2:455.

944. Baden Powell, *Essays and reviews*, 1860, p. 139. Italics in the original. See Shea & Whitla, *Essays and reviews*, 2000 and Corsi, *Science and religion*, 1988.

945. Wallace to Silk 22 Dec. 1861 NHM WP1/3/53.

946. Wallace to Sclater 7 Feb. 1862 ZSL GB 0814 BADW.

947. Wallace, Bucerotidae, or Hornbills, 1863, p. 314. MA1:213–214.

948. *The Straits Times*, 25 Jan. 1862, p. 3.

949. *Singapore Free Press*, 19 July 1904.

950. Rookmaaker & van Wyhe, Charles Allen, 2012. Sherry, *Conrad's eastern world*, 1966, p. 143 suggests that Allen was the model for the character Stein in Joseph Conrad's novel *Lord Jim* (1900).

951. *Singapore Free Press*, 2 Apr. 1906, p. 5.

952. *The Straits Times*, 25 Jan. 1862, p. 3. Yet, as Raby, *Wallace*, 2001 pointed out, Wallace stated in MA1:209 that the siamang "died just before I started".

953. ML1:383.

954. *The Straits Times*, 18 May 1861, p. 4.

955. Barbour, *Naturalist at large*, 1943, p. 42.

956. Note kept with Wallace to Sclater of 7 Feb. 1862 ZSL GB 0814 BADW. The account in Raby, *Wallace*, 2001 on the acquisition, sale and display of the Birds of Paradise in London is excellent. Moss and Waterworth are listed in the *Straits Calendar and Directory* for 1862.

957. *The Straits Times*, 8 Feb. 1862, p. 3.

958. Wallace to Sclater 7 Feb. 1862 ZSL GB 0814 BADW.

959. *The Straits Times*, 15 Feb. 1862, p. 4. The ticket for Marseilles was $528 which was where he eventually alighted, but not according to his original plan. Wallace to Sclater 4 Apr. 1862 ZSL.

960. Marchant, A man of the time, 1905.

961. In his notes on the localities in Pascoe's Longicornia Malayana, Wallace wrote (1869: 691): "The small collection from Penang consists of a few insects given me by Mr. Lamb on my way home, and of a few more collected by a native sent there by a friend". [Lamb's "Penang" was the mainland opposite Penang I. (Province Wellesley), *cf.* Wallace, Catalogue of the Cetoniidae, 1868, p. 522.]

962. Wallace to Stevens 16 Feb. 1862 ZSL GB 0814 BADW.

963. ML1:384. The Elephanta caves are located on an island in the harbour.

964. *Popular overland guide*, 1861, p. 56.

965. ML1:384.

966. *Times*, 27 Mar. 1862, p. 12.

967. Wallace to Sclater 18 Mar. 1862 ZSL GB 0814 BADW.

968. ML2:396.

969. *The Straits Times*, 1 Feb. 1862, p. 4.

970. *Allen's Indian Mail*, 31 Mar. 1862 [p. 1] Passengers by the present mail: from Malta "Mr. Wallace", the only passenger listed. See *Notebook 5*, p. 44.

971. ML1:384.

972. MA1:xiii.

973. ML2:360.

974. Wallace to F. Bates 2 Mar. 1858 and Wallace to Pascoe 28 Nov. 1858 NHM WP1/8/262.

975. See Baker, Wallace's record of his consignments, 2001, p. 178.

976. Baker, Wallace's record of his consignments, 2001, pp. 255, 257.

977. MA1:xiv.

978. George, Alfred Wallace, the gentle trader, 1979.

979. Rookmaaker & van Wyhe, Charles Allen, 2012, p. 32: at least 28,483 were collected by Allen during 1860–2 and considering that at least 10,000 would have been collected by him during Apr. 1854–Jan. 1856.

980. *Notebook 4*, p. 63b.

981. MA1:476.

982. MA1:476.

983. Apr.–Oct. 1862 *Notebook 5*. See Raby, *Wallace*, 2001. Fagan, Theory and practice in the field, 2007 claims that Darwin acquired more of his specimens from others than Wallace did. This seems impossible to verify and probably incorrect.

984. Marchant, A man of the time, 1905.

985. I am grateful to Anna Mayer for informing me of Lorquin.

986. On 11 Dec. 1866 Bunbury met "the great naturalist traveller" *Life of Sir Charles J.F. Bunbury*, 1906, 2:211.

987. Edward Clodd's memoir introducing Bates' *Naturalist on the River Amazons*, 1892; quoted in Raby, *Wallace*, 2001, p. 167.

988. Darwin to Wallace 20 Aug. [1862] CCD10:371 and Emma Darwin's diary for 1862 In John van Wyhe, *The complete work of Charles Darwin online*, http://darwin-online.org.uk/EmmaDiaries.html). Lenny was ill on 12 June and Etty Darwin left on 16 July.

989. ML2:1.

990. "A label in [Darwin]'s hand on a specimen box now on display in his old study in Down House reads: 'Bees: Timor Wallace of which I have Comb'." CCD7:242. See MA1:311–315.

991. Anon., *The hand-book of etiquette*, 1860, p. 38.

992. Kottler, Wallace, the origin of man, and Spiritualism, 1974, see Milner, Darwin for the prosecution, 1990.

993. Raby, *Wallace*, 2001, p. 181. Wallace only identified her as "Miss L—" in ML1:409. She was first identified by Peter Raby.

994. Crawfurd, *History of the Indian Archipelago*, 1820.

995. Darwin to Wallace 22 Mar. [1869] CCD17:149.

996. Wallace, *Narrative of travels on the Amazon and Rio Negro*, 1853, p. 60.

997. Wallace, *On miracles and modern spiritualism,* 1875. See also Wallace, The facts beat me, 1908.

998. Wallace, *Insulinde: het land van den orang-oetan en den paradijsvogel*, 1870–1.

999. H.O. Forbes, *A naturalist's wanderings*, 1885, p. [v]. See also A. Forbes, *Insulinde: experiences of a naturalist's wife*, 1887.

1000. Wallace, Spiritualism in Java, 1868.

1001. See Garwood, *Flat Earth*, 2007.

1002. John Hampden to Wallace in ML2:371. I empathise with Wallace having been similarly persecuted for publishing an academic paper on Darwin in 2007.

1003. Wallace, Debt of science to Darwin, 1883.

1004. Wallace, The centenary of Darwin, 1909.

1005. Romanes, Darwin's latest critics, 1890, p. 831. See ML2:317ff.

1006. Bettany, Biographical introduction, 1889, p. ix.

1007. Sahlins, *Waiting for Foucault, still*, 2002, p. 7.

1008. Henry, *The scientific revolution*, 1997.

1009. Elton, *The practice of history*, 1967, p. 27.

1010. Lieberson, *A matter of taste*, 1985.

1011. Richards, *The meaning of evolution*, 1992, p. 2.

1012. I made this argument for pluralism in: The history of science is dead. Long live the history of science!, 2005. See also the introduction to Rudwick, *Bursting the limits of time*, 2005. For an accessible overview of the history of science see Johnston, *History of science*, 2009.

1013. See Secord, *Victorian sensation*, 2000.

1014. ML1:232.

REFERENCES

Wallace's publications are followed by the numbers assigned to them in the bibliographical list by Charles H. Smith. A few periodical articles cited in the notes are not relisted here.

Allingham, E.G. 1924. *A romance of the rostrum being the business life of Henry Stevens and the history of thirty-eight King Street*. London: Witherby.

Anon. 1853. *Accounts and papers of the House of Commons*, vol. xcv.

Anon. 1812. Huber on ants. *Edinburgh Review* 20:143–144.

Anon. 1823. *The youthful travellers*. London: Darton.

Anon. 1845. Popular works on natural history. *Westminster Review* 44.

Anon. 1852. A visit to Kew Gardens. *Hogg's Instructor* 9:216.

Anon. 1853. *Extracts from the diary of a field officer of the Bengal army*. London: Allen.

Anon. 1853. The French in the South Seas. *New Monthly Magazine* 1853:48–63.

Anon. 1854. *Etiquette: social ethics, and the courtesies of society*. London: Orr.

Anon. 1854. Ocean routes. *Chambers's papers for the people* 8(57).

Anon. 1859. On the origin of species. *Saturday Review* (24 Dec.):775–776.

Anon. 1860. *Catalogue of books in the Singapore library, with regulations and by-laws. September 1860*. Singapore: Mission Press.

Anon. 1860. Science. *Westminster Review* 73:295.

Anon. 1860. *The hand-book of etiquette: being a complete guide to the usages of polite society*. London: Cassell.

Anon. 1861. *Popular overland guide*. London: Ward & Lock.

Anon. 1862. *Nieuw Guinea: ethnographisch en natuurkundig onderzocht en beschreven in 1858 door een Nederlandsch Indische commissie, uitgegeven door het Koninklijk Instituut voor Taal-, Land-en Volkenkunde van Nederlandsch Indie.* Amsterdam: Muller.

Anon. 1870. The ocean steamer. *Harper's New Monthly Magazine* 41(242).

Anon. 1908. The *Darwin–Wallace celebration held on Thursday, 1st July, 1908 by the Linnean Society of London.* London: Linnean Society.

Aritonang, J.S. & Steenbrink, K.A. eds. *A history of Christianity in Indonesia.* Leiden: Koninklijke Brill.

Armstrong, P. 1991. *Under the blue vault of heaven: A study of Charles Darwin's sojourn in the Cocos (Keeling) Islands.* Nedlands: Indian Ocean Centre for Peace Studies.

Ashton, E.R. 1930. Memoirs of a photographic pioneer. *British Journal of Photography* (13 June):353–355.

Baden Powell, B.H. *et al.* 1860. *Essays and reviews.* London: Parker & Son.

Baker, D.B. 1995. Pfeiffer, Wallace, Allen and Smith the discovery of the Hymenoptera of the Malay archipelago, *Archives of Natural History* 23(2):153–200.

——. 2001. Alfred Russel Wallace's record of his consignments to Samuel Stevens, 1854–1861. *Zoologische Mededelingen, Leiden* 75(16):251–341.

Barlow, N. ed. 1958. *The autobiography of Charles Darwin 1809–1882. With the original omissions restored. Edited and with appendix and note.* London: Collins.

——. ed. 1963. Darwin's ornithological notes. *Bulletin of the British Museum (Natural History), Historical Series* 2(7):201–278.

Barrett, P.H., Gautrey, P.J., Herbert, S., Kohn, D. & Smith, S. eds. 1987. *Charles Darwin's notebooks, 1836–1844: geology, transmutation of species, metaphysical enquiries.* London: British Museum.

Bastin, J. 1986. Introduction to Wallace, *The Malay Archipelago.* Oxford: U. Press.

Bates, H.W. 1862. Contributions to an insect fauna of the Amazon Valley. *Lepidoptera: Heliconidae.* [Read 21 Nov. 1861] *Transactions of the Linnean Society of London* 23:495–566.

——. 1864. Hints on the collection of objects of natural history. *Journal of the Royal Geographical Society* 34:306–307.

Beccaloni, G. 2008. Homes sweet homes: a biographical tour of Wallace's many places of residence. In: Smith, C.H. & Beccaloni, G. eds. *Natural selection and beyond*, p. 7–43.

Beccaloni, G. & Smith, C.H. 2008. Celebrations for Darwin downplay Wallace's role. *Nature* 451(28 Feb.):1050.

Beddall, B.G. 1968. Wallace, Darwin, and the theory of natural selection: a study in the development of ideas and attitudes. *Journal of the History of Biology* 1(2):261–323.

——. 1988. Darwin and divergence: the Wallace connection. *Journal of the History of Biology* 21(1):1–68.

——. 1988. Wallace's annotated copy of Darwin's *Origin of species. Journal of the History of Biology* 21(2):265–289.

Bell, T. 1859. Presidential address. *Journal of the Linnean Society of London: Zoology* 4.

Benton, T. 2009. Darwin and Wallace as environmental philosophers. *Environmental Values* 18:487–502.

———. 2009. Race, sex and the 'earthly paradise': Wallace versus Darwin on human evolution and prospects. *Sociological* Review (s2)57:23–46.

Berry, A. ed. 2002. *Infinite tropics: an Alfred Russel Wallace anthology*. London & New York: Verso.

———. 2008. "Ardent beetle-hunters": natural history, collecting, and the theory of evolution. In: Smith, C.H. & Beccaloni, G. eds. *Natural selection and beyond*, p. 47–65.

Berry, A. & Browne, J. 2008. The other beetle-hunter. *Nature* 453:1188–1190.

Berthoud, J.A., Davis L.L. & Reid, S.W. eds. 2008. *Joseph Conrad: 'Twixt land and sea: tales*. Cambridge: U. Press.

Bettany, G.T. 1889. Biographical introduction. In: Wallace, A.R. *A narrative of travels on the Amazon and Rio Negro*, 2d edn. London: Ward, p. iii–ix.

———. 1887. *Life of Charles Darwin*. London: Walter Scott.

Bickmore, A.S. 1869. *Travels in the East Indian Archipelago*. London: Murray.

Bleeker, P. 1856. *Reis door de Minahassa en den Molukschen archipel gedaan in 1855*. 2 vols. Batavia.

Blyth, E. 1835. An attempt to classify the "varieties" of animals, with observations on the marked seasonal and other changes which naturally take place in various British species, and which do not constitute varieties. *Magazine of Natural History* 8:40–53.

Bonaparte, C.L. 1850. *Conspectus generum avium*, [T. 1]. Lugduni Batavorum.

Boomgaard, P. 2001. *Frontiers of fear: tigers and people in the Malay world, 1600–1950*. Yale U. Press.

Bosma, U. & Raben, R. 2008. *Being "Dutch" in the Indies: A History of Creolisation and Empire, 1500–1920*, Singapore, Athens: NUS Press & Ohio U. Press.

Bowlby, J. 1990. *Charles Darwin: a biography*. London: Hutchinson.

Bowler, P.J. 1976. Alfred Russel Wallace's concepts of variation. *Journal of the History of Medicine* 31(1):17–29.

———. 1976. Malthus, Darwin, and the concept of struggle. *Journal of the History of Ideas* 37(4):631–650.

———. 1983. *The eclipse of Darwinism: anti-Darwinian evolution theories in the decades around 1900*. Baltimore & London: Johns Hopkins U. Press.

———. 1984. [Review of Brooks 1984] Wallace and Darwinism. *Science* 224:277–278.

———. 1984. *Evolution history of an idea*. Berkeley & London: U. of California Press.

———. 1990. *Charles Darwin: the man and his influence*. Cambridge, MA: Blackwell.

Boyer, P. 1994. Cultural transmission and the biology in history. In: Boyer, P. *The naturalness of religious ideas: a cognitive theory of religion*. Berkeley: U. of California Press.

Brackman, A. 1980. *A delicate arrangement: the strange case of Charles Darwin and Alfred Russel Wallace*. New York: Times Books.

Bronowski, J. 1973. The ascent of man (BBC 1973), episode 8.

Brooks, J.L. 1984. *Just before the origin: Alfred Russel Wallace's theory of evolution*. New York and Guildford: Columbia U. Press.

Browne, J. 1980. Darwin's botanical arithmetic and the "principle of divergence," 1854–1858. *Journal of the History of Biology* 13:53–89.

——. 1983. *The secular ark: studies in the history of biogeography*. New Haven & London: Yale U. Press.

——. 1995. *Charles Darwin, vol. 1: Voyaging*. London: Jonathan Cape.

——. 2002. *Charles Darwin, vol. 2: The power of place*. London: Jonathan Cape.

——. 2006. *Darwin's Origin of species: A biography*. London: Atlantic.

——. 2006. Science in culture: a bigger picture of apes. *Nature* 439(142, 12 Jan.):142.

——. 2010. Making Darwin: biography and the changing representations of Charles Darwin. *Journal of Interdisciplinary History* 40(3):347–373.

Bryant, W. 2003. *Naturalist in the river: the life and early writings of Alfred Russel Wallace*. New York: iUniverse.

Bryson, B. 2003. *A short history of nearly everything*. London: Doubleday.

——. 2010. *At home*. London: Random House.

Buckley, C.B. 1902. *An anecdotal history of old times in Singapore 1819–1867*. Singapore: Fraser.

Bulmer, M. 2005. The theory of natural selection of Alfred Russel Wallace FRS. *Notes and Records of the Royal Society of London* 59(2):125–136.

Bunbury, C.J.F. 1906. *Life of Sir Charles J. F. Bunbury*. 2 vols. London: Murray.

Bunyon, C.J. 1889. *Memoirs of Francis Thomas McDougall, D.C.L., F.R.C.S., sometime Bishop of Labuan and Sarawak, and of Harriette, his wife*. London: Longmans.

Burkhardt, F. & Smith, S. eds. 1985. *A calendar of the correspondence of Charles Darwin, 1821–1882*. New York & London: Garland.

Burkhardt, F. *et al.* eds. 1985–. *The correspondence of Charles Darwin*. 19 vols. Cambridge: U. Press.

[Burton, J.W.] 1828. Stephens's British insects. *Monthly Review* 39:370.

——. 1837. *Lectures on entomology*. London: Simpkin & Marshall.

Calendar: Burkhardt, F. and Smith, S. eds. 1985. *A calendar of the correspondence of Charles Darwin, 1821–1882*. New York & London: Garland (superseded by the online database http://www.darwinproject.ac.uk/).

Camerini, J. 1993. Evolution, biogeography, and maps: an early history of Wallace's line. *Isis* 84:700–727.

——. 1996. Wallace in the field. *Osiris* (2d Series):44–65.

——. ed. 2002. *The Alfred Russel Wallace reader: a selection of writings from the field*. Baltimore: John Hopkins U. Press.

Cameron, J. 1865. *Our tropical possessions in Malayan India*. London: Smith Elder.

Campo, J.N.F.M. à. 2002. *Engines of empire*. Hilversum: Verloren.

Cannadine, D. 1999. *Rise and fall of class in Britain*. New York: Columbia U. Press.

Carneri, B. 1871. *Sittlichkeit und Darwinismus*. Vienna: Braumüller.

CCD: Burkhardt, F. *et al.* eds. 1985–. *The correspondence of Charles Darwin*. 19 vols. Cambridge: U. Press.

[Chambers, R.] 1844. *Vestiges of the natural history of creation*. London: John Churchill.

Chancellor, G. & Wyhe, J. van eds. (with Rookmaaker, K.) 2009. *Charles Darwin's notebooks from the voyage of the Beagle*. Cambridge: U. Press.

Claeys, G. 2008. Wallace, women, and eugenics. In: Smith, C.H. & Baccaloni, G. eds. *Natural selection and beyond*, p. 263–378.

Clodd, E. 1870. *Memories*. London: Chapman & Hall.

Colp, R. 1977. *To be an invalid: the illness of Charles Darwin*. Chicago & London: U. Press.

Collingwood, Cuthbert. 1868. *Rambles of a naturalist on the shores and waters of the China sea... 1866 and 1867*. London: John Murray.

Cooter, R. 1989. *Phrenology in the British Isles*. Metuchen & London: Scarecrow.

Corsi, P. 1988. *Science and religion: Baden Powell and the Anglican debate, 1800–1860*. Cambridge: U. Press.

Costa, J.T. 2009. The Darwinian revelation: tracing the origin and evolution of an idea. *BioScience* 59(10):886–894.

Cox, H. & Metcalfe, S. 1998. The Borneo Company Limited: the origins of a nineteenth century networked multinational. *Asia Pacific Business Review* 4(4):53–69.

Cranbrook, Earl of, Hills, D.M., McCarthy, C.J. & Prys-Jones, R. 2005. Wallace, collector: tracing his vertebrate specimens. In: Tuen, A.A., & Das, I. eds. *Wallace in Sarawak*, p. 8–34.

Crawfurd, J. 1820. *History of the Indian Archipelago*. 2 vols. Edinburgh: Constable.

——. 1852. *A grammar and dictionary of the Malay language, with a preliminary dissertation*, vol. 2 Malay and English, and English Malay dictionaries. London: Smith, Elder & Co.

Creswell, M. 1878. From Dudley colliery to Borneo. *Newcastle Courant*, 18 January to 12 April 1878. Transcribed and annotated by Martin Laverty. Nov. 2010: http://g8fight.blogspot.sg/2011/07/from-dudley-colliery-to-borneo-by.html

Dana, J.D. 1857. Thoughts on species. *Annals and Magazine of Natural History*, Supplement to (2ds.) 20(121):485–497.

Daniels, B. 2001. *Usk: origin of a thinker: The story of Alfred Russel Wallace*. Usk: Poetry Monthly Press.

Darwin, C.R. [1835]. [Extracts from letters addressed to Professor Henslow]. Cambridge: [privately printed].

——. 1839. *Journal and remarks. 1832–1836*. London: Colburn.

——. ed. 1841. *Birds Part 3 of The zoology of the voyage of H.M.S. Beagle. by John Gould*. London: Smith Elder and Co.

——. 1845. *Journal of researches into the natural history and geology of the countries visited during the voyage of H.M.S. Beagle round the world, under the Command of Capt. Fitz Roy, R.N.*, 2d edn. London: Murray.

——. 1851. *A monograph of the sub-class Cirripedia, vol. 1. The Lepadidæ; or, pedunculated cirripedes*. London: Ray Society.

——. 1851. *A monograph on the fossil Lepadidæ, or pedunculated cirripedes of Great Britain*. London: Palæontographical Society.

——. 1854. *A monograph on the sub-class Cirripedia, vol. 2. The Balanidæ, (or sessile cirripedes); the Verrucidæ*. London: Ray Society.

——. 1859. *On the origin of species by means of natural selection, or the preservation of favoured races in the struggle for life*. London: Murray.

——. 1868. *The variation of animals and plants under domestication*. 2 vols. London: Murray.

——. 1872. *The expression of the emotions in man and animals*. London: Murray.

——. 1872. *The origin of species by means of natural selection, or the preservation of favoured races in the struggle for life*, 6th edn. London: Murray.

Darwin, F. ed. 1887. *The Life and letters of Charles Darwin*, 3 vols. London: Murray.

———. ed. 1892. *Charles Darwin: his life told in an autobiographical chapter* [abridged edn.]. London: Murray.

———. ed. 1909. *Foundations of the Origin of species*. Cambridge: U. Press.

Davies, R. 2008. *The Darwin conspiracy: origins of a scientific crime*. London: Golden Square books.

———. 2012. How Charles Darwin received Wallace's Ternate paper 15 days earlier than he claimed: a comment on van Wyhe and Rookmaaker (2012). *Biological Journal of the Linnean Society* 105:472–477.

Daws, G. & Fujita, M. 1999. *Archipelago: The islands of Indonesia*. U. of California Press.

Dawson, G. 2007. *Darwin, literature and Victorian respectability*. Cambridge: U. Press.

De Beer, G. 1963. *Charles Darwin: evolution by natural selection*. London: Nelson.

De Clercq, F.S.A. 1890. *Bijdragen tot de kennis der residentie Ternate*. Leiden: Brill. (Trans. by Taylor, P.M.& Richards, M.N.: Ternate: The Residency and its Sultanate, 2010).

Decroix, P. 2005. *History of the Church and churches in Malaysia and Singapore (1511–2000)*. Penang: self-published.

Desmond, A. 1989. *Politics of evolution*. Chicago: U. Press.

———. 1997. *Huxley: from devil's disciple to evolution's high priest*. Harmondsworth: Penguin.

Desmond, A., Browne, J. & Moore, J.R. 2004. Darwin, Charles Robert. *Oxford Dictionary of National Biography*. U. Press.

Desmond, A. & Moore, J.R. 1992. *Darwin*. Harmondsworth: Penguin.

———. 1998. Transgressing boundaries. *Journal of Victorian Culture* 3:147–168.

———. 2009. *Darwin's sacred cause*. London: Penguin.

Dickenson, J.T. 1838. Indian Archipelago: Journal of Mr. Dickenson on a missionary voyage. *Missionary Herald* 34:227–233.

Dixon, M. & Radick, G. 2009. *Darwin in Ilkley*. Stroud: History Press.

Douglas, J.W. 1856. *The world of insects: a guide to its wonders*. London: Van Voorst.

Dufour, P. 1851. *Histoire de la Prostitution*. Paris: Seré.

Durant, J. 1979. Scientific naturalism and social reform in the thought of Alfred Russel Wallace. *British Journal for the History of Science* 12(1):31–58.

Eames, J.A. 1855. *Another budget; or things which I saw in the East*, 2d edn. Boston: Ticknor.

Earl, G.W. 1845. On the physical structure and arrangement of the islands in the Indian Archipelago. *Journal of the Royal Geographical Society* 15:358–365.

Eiseley, L. 1979. *Darwin and the mysterious Mr X*. London: J.M. Dent.

Elton, G.E. 1967. *The practice of history*. London: Fontana.

Ely, A. 1853. The East India islands. *De Bow's Review* 15:14–36, 243–254.

Engels, F. 1954. *Dialectics of nature*. Moscow: Foreign Languages Publishing House.

England, R. 1997. Natural selection before the *Origin*: public reactions of some naturalists to the Darwin–Wallace papers (Thomas Boyd, Arthur Hussey, and Henry Baker Tristram). *Journal of the History of Biology* 30:267–290.

Evenhuis, N.L. 2003. Publication and dating of the journals forming the Annals and Magazine of Natural History and the Journal of Natural History. *Zootaxa* 385:1–68.

Fagan, M.B. 2007. Wallace, Darwin, and the practice of natural history. *Journal of the History of Biology* 40(4):601–635.

——. 2008. Theory and practice in the field: Wallace's work in natural history (1844–1858). In: Smith, C.H. & Beccaloni, G. eds. *Natural selection and beyond*, p. 66–90.

Fallows, S. 1835. *Progressive dictionary of the English language*. Chicago: Nabu.

Fichman, M. 1981. *Alfred Russel Wallace*. Boston: Twayne.

——. 1997. Biology and politics. Defining the boundaries. In: Lightman, B. ed. *Victorian science in context*. Chicago: U. Press, p. 94–118.

——. 2004. *An elusive Victorian: the evolution of Alfred Russel Wallace*. Chicago: U. Press.

Fish, D.T. 1882. Charles Darwin. *Garden* 21(29 Apr.):302.

FitzRoy, R. 1839. *Narrative of the surveying voyages of his Majesty's ships Adventure and Beagle*. London: Colburn.

Forbes, A. 1887. *Insulinde: experiences of a naturalist's wife in the Eastern Archipelago*. Edinburgh & London: Blackwood & Sons.

Forbes, E. 1854. Anniversary address of the president (17 Feb. 1854). *Quarterly Journal of the Geological Society of London* 10:xxii–lxxxi.

——. 1854. On the manifestation of polarity in the distribution of organized beings in time. *Notices of the Proceedings of the Royal Institution of Great Britain* 1:428–433.

Forbes, H.O. 1885. *A naturalist's wanderings in the eastern archipelago*. London: Sampson.

Foxcroft, J. 1857. To entomologists. *Substitute; or, entomological exchange facilitator* 12.

Freedley, E.T. 1853. *How to make money: a practical treatise on money; with inquiry into the chances of success and causes of failure*. London: Kessinger.

Freeman, R.B. 2010. *Charles Darwin: a companion*, 2d edn. Edited by John van Wyhe. (*Darwin Online*, http://darwin-online.org.uk/content/frameset?itemID=A27b&viewtype=text&pageseq=1).

Fry, E. 1846. On the relation of the Edentata to the reptiles, especially of the armadillos to the tortoises. *Annals and Magazine of Natural History* 18:278–280.

Fyfe, A. & Lightman, B. eds. 2007. *Science in the marketplace: nineteenth-century sites and experiences*. Chicago: U. Press.

Fyfe, A. 2004. *Science and salvation: evangelical popular science publishing in Victorian Britain*. Chicago & London: U. Press.

Gale, B.G. 1972. Darwin and the concept of a struggle for existence: a study in the extrascientific origins of scientific ideas. *Isis* 63(3):321–344.

Gardiner, B.G. 1995. The joint essay of Darwin and Wallace. *Linnean* 11(1):13–24.

Garwood, C. 2007. *Flat Earth*. London: Macmillan.

George, W. 1964. *Biologist philosopher: a study of the life and writings of Alfred Russel Wallace*. London: Abelard-Schuman.

——. 1979. Alfred Wallace, the gentle trader: collecting in Amazonia and the Malay Archipelago 1848–1862. *Journal of the Society for the Bibliography of Natural History* 9(4):503–514.

Ghiselin, M.T. 1984. *The triumph of the Darwinian method*. Chicago & London: U. Press.

Glaubrecht, M. & Kotrba, M. 2004. Alfred Russel Wallace's discovery of "curious horned flies" and the aftermath. *Archives of Natural History* 31:275–299.

Glick, T.F. & Kohn, D. 1996. *On evolution*. Indianapolis & Cambridge: Hackett.

Gordon, S. 1989. Darwin and political economy: the connection reconsidered. *Journal of the History of Biology* 22(3):437–459.

Gould, S.J. 1977. *Ever since Darwin*. Harmondsworth: Penguin.

Gray, J.E. 1859. [Notes on a new bird-of-paradise discovered by Mr. Wallace.] *Proceedings of the Zoological Society of London* 27:130.

Greene, J.C. 1959. *The death of Adam*. Ames: Iowa State U. Press.

Gribbin, J. 1997. *Darwin in 90 minutes*. London: Constable.

Gruber, H.E. 1974. *Darwin on man*. London: Wildwood House.

Haenen, P. & Huizinga, F. eds. 2001. *Sources on Netherlands New Guinea in the Indonesian National Archives: the general and political reports of Ternate residency (1824–1889)*. Jakarta: Arsip Nasional RI.

Hall-Jones, J. & Hooi, C. 1979. *An early surveyor in Singapore: John Turnbull Thomson in Singapore, 1841–1853*. Singapore: National Museum.

Hamilton, W.J. 1856. The anniversary address of the president. *Quarterly Journal of the Geological Society of London* 12:xxvi–cxix.

Hands, G. 2001. *Darwin: a beginner's guide*. London: Hodder & Stoughton.

Harrisson, T. 1958. Alfred Russel Wallace and a century of evolution in Borneo. *Proceedings of the Centenary and Bicentenary Congress of Biology*, p. 25–38.

Haughton, S. 1860. Presidential address. *Journal of the Geological Society of Dublin* 8:137–156.

Heij, C.J. 2010. *Biographical notes of Antonie Augustus Bruijn (1842–1890)*. Bogor: PT Penerbit IPB.

Henry, J. 1997. *The scientific revolution and the origins of modern science*. Basingstoke: Macmillan.

Herbert, S. ed. 1980. The red notebook of Charles Darwin. *Bulletin of the British Museum (Natural History) Historical Series* 7:1–164.

———. 2010. "A Universal Collector": Charles Darwin's extraction of meaning from his Galápagos experience. *Proceedings of the California Academy of Sciences* (s4)61, Supplement II (No. 5):45–68.

Hodge, M.J.S. 1983. Darwin and the laws of the animate part of the terrestrial system (1835–1837): On the Lyellian origins of his zoonomical explanatory program. *Studies in History of Biology* 6:1–106.

———. 1992. Review of Desmond & Moore, *Darwin*. *History and Philosophy of the Life Sciences* 14(2):329–342.

Holmberg, E. 2009. A community of prestige: a social history of the cosmopolitan elite class in colonial Singapore. Singapore: NUS PhD thesis.

Homans, I.S. 1860. *A cyclopedia of commerce and commercial navigation*, 2d edn. New York: Harpers.

Hopkins, R.S. 1969. *Darwin's South America*. New York: John Day.

Howes, P.G. 1919. *Insect behavior*. Boston: Gorham Press.

Huc, E.R. 1854. *L'Empire chinois: faisant suite à l'ouvrage intitulé Souvenirs d'un voyage dans la Tartarie et le Thibet*. 2 vols. Paris: L'imprimerie Impériale.

Hughes, R.E. 1989. Alfred Russel Wallace: some notes on the Welsh connection. *British Journal for the History of Science* 22:401–418.

Hull, D.L. 2005. Deconstructing Darwin: evolutionary theory in context. *Journal of the History of Biology* 38:137–152.

Hume, D. 1793. *Essays and treatises on several subjects*, vol. 4. Edinburgh: Cadell.

Huxley, J. & Kettlewell, H.B.D. 1965. *Charles Darwin and his world*. London: Thames & Hudson.

Irvine, W. 1956. *Apes, angels & Victorians: a joint biography of Darwin and Huxley*. London: Weidenfeld & Nicolson.

Jagor, F. 1866. *Singapore Malacca Java Reiseskizzen*. Berlin: Springer.

James, F. 2005. An 'open clash between science and the church'? Wilberforce, Huxley and Hooker on Darwin at the British Association, Oxford, 1860. In: Knight, D.M. & Eddy, M.D. eds. *Science and beliefs*. Aldershot: Ashgate, p. 171–194.

Jardine, N., Secord, J. & Spary, E. eds. 1997. *The cultures of natural history*. Cambridge: U. Press.

Johnson, C.N. 2007. The preface to Darwin's *Origin of species*: the curious history of the "historical sketch". *Journal of the History of Biology* 40(3):529–556.

Johnston, S.F. 2009. *History of science*. Oxford: Oneworld.

Jones, E. 1959. *Free associations: memoirs of a psychoanalyst*. New York: Basic Books.

Jones, M.L. 1982. The orangutan in captivity. In: de Boer, L.E.M. ed. *The Orang utan*. The Hague: Junk, p. 17–37.

Jones. G. 2002. Alfred Russel Wallace, Robert Owen and the theory of natural selection. *British Journal of the History of Science* 35:73–96.

Jukes, J.B. 1850. *A sketch of the physical structure of Australia*. London: T. & W. Boone.

Keynes, R. ed. 2000. *Charles Darwin's zoology notes & specimen lists from H.M.S. Beagle*. Cambridge: U. Press.

[Kinloch, C.W.] 1853. *De Zieke Reiziger: or, rambles in Java and the Straits* [in 1852]. London: Simpkin & Marshall.

Knapp, S. 1999. *Footsteps in the forest: Alfred Russel Wallace in the Amazon*. London: NHM.

Knight, C. ed. 1854. *The English cyclopaedia: a new dictionary of universal knowledge: natural history*. London: Bradbury & Evans.

Knight, D. 1981. *Ordering the world: a history of classifying man*. London: Burnett.

Knighton, W. 1855. *Tropical sketches; or reminiscences of an Indian journalist*. London: Hurst & Blackett.

Kohn, D. 1980. Theories to work by: rejected theories, reproduction, and Darwin's path to natural selection. *Studies in History of Biology* 4:67–170.

———.1981. On the origin of the principle of diversity. *Science* 213(4512):1105–1108.

———. ed. 1985. *The Darwinian heritage*. Princeton: U. Press.

———. 1985. Internal dialogue. In: Kohn, D. ed. *The Darwinian heritage*, p. 245–257.

Kohn, M. 2004. *A reason for everything*. London: Faber.

Kottler, M.J. 1974. Alfred Russel Wallace, the origin of man, and spiritualism. *Isis* 65:144–192.

———. 1985. Charles Darwin and Alfred Russel Wallace. In: Kohn, D. ed. *The Darwinian heritage*, p. 367–432.

Kraan, Alfons van der. 1980. *Lombok: Conquest, colonization, and underdevelopment, 1870–1940*. Singapore.

Lamarck, J.B. 1802. *Recherches sur l'organisation des corps vivants*. Paris: Maradan.

——. 1809. *Philosophie zoologique*. Paris: Dentu.

——. 1815–1822. *Histoire naturelle des animaux sans vertèbres*. 7 vols. Paris: Verdière.

Larkum, A. 2009. *A natural calling: life, letters and diaries of Charles Darwin and William Darwin Fox*. Dordrecht & New York: Springer.

Larsen, A. 1996. *Equipment for the field*. In: Jardine, N., Secord, J.A. & Spary, E.C. eds. *Cultures of natural history*. Cambridge: U. Press, p. 358–377.

Larson, E.J. 2001. *Evolution's workshop: God and science on the Galápagos Islands*. London: Allen Lane.

——. 2004. *Evolution: The remarkable history of a scientific theory*. New York: Random House.

LePoer, L.B. 1991. *Singapore: A country study*. Washington: Library of Congress.

Lesson, R.P. 1834–5. *Histoire naturelle des oiseaux de paradis et des épimaques*. Paris: Bertrand.

Leveson, H.A. 1865. *The hunting grounds of the Old World: Asia*, 3d edn. London: Longman.

Lieberson, S. 1985. *A matter of taste: how names, fashions, and culture change*. New Haven & London: Yale U. Press.

Liew, C. 1993. The Roman Catholic Church of Singapore, 1819–1910: from mission to church. Singapore: NUS PhD.

Lindley, J. 1832. *An introduction to botany*. London: Ridgway.

Liu, G. 2001. *Singapore a pictorial history: 1819–2000*. Richmond: Curzon.

Lloyd, D., Wimpenny, J. & Venables. A. 2010. Alfred Russel Wallace deserves better. *Journal of Biosciences* 35:339–349.

Loritz, D., 1999. *How the brain evolved language*. New York & Oxford: U. Press.

Loudon, J.C. 1836. *An encyclopedia of plants*. London: Longman.

Love, A.C. 2002. Darwin and Cirripedia prior to 1846, exploring the origins of the barnacle research. *Journal of the History of Biology* 35:251–289.

Lowrey, K.B. 2010. Alfred Russel Wallace as ancestor figure: reflections on anthropological lineage after the Darwin bicentennial. *Anthropology Today* 26(4):18–21.

Lyell, C. 1835. *Principles of geology: being an inquiry how far the former changes of the Earth's surface are referable to causes now in operation*, 4th edn. 4 vols. London: Murray.

——. 1863. *The geological evidences of the antiquity of man: with remarks on theories of the origin of species by variation*. London: Murray.

MA: Wallace, A.R. 1869. *The Malay Archipelago: The land of the orang-utan, and the bird of paradise. A narrative of travel, with studies of man and nature*. 2 vols. London: Macmillan & Co.

Malthus, T. 1826. *An essay on the principle of population*. 2 vols. London: Murray.

Marchant, J. 1905. A man of the time: Dr. Alfred Russel Wallace and his coming autobiography. *Book Monthly* 2(8):545–549 [S743].

——. ed. 1916. *Alfred Russel Wallace letters and reminiscences*. 2 vols. London: Cassell.

Matthew, P. 1831. *On naval timber and arboriculture*. London: Longman.

——. 1860. Nature's law of selection. *Gardeners' Chronicle* (7 Apr.):312–313.

Maunder, S. 1853. *The scientific and literary treasury*, new edn. London: Longman.

Mayr, E. 1982. *The growth of biological thought*. Cambridge, Mass.; London: Belknap.

McCalman, I. 2010. *Darwin's armada*. London: Pocketbook.

McDougall, H. 1854. *Letters from Sarawak*. London: Grant & Griffith.

McKinney, H.L. 1972. *Wallace and natural selection*. New Haven: U. Press.

Mereweather, J.D. 1859. *Diary of a working clergyman in Australia and Tasmania*. London: Hatchard.

Messer, A.C. 1984. Chalicodoma pluto: The world's largest bee rediscovered living communally in termite nests (Hymenoptera: Megachilidae). *Journal of the Kansas Entomological Society* 57(1):165–168.

Meyer, A.B. 1895. How was Wallace led to the discovery of natural selection? *Nature* 52(1348):415 [S516].

Milner, R. 1990. Darwin for the prosecution, Wallace for the defense. *North Country Naturalist* 2:19–50.

ML: Wallace, A.R. 1905. *My life: A record of events and opinions*. 2 vols. London: Chapman & Hall.

Mohnike, O.G. 1883. *Blicke auf das Pflanzen- und Thierreich in den niederländischen Malaienländern*. Münster: Aschendorff.

Moody, J.W.T.J. 1971. The reading of the Darwin and Wallace papers: a historical "non-event". *Journal of the Society for the Bibliography of Natural History* 5(6):474–476.

Moore, J.R. 1979. *The post-Darwinian controversies*. Cambridge: U. Press.

———. 1997. Wallace's Malthusian moment: the common context revisited. In: Lightman, B. ed. *Victorian science in context*. Chicago: U. Press, p. 290–311.

———. 2008. Wallace in Wonderland. In: Smith, C.H. & Beccaloni, G. eds. *Natural selection and beyond*, p. 353–367.

Moorehead, A. 1969. *Darwin and the Beagle*. New York: Harper & Row.

Morrell, Jack. 1997. *Science, culture and politics in Britain, 1750–1870*. Aldershot & Brookfield: Variorum.

Munz, P. 1985. *Our knowledge of the growth of knowledge*. London: Routledge.

[Neale, F.A.] 1854. *The old arm-chair, a retrospective panorama of travels by land and sea*. London: SPCK.

Newton, A. 1888. The early days of Darwinism. *Macmillan's Magazine* 57:244.

Ng, P.K.L, Corlett, R.T. *et al.* eds. 2011. *Singapore biodiversity: an encyclopedia of the natural environment and sustainable development*. Singapore: Editions Didier Millet with Raffles Museum of Biodiversity Research.

Nolan, E.H. ed. 1859. *The illustrated history of the British Empire in India*. London: Virtue.

Olby, R. 1967. *Charles Darwin*. Oxford: U. Press.

Origin: Darwin, C.R. 1859. *On the origin of species by means of natural selection, or the preservation of favoured races in the struggle for life*. London: Murray.

Ospovat, D. 1995. *The development of Darwin's theory: natural history, natural theology, and natural selection, 1838–1859*. Cambridge: U. Press.

Overweel, J.A. ed. 1995. *Irian Jaya source materials. no. 13. Topics relating to Netherlands New Guinea in Ternate Residency Memoranda of Transfer and other assorted documents*. Leiden: DSALCUL/IRIS.

Owen, R.D. 1840. *A lecture on consistency as delivered in New York, Boston, and London*. London: B. D. Cousins.

Owen, R. 1856. Description of the skull of a large species of *Dicynodon* (*D. tigriceps, Ow.*). *Transactions of the Geological Society of London* 1856:233–240.

Palmer, A.H. 1895. *The life of Joseph Wolf, animal painter.* London & New York: Longmans.

Pearce, T. 2010. "A great complication of circumstances" — Darwin and the economy of nature. *Journal of the History of Biology* 43:493–528.

Pearson, M.B. 2005. A.R. Wallace's Malay archipelago journals and notebook (privately printed). London: Linnean Society.

Pfeiffer, I.L. [1852]. *A woman's journey round the world*, 2d edn. London: National Illustrated Library.

——. 1855. *A lady's second journey round the world.* London: Spottiswoode.

Phillips, J.A. & Wetherell, C. 1995. The great reform act of 1832 and the political modernization of England. *American Historical Review* 100:411–436.

Pickering, C. & Hall, J.C. 1854. *The races of man.* London: Bohn.

Pietsch, T.W. 2012. *Trees of life: A visual history of evolution.* Baltimore: Johns Hopkins U. Press.

Pike, E.R. 1962. *The true book about Charles Darwin.* London: Muller.

Porter, D. 1993. On the road to the Origin with Darwin, Hooker, and Gray. *Journal of the History of Biology* 26(1):1–38.

Prichard, J.C. 1843. *The natural history of man.* London: Bailliere.

Quammen, D. 1997. *The song of the dodo.* London: Pimlico.

——. 2008. The man who wasn't Darwin. *National Geographic* 106:33.

Raby, P. 2001. *Alfred Russel Wallace: a life.* Princeton: U. Press.

Reeve, T. 2009. *Charles Darwin's home: Down House.* London: English Heritage.

Richards, E. 1983. Will the real Charles Darwin please stand up. *New Scientist* 22:884–887.

——. 1989. Huxley and woman's place in science: the 'woman question' and the control of Victorian anthropology. In: Moore, J.R. ed. *History, humanity, and evolution.* Cambridge: U. Press, p. 253–284.

Richards, E. 1992. *The meaning of evolution.* Chicago & London: U. Press.

[Roberts, D.] [1850.] *The route of the Overland Mail to India.* London: Atchley.

[Robertson, J.] 1868. The variation of plants and animals under domestication. *Athenaeum* (15 Feb.):243–244.

Rockell, F. 1912. The last of the great Victorians: special interview with Dr. Alfred Russel Wallace. *Millgate Monthly* 7, part 2(83):657–663.

Romanes, G.J. 1890. Darwin's latest critics. *Nineteenth Century* 27(May):823–832.

Rookmaaker, K. & Wyhe, J. van eds. Litchfield, Henrietta. [On plagiarism and scientific jealousy] (*Darwin Online*, http://darwin-online.org.uk/content/frameset?pageseq=1&itemID=CUL-DAR262.23.3&viewtype=text).

——. eds. 2012. Wallace 'Direction for Collecting in the Tropics by A.R. Wallace'. http://wallace-online.org/content/frameset?pageseq=1&itemID=CUL-DAR270.1.2&viewtype=text

——. 2012. In Wallace's shadow: the forgotten assistant of Alfred Russel Wallace, Charles Allen (1839–1892). *Journal of the Malayan Branch of the Royal Asiatic Society* 85(2):17–54.

Rose, K. & Rose, S. 2000. *Alas, poor Darwin.* London: Cape.

Rosen, B.R. & Darrell, J. 2011. A generalized historical trajectory for Charles Darwin's specimen collections, with a case study of his coral reef specimen list in the Natural History Museum, London. In: Stoppa, F. & Veraldi, R. eds. *Darwin tra scienza, storia e società*. Rome: Edizioni Universitairie Romane, p. 133–198.

Rudwick, M.J.S. 1985. *The meaning of fossils: episodes in the history of palaeontology*, 2d edn. Chicago: U. Press.

———. 2005. *Bursting the limits of time: the reconstruction of geohistory in the age of revolution*. Chicago: U. Press.

———. 2008. *Worlds before Adam: The reconstruction of geohistory in the age of reform*. Chicago: U. Press.

Runciman, S. 2010. *The white Rajah: a history of Sarawak from 1841 to 1946*. Cambridge: U. Press.

Rupke, N. 2009. *Richard Owen: biology without Darwin*, revised edn. Chicago: U. Press.

Ruse, M. 1979. *Darwinian revolution: nature red in tooth and claw*. Chicago: U. Press.

———. 1996. *Monad to man*. Cambridge, Mass. & London: U. Press.

Sahlins, M. 1996. *How "natives" think: about Captain Cook, for example*. Chicago: U. Press.

———. 2002. *Waiting for Foucault, still*. Chicago: Prickly Paradigm.

St. John, S. 1862. *Life in the forests of the Far East*. London: Smith, Elder, & Co.

———. 1879. *The life of Sir James Brooke, Rajah of Sarawak*. Edinburgh & London: Blackwood.

Schwartz, J.S. 1990. Darwin, Wallace, and Huxley, and "Vestiges of the natural history of creation". *Journal of the History of Biology* 23(1):127–153.

Sclater, P.L. 1858. On the general geographical distribution of the members of the class Aves. *Journal of Proceedings of the Linnean Society: Zoology*, 2:130–145.

———. 1860. Note on Wallace's Standard-wing, Semioptera wallacii. *Ibis* 2(5 Jan.):26–28, pl. II.

Secord, J. 1981. Nature's Fancy: Charles Darwin and the breeding of pigeons, *Isis* 72: 162–186.

———. 1997. Introduction to Lyell, *Principles of geology*. Harmondsworth: Penguin.

———. 2000. *Victorian sensation: the extraordinary publication, reception, and secret authorship of Vestiges of the natural history of creation*. Chicago: U. Press.

———. ed. 2004. *Collected works of Mary Somerville*. London: Thoemmes.

Severin, T. 1997. *The Spice Islands voyage: in search of Wallace*. London: Abacus.

Shea, V. & Whitla, W. eds. 2010. *Essays and reviews: the 1860 text and its reading*. Charlottesville: U. Press of Virginia.

Shermer, M. 2002. *In Darwin's shadow: the life and science of Alfred Russel Wallace*. Oxford: U. Press.

Sherry, N. 1966. *Conrad's eastern world*. Cambridge: U. Press.

Sidebottom, J.K. 1948. *The overland mail*. London: Allen & Unwin.

[Simonides, A.] 1854. *The Borneo question*. Singapore: Simonides.

Simpson, A.W.B. 1984. *Cannibalism and the common law*. Chicago: U. Press.

Sloan, P.R. 1985. Darwin's invertebrate program, 1826–1836. In: Kohn, D. ed. *The Darwinian heritage*. Princeton: U. Press, p. 71–120.

Slotten, R.A. 2004. *The heretic in Darwin's court: the life of Alfred Russel Wallace.* New York: Columbia U. Press.

Smith, A.Z. 1986. *A history of the Hope Entomological Collections in the University Museum Oxford with lists of archives and collections.* Oxford: U. Museum (Publication 2).

Smith, S. 1987. Historical preface. In: Barrett, P.H. *et al.* eds. *Charles Darwin's notebooks, 1836–1844.* Cambridge: U. Press.

Smith, C.H. 1991. *Alfred Russel Wallace: an anthology of his shorter writings.* Oxford: U. Press.

———. ed. 1998. *The Alfred Russel Wallace page* (http://people.wku.edu/charles.smith/index1.htm).

———. 2004. Wallace, Alfred Russel. *Oxford Dictionary of National Biography,* (http://www.oxforddnb.com/view/article/36700).

———. 2004. *Wallace's unfinished business. Complexity* 10(2):25–32.

———. 2005. Alfred Russel Wallace, past and future. *Journal of Biogeography* 32:1509–1515.

Smith, C.H. & Beccaloni, G. eds. 2008. *Natural selection and beyond: the intellectual legacy of Alfred Russel Wallace.* Oxford: U. Press.

Smith, F. 1860. Catalogue of hymenopterous insects collected by Mr. A.R. Wallace in the islands of Bachian, Kaisaa, Amboyna, Gilolo, and at Dory in New Guinea. *Journal of the Proceedings of the Linnean Society: Zoology* 5(17b)[Supplement to vol. 4]:93–143, pl.1.

Somerville. M. 1849. *Physical geography,* new edn, vol. 2. London: Murray.

Sperber, D. 1985. Anthropology and psychology: towards an epidemiology of representations. *Man* 20:73–89.

Stagl, V. 1999. Carl Ludwig Doleschall: Arzt, Forscher und Sammler. *Quadrifina* 2:195–203.

Staten Generaal van Nederland. 1853. Verslag van het beheer en den staat der Oost-Indische bezittingen over 1853. Kamerstuk Tweede Kamer 1855–1856 kamerstuknummer XIII ondernummer 2. *Bijblad van de Nederlandsche Staatscourant* 1855–1856:77–124.

———. 1858. *Verslag van het beheer en den staat der Nederlandsche bezittingen en kolonien in Oost- en West-indie en ter kust van Guinea. over 1854.* Utrecht: Kemink.

Stauffer, R.C. ed. 1975. *Charles Darwin's natural selection: being the second part of his big species book written from 1856 to 1858.* Cambridge: U. Press.

Stevens, S. [1850.] Directions for collecting and preserving specimens of natural history in tropical climates. In: Leveson, H.A. 1865. *The hunting grounds of the Old World: Asia,* 3d edn. London: Longman, p. 639–653.

Stocqueler, J.H. 1850. *Overland companion.* London: Allen.

———. 1854. *The hand-book of British India,* 3d edn. London: Allen.

Stott, R. 2003. *Darwin and the barnacle.* London: Faber.

———. 2012. *Darwin's ghosts.* London: Bloomsbury.

Strickland, H.E. 1840. On the true method of discovering the natural system in zoology and botany. *Annals of Natural History* 6(36):184–194.

———. 1844. Description of a chart of the natural affinities of the insessorial order of birds. *Report of the meeting of the British Association for the Advancement of Science* 13(Sections):69.

Sulloway, F.J. 1982. Darwin and his finches: the evolution of a legend. *Journal of the History of Biology* 15(1):1–53.

——. 1982. The *Beagle* collections of Darwin's finches. *Bulletin of the British Museum of Natural History: Zoology* 43:49–94.

——. 1996. *Born to rebel: birth order, family dynamics, and creative lives.* London: Brown.

Sunquist, M.E. & Sunquist, F. 2002. *Wild cats of the world.* Chicago: U. Press.

[Thackeray, W.M.] 1846. *Notes of a journey from Cornhill to Grand Cairo by way of Lisbon, Athens, Constantinople and Jerusalem.* London: Chapman & Hall.

Thomson, J.T. 1852. *Account of the Horsburgh Light-house, Erected on Pedra Branca, near Singapore.* Singapore: G.M. Fredrick. [Reprinted from *Journal of the Indian Archipelago and Eastern Asia* 6:376–544, 1852.]

——. 1864. *Some glimpses into life in the Far East.* London: Richardson.

Thomson, K.S. 1995. *HMS Beagle: the story of Darwin's ship.* New York: Norton.

——. 2009. *The young Charles Darwin.* New Haven & London: Yale U. Press.

Train, G.F. 1857. *An American merchant in Europe, Asia and Australia.* New York: Putnam.

Turnbull, C.M. 2009. *A history of Singapore: 1819–2005.* Singapore: NUS Press.

Turner, S.P. 2002. *Brains, practices, relativism: social theory after cognitive science.* Chicago & London: U. Press.

Turnor, L. 1830. *History of the ancient town and borough of Hertford.* Hertford: Austin.

Tyndall, J. 1874. Address Delivered Before the British Association Assembled at Belfast, With Additions. London: Longmans.

Veth, P.J. 1870–1. Annotations. In: Wallace, *Insulinde: het land van den orang-oetan en den paradijsvogel.* 2 vols. Amsterdam: Kampen.

Vetter, J. 2006. Wallace's other line: human biogeography and field practice in the eastern colonial tropics. *Journal of the History of Biology* 39(1):89–123.

——. 2010. The unmaking of an anthropologist: Wallace returns from the field, 1862–70. *Notes and Records of the Royal Society* 64(1):25–42.

Vickers, A. 2005. *A history of modern Indonesia.* Cambridge: U. Press.

Von Buch, L. 1836. *Description physique des Iles Canaries suivie d'une indication des principaux volcans du globe.* Paris & Strasbourg: Levrault.

Von Rosenberg, H. 1878. *Der malayische Archipel: Land und Leute.* Leipzig: Weigel.

Wahrman, D. 1995. *Imagining the middle class: the political representation of class in Britain c. 1780–1840.* Cambridge: U. Press.

Wallace, A.R. 1845. An essay, on the best method of conducting the Kington Mechanic's Institution. In: Parry, R. ed. *The History of Kington.* Kington: Humphreys, p. 66–70. [S1a]

——. 1847. [*Trichius fasciatus*]. *Zoologist* 5: 1676. [S2]

——. 1853. *Narrative of travels on the Amazon and Rio Negro, with an account of the native tribes, and observations on the climate, geology, and natural history of the Amazon valley.* London: Reeve & Co. [S714]

——. 1854. [Letter]. *Zoologist* 12(142):4395–4397. [S14]

——. 1854. Letters from the Eastern Archipelago. *Literary Gazette* 1961:739. [S13a]

——. 1855. Borneo. *Literary Gazette* 2023:683–684. [S22]

——. 1855. Description of a new species of Ornithoptera: Ornithoptera brookiana, Wallace. *Proceedings of the Entomological Society of London* 1855(2 April):104–105. [S16]

——. 1855. Extracts of a letter from Mr. Wallace. *Hooker's Journal of Botany* 7(7):200–209. [S19]

——. 1855. [Letter]. *Zoologist* 13(154):4803–4807. [S21]

——. 1855. On the law which has regulated the introduction of new species. *Annals and Magazine of Natural History* (ser. 2) 16(93):184–196. [S20]

——. 1855. On the ornithology of Malacca. *Annals and Magazine of Natural History* (ser. 2) 15(86, February):95–99. [S15]

——. 1855. The entomology of Malacca. *Zoologist* 13(149):4636–4639. [S17]

——. 1856. A new kind of baby. *Chambers's Journal* (ser. 3) 6(151):325–327. [S30]

——. 1856. Attempts at a natural arrangement of birds. *Annals and Magazine of Natural History* (ser. 2) 18:193–216. [S28]

——. 1856. Observations on the zoology of Borneo. *Zoologist* 14(164):5113–5117. [S25]

——. 1856. On the bamboo and durian of Borneo. *Hooker's Journal of Botany* 8(8):225–230. [S27]

——. 1856. On the habits of the Orang-utan of Borneo. *Annals and Magazine of Natural History* (ser. 2) 18(103):26–32. [S26]

——. 1856. On the Orang-utan or Mias of Borneo. *Annals and Magazine of Natural History* (ser. 2) 17(102):471–476. [S24]

——. 1857. [Letter]. *Proceedings of the Entomological Society of London*, p. 91–93. [S35]

——. 1857. Notes of a journey up the Sadong River, in North-West Borneo. *Proceedings of the Royal Geographical Society of London* 1(6):193–205. [S29]

——. 1857. On the Great Bird of Paradise, Paradisea apoda, Linn.; 'Burong mati' (Dead Bird) of the Malays; 'Fanéhan' of the natives of Aru. *Annals and Magazine of Natural History* (ser. 2) 20(120):411–416. [S37]

——. 1857. On the natural history of the Aru Islands. *Annals and Magazine of Natural History*, Supplement to vol. 20 (2d s.) 121:473–485. [S38]

——. 1858. [Letter]. *Zoologist* 16(191–192):6120–6124. [S44]

——. 1858. Note on the theory of permanent and geographical varieties. *Zoologist* 16(185–186):5887–5888. [S39]

——. 1858. On the Arru Islands. *Proceedings of the Royal Geographical Society of London* 2(3):163–170. [S41]

——. 1859. [Letter]. *Proceedings of the Zoological Society of London* 27: 129. [S48]

——. 1859. Letter from Mr. Wallace concerning the geographical distribution of birds. *Ibis* 1(4):449–454. [S52]

——. 1860. [Letter]. *Ibis* 2(6):197–199. [S58]

——. 1860. On the zoological geography of the Malay archipelago. *Journal of the Proceedings of the Linnean Society: Zoology* 4:172–184. [S53]

——. 1860. The ornithology of Northern Celebes. *Ibis* 2(6):140–147. [S57]

——. 1862. Descriptions of three new species of Pitta from the Moluccas. *Proceedings of the Zoological Society of London* (24 June):187–188. [S69]

——. 1863. On the physical geography of the Malay Archipelago. *Journal of the Royal Geographical Society* 33:217–234. [S78]

——. 1863. The Bucerotidae, or Hornbills. *Intellectual Observer* 3(5):309–317. [S79]

——. 1864. [Discussion.] In: Guppy, H.F.J. Notes on the capabilities of the Negro for civilisation. *Journal of the Anthropological Society of London* 2:ccxiii–ccxiv. [S95]

——. 1864. The origin of human races and the antiquity of man deduced from the theory of "natural selection". *Journal of the Anthropological Society of London* 2:clviii–clxx. [S93]

——. 1865. [Discussion] In: Donovan, C. On craniology and phrenology in relation to ethnology. *Ethnological Journal* 2:97. [S113a]

——. 1865. List of the land shells collected by Mr. Wallace in the Malay Archipelago, with descriptions of the new species by Mr. Henry Adams. *Proceedings of the Zoological Society of London* (25 April):405–416, pl. 21. [S109]

——. 1865. On the progress of civilization in Northern Celebes. *Report of the British Association for the Advancement of Science* 34(part 2: Sections):149–150. [S104]

——. 1867. Anthropology address. *Report of the British Association for the Advancement of Science* 36(Sections):93–94. [S119]

——. 1867. [Discussion.] In: Hunt, J. On physio-anthropology, its aim and method. *Journal of the Anthropological Society of London* 5:cclvi–cclvii. [S133]

——. 1867. Mimicry, and other protective resemblances among animals. *Westminster Review* (n.s.) 32(173, 1 July):1–43. [S134]

——. 1867. The disguises of insects. *Hardwicke's Science-Gossip* 3:193–198. [S138]

——. 1867. The philosophy of birds' nests. *Intellectual Observer* 11(6):413–420. [S136]

——. 1868. A catalogue of the Cetoniidae of the Malayan Archipelago, with descriptions of the new species. *Transactions of the Entomological Society of London* (ser. 3) 4(part V):519–601, pls. XI–XIV. [S135]

——. 1868. [Discussion.] In: Morris, F.O. On the difficulties of Darwinism. *Athenaeum* 2134:373. [S142a]

——. 1868. Spiritualism in Java. *Spiritual Magazine* (n.s.) 3:92. [S141a]

——. 1869. Notes on the localities given in Longicornia Malayana, with an estimate of the comparative value of the collections made at each of them. *Transactions of the Entomological Society of London* (ser. 3) 3(part VII):691–696. [S154]

——. 1869. *The Malay Archipelago: The land of the orang-utan, and the bird of paradise. A narrative of travel, with studies of man and nature.* 2 vols. London: Macmillan & Co.

——. 1870. *Contributions to the theory of natural selection. A series of essays.* London & New York: Macmillan & Co. [S716]

——. 1870–1. *Insulinde: het land van den orang-oetan en den paradijsvogel; uit het Engelsch vertaald en van aanteekeningen voorzien door P. J. Veth.* 2 vols. Amsterdam: Kampen.

——. 1873. Introduction. In: Smith, F. A catalogue of the Aculeate Hymenoptera and Ichneumonidæ of India and the Eastern Archipelago. *Journal of the Linnean Society: Zoology* 11:285–302. [S156]

——. 1875. *On miracles and modern spiritualism. Three essays.* London: Burns. [S717]

——. 1880. Popular natural history. *Nature* 21(532):232–235. [S321]

——. 1883. The debt of science to Darwin. *Century Magazine* 25(3 Jan.):420–432. [S358]

——. 1891. *Natural selection and tropical nature; essays on descriptive and theoretical biology.* London & New York: Macmillan & Co. [S725]

———. 1898. *The wonderful century; its successes and its failures.* London: Sonnenschein; New York: Dodd, Mead; Toronto: Morang. [S726]

———. 1903. [Letter to Cockerell]. In: Cockerell, T.D.A. 1903. The making of biologists. *Popular Science Monthly* 62(6):517. [S603a]

———. 1903. The dawn of a great discovery: "My relations with Darwin in reference to the theory of natural selection." *Black and White* 25(17 Jan.):78. [S599]

———. 1905. *My life: A record of events and opinions.* 2 vols. London: Chapman & Hall.

———. 1906. The native problem in South Africa and elsewhere. *Independent Review* 11(38):174–182 [S630].

———. 1908. [Acceptance speech]. In: *The Darwin–Wallace celebration held on Thursday, 1st July 1908, by the Linnean Society of London.* London: Linnean Society, p. 5–11. [S656]

———. 1908. The facts beat me. *Delineator* 72:542. [S662]

———. 1909. The centenary of Darwin. *Clarion* 897:5. [S673]

———. 1910. *The world of life: a manifestation of creative power, directive mind and ultimate purpose.* London: Chapman & Hall. [S732]

[Whewell, W.] 1853. *A dialogue on the plurality of worlds.* London: Parker.

Wiersma, J.N. 1885. Duijvenbode. *Navorscher* 36:132–136.

Wilkes, C. 1845. *Narrative of the United States exploring expedition*, vol. 5. London: Putnam.

Williams-Ellis, A. 1966. *Darwin's moon: a biography of Alfred Russel Wallace.* London: Blackie.

Wilson, J.G. 2000. *The forgotten naturalist: Alfred Russel Wallace.* [Tennyson, S.A.]: Arcadia.

Wilson, L.G. ed. 1970. *Sir Charles Lyell's scientific journals on the species question.* New Haven: Yale U. Press.

Winch, D. 1987. *Malthus.* Oxford: U. Press.

———. 2001. Darwin fallen among political economists. *Proceedings of the American Philosophical Society* 145(4):415–437.

Winchester, S. 2003. *Krakatoa, the day the world exploded.* London & New York: Viking.

Winter, A. 1998. *Mesmerized: powers of mind in Victorian Britain.* Chicago: U. Press.

Wollaston, T.V. 1856. *On the variation of species with special reference to the Insecta; followed by an inquiry into the nature of genera.* London: Van Voorst.

Wouters, C. 2007. *Informalization: manners and emotions since 1890.* London: SAGE.

Wyhe, J. van 2002. The authority of human nature: the *Schädellehre* of Franz Joseph Gall. *British Journal for the History of Science* 35:17–42.

———. ed. 2002–. *The complete work of Charles Darwin online.* (http://darwin-online.org.uk/)

———. 2003. George Combe's law of hereditary descent. In: A cultural history of heredity II: 18th and 19th Centuries. *Max Planck Institut für Wissenschaftsgeschichte, Preprint* 247:165–174.

———. 2004. *Phrenology and the origins of Victorian scientific naturalism.* Aldershot: Ashgate.

———. 2005. The history of science is dead. Long live the history of science! *Viewpoint: Newsletter of the British Society for the History of Science* 80(June):1–2.

———. ed. 2006. Darwin's "Journal" (1809–1881) (*Darwin Online*, http://darwin-online.org. uk/content/frameset?viewtype=side&itemID=CUL-DAR158.1–76&pageseq=3).

———. 2007. Mind the gap: Did Darwin avoid publishing his theory for many years? *Notes and Records of the Royal Society* 61:177–205.

———. 2007. The diffusion of phrenology through public lecturing. In: Fyfe, A. & Lightman, B. eds. *Science in the marketplace: nineteenth-century sites and experiences.* Chicago: U. Press, p. 60–96.

———. 2009. *Darwin in Cambridge.* Cambridge: Christ's College.

———. 2012. Where do Darwin's finches come from? *Evolutionary Review* 3(1):185–195.

———. 2012–. *Wallace Online* (http://wallace-online.org/).

———. 2013. Alfred Russel Wallace. In: Ruse, M. ed. *Cambridge encyclopedia of Darwin and evolutionary thought.* Cambridge: U. Press.

———. 2013. "My appointment received the sanction of the Admiralty": Why Charles Darwin really was the naturalist on HMS *Beagle. Studies in History and Philosophy of Biological and Biomedical Sciences* 44(3) (September): 316–326.

Wyhe, J. van & Rookmaaker, K. 2012. A new theory to explain the receipt of Wallace's Ternate essay by Darwin in 1858. *Biological Journal of the Linnean Society* 105(1):249–252.

———. eds. 2013. *Alfred Russel Wallace: Letters from the Malay Archipelago.* Oxford: U. Press.

———. 2013. Wallace's mystery flycatcher at the Raffles Museum. *The Raffles Bulletin of Zoology* 61(1):1–5.

Young, D. 1992. *The discovery of evolution.* Cambridge & London: U. Press.

INDEX

fossil   2, 6, 7, 16, 17, 23, 37, 82, 104,
   105, 109, 139, 206, 213, 289, 318,
   331, 335, 353
Fox, William Darwin   21, 95, 121,
   237, 251, 252, 254, 325, 333, 343,
   365
Foxcroft, James   260, 366
France   76, 305
Fuegians   219, 220

Gading   92
Galápagos   6, 7, 18, 104, 118, 148,
   318
Gall, Franz Joseph   27–29
Galle, *see* Ceylon   56, 57, 225, 304,
   329
Gam Island   285
gambier   65, 68, 77, 78, 84, 87
Geach, Frederick F.   288, 289, 301,
   302, 304, 309, 370
Geographical distribution (1876)   270
gibbon   298, 301
Gibraltar   45, 50
Gilolo (Halmaheira)   161, 197,
   202–205, 217, 218, 221, 223, 229,
   259, 260, 263, 264, 266, 268, 277,
   278, 287, 317, 350
giraffe   17, 130
Goa   51, 157
Goldman, Carel Frederik   161, 172,
   191, 192, 275
Goram (Gorong)   169, 282, 369
Gray, Asa   214, 227, 238, 240–242,
   246, 251, 253, 254, 257, 358, 360,
   361, 366
Gray, George   266, 368
Gunung Api   189
Gunung Malam   295
Guthrie, James   76

Hall, Spencer   28
Hamadryas   167
Hamilton, Gray and Co.   68, 73, 88,
   96, 145, 280
Hamilton, William John   110, 111,
   337
Hart, Alfred Edward   288, 370
Hatosua   277
Haughton, Samuel   227, 241, 358
Heliconidae   82, 167
Helms, Ludvig Verner   99, 113, 114
Hertford   11–13, 26, 324
Hertford Grammar School   12
Hicks, Edwin Thomas   29, 30
higher law   108, 113, 142, 211, 212,
   217
Hoddesdon   13
Hodge, M. J. S.   105, 336, 347, 363,
   364, 366
Holmberg, Erik   65, 329, 343
Hong Kong   44–46, 57, 61, 303
Hooker, Joseph Dalton   144, 214,
   227, 228, 236, 238, 240–242,
   245–248, 251–256, 260–262, 266,
   271, 272, 313, 326, 327, 354, 356,
   360–362, 364–367
hornbill   121, 160, 300, 371
Horsburgh Lighthouse   67
Hughes, R. E.   29, 30, 326
Huguenin, Otto Fredrik Ulrich
   Jacobus   268, 309, 368
Humboldt, Alexander von   22, 23,
   318
Hume, David   15, 324
Hussein Shah   63
Huxley, Thomas Henry   2, 5, 151,
   227, 251, 284, 313, 314, 324, 336,
   344, 352, 354, 363, 364, 366
Hymenoptera   115, 327, 338